AF531346

UNDERSTANDING GENETICS

UNDERSTANDING GENETICS

By

Dr. Rajiv Tyagi

Department of Zoology

M.M. College

Modi Nagar (U.P.)

(India)

DISCOVERY PUBLISHING HOUSE PVT. LTD.

NEW DELHI-110 002

First Published-2009

ISBN 978-81-8356-458-8

Published by:

DISCOVERY PUBLISHING HOUSE PVT. LTD.

4831/24, Ansari Road, Prahlad Street,
Darya Ganj, New Delhi-110002 (India)
Phone: 23279245 • Fax: 91-11-23253475
E-mail: dphbooks@rediffmail.com
dphtemp@indiatimes.com
Website: www.discoverypublishinghouse.com

Printed at:

Sachin Printers
Delhi

Preface

The present title " Understanding Genetics" has been written for those students interested in careers in diverse fields of biological sciences. It provides a structured approach to learning by covering all the important topics in a uniform, systematic format. The book has been comprehensively designed incorporating recent advances in this fast moving field. It also provides accessible information on genetics in compact form for undergraduate students in biology and related life sciences. It is intelligible to the educated layman, though it deals with some complex ideas. It is an adequate text for all the requirements of students in this area. In addition, busy lecturers who require a quick reference compendium will find it useful, particularly for tuitional planning. Simple, yet hopefully clear figures and tables are provided throughout the book.

The over-riding goal of this book, and indeed of the whole *Understanding series,* is to present the essential information concerning genetics in a compact, readily accessible form which leads itself to student learning and revision. The convergence of various approaches has generated a rich panorama of detail, the significance of which we are still attempting to unraval. The present text has been written as an introduction to this rapidly growing field.

To make the work more comprehensive and informative, the author has consulted many authoritative books, research journals, abstracts, monographs etc., so there can be no claim to originality except in the manner of treatment.

The author expresses his thanks to his friends and colleagues whose continue inspirations have initiated him to bring out this book.

The author expresses his gratitude to Mr. Wasan and staff of M/s Discovery Publishing House Pvt. Ltd. for their whole hearted cooperation in the publication of this book.

In the mean time, the author will remain sincerely responsible for any shortcomings of the book and be grateful to the readers for their suggestions and constructive criticism for the continuous betterment of the book. He takes this opportunity to appeal to the readers to send their suggestions straightaway to his Publisher.

Author

CONTENTS

1. Mendelian Principles 1—24

Mendel's Experiments, Plants Differing in One Character, Plants Differing in Two Characters, Rules of Probability, Methods for Working Problems, Simple Mendelian Genetics in Humans, Simple Mendelian Genetics in Agriculture, Genetic Dissection.

2. Mendelian Analysis 25—42

Variations on Dominance Relations, Multiple Alleles, ABO Blood Group in Humans, C Gene in Rabbits, Operational Test of Allelism, Clover Chevrons, Incompatibility Alleles in Plants, Tissue Incompatibility in Humans, Lethal Genes, Several Genes Affecting the Same Character, Coat Colour in Mammals, Examples of Gene Interaction in Other Organisms, Penetrance and Expressivity.

3. Genetic Material 43—65

Requirements for the Genetic Material, Presence of Genetic Material, The Independence of Different Genes, Chromosomes in Heredity, Arrangement of Genes, Linear, Identification of DNA as the Genetic Material, Transformation Experiments, The Blendor Experiment, Transfer Experiments, Properties of a Genetic Material, Control of the Enzymes, Replication, Location, Storage and Transmission of Genetic Information by DNA, Transmission from Parent to Progeny, Chemical Stability of DNA, Mutation, RNA as Genetic Material in Small Viruses.

4. Chromosome 66—87

Eukaryotic Cell, Eukaryotic Chromosome, DNA Arrangement, Nucleoprotein Composition, Chromosomal Banding, Centromeres and Telomeres, C-Value Paradox, DNA-DNA Hybridization.

5. Chromosome in Inheritance **88—119**

Variation in Chromosome Structure, Natural Variation Exists in Chromosome Structure, Mutations can Alter Chromosome Structure, Loss of Genetic Material in a Deficiency Tends to be Detrimental to an Organism, Deficiencies can be Detected Using Cytological, Genetic, and Molecular Techniques, Duplications Tend to be Less Harmful than Deletions, In 1936, Bridges Found that Gene Duplications Produced the Bar and Ultra-bar Eyes Phenotype in Drosophila, Duplications Provide Additional Material for Gene Evolution, Sometimes Leading to the Development of Gene Families, Inversions Often Occur Without Phenotypic Consequences, Inversion Heterozygotes may Produce Abnormal Gametes Due to Crossing Over, Translocations can be Balanced or Unbalanced; Unbalanced Translocations Usually have Detrimental Phenotypic Effects, Individuals with Balanced Translocations may Produce Abnormal Gametes due to the Segregation of Chromosomes, Variation in Chromosome Number, Aneuploidy Causes an Imbalance in Gene Expression that is often Detrimental to the Phenotype of the Individual, Aneuploidy in Humans causes Abnormal Phenotypes, Variations in Euploidy Occur Naturally in a few Animal Species, Variations in Euploidy can Occur in Certain Tissues within an Animal; Polytene Chromosomes are an Extreme example of Repeated Chromosome Doubling, Variations in Euploidy are Common in Plants; this Variation has allowed the Development of Many Agricultural Crops, Natural and Experimental Ways to Produce Variations in Chromosome Number, Meiotic Nondisjunction can Produce Aneuploidy; Complete Nondisjunction can Produce Polyploidy, Mitotic Nondisjunction or Chromosome loss can Produce a Patch of Tissue with an Altered Chromosome Number, Changes in Euploidy can Occur by Autopolyploidy, Alloploidy, and Allopolyploidy, Experimental Treatments can Promote Polyploidy, Alloploids are Often Sterile, but Allopolyploids are More Likely to be Fertile, Cell Fusion Techniques can be Used to Make Hybrid Plants, Monoploids Produced in Agricultural and Genetic Research can be Used to Create Homozygous and Hybrid Strains.

6. Chromosome Aberrations **120—136**

Deletions, Genetic Significance of Deficiencies, Occurrence of Deletions in Human Beings, Duplication, Duplication of Complete Gene System, Gene Duplication as a source of New Genetic

Material, Inversions, Types of Inversions, Pericentric Inversion, Paracentric Inversions, Cytological and Genetical Methods for Detection and Identification of Inversions, Translocations, Cytology of a Translocation Heterozygote, Breeding Behaviour of a Translocation Heterozygote, Interchange Heterozygosity in Oenothera, Chromosome Aberrations in Humans.

7. **Polyploidy** **137—159**

Aneuploidy, Nullisomics, Monosomics, Trisomics, Somatic Aneuploids, Aneuploidy in Man, The Turner Syndrome (45,X), The Klinefelter Syndrome (47,XXY), Aneuploidy of X Chromosomes and Mental Deficiency, 47,XYY and Behaviour, Chromosomal Mosaics, Abnormal Euploidy, Monoploids, Polyploids, Triploids, Autotetraploids, Allopolyploids, Somatic Allopolyploids from Cell Hybridization, Application of Polyploidy, New World Cotton, Primrose Hybridization, Tobacco Resistance, Polyploid Fruits, Flowers, and Wheat, Chromosome Anomalies in Spontaneous Abortions in Humans.

8. **Cytological Genetics** **160—183**

Chromosome Banding, Structure and Functions of Special Types of Chromatin, Detection of Changes in Chromosome Number and Structure, Change in Number Chromosomes, Detection of Chromosomal Rearrangements by Different Chromosome Shapes, Sizes, or Banding Patterns, Detection of Rearrangements by Meiotic Synapsis or Interphase Synapsis of Polytene Chromosomes, Ultrastructure of Mitotic and Meiotic Chromatin, Mitotic Chromatin, Meiotic Chromatin, DNA Replication and Transcription at the Cytological Level, Sister Chromatid Exchange.

9. **Quantitative Genetics** **184—213**

Quantitative Traits, Quantitative Traits Exhibit a Continuum of Phenotypic Variation that may Follow a Normal Distribution, Statistical Methods are Used to Evaluate a Frequency Distribution Quantitatively, Some Statistical Methods Compare two Variables with Each Other, Polygenetic Inheritance, Polygenic Inheritance Creates Overlaps between Genotypes and Phenotypes, Quantitative Trait Loci (QTLs) can now be Mapped by RFLP Analysis, Heritability, Phenotypic Variance is Due to the Additive Effects of Genetic Variance and Environmental Variance, Heritability is the Relative Amount of Phenotypic Variation that is due to Genetic Factors, Holt Found that the Heritability of Dermal Ridge Count in Human Fingerprints is very High, Heritability Values are only Relevant to Particular Groups Raised in a Particular Environment, Selective Breeding of Species can alter Quantitative Traits Dramatically, Heterosis may be Explained by Dominance or Overdominance.

10. Developmental Genetics **214—243**

Invertebrate Development, Early Stages of Embryonic Development Determine the Pattern of Structures in the Adult Organism, Study of Drosophila Mutants with Disrupted Development Patterns has Identified Genes that Control Development, Generation of a Body Pattern Depends on the Positional Information that each Cell Receives during Development, Gene Products of Maternal Effect Genes are Deposited Asymmet-rically into the Oocyte and Establish the Anteroposterior and Dorsoventral Axes at a Very Early Stage of Development, Gap, Pair-rule, and Segment-polarity Genes act Sequentially to Divide the Drosophila Embryo into Segments, Expression of Homeotic Genes Controls the Phenotypic Characteristics of Segments, Developmental Fate of each Cell in the Nematode Caenorhabditis elegans is Known, Horvitz and Ambros Found that Heterochronic Mutations Disrupt the Timing of Developmental Changes in C. elegans, Vertebrate Development, Researchers have Identified Homeotic Genes in Vertebrates, Genes that Encode Transcription Factors also Play a Key Role in Cell Differentiation, Plant Development, Plant Growth occurs from Meristems that are Formed during Embryonic Development, Plant Homeotic Genes Control Flower Development.

11. Population Genetics **244—276**

Genes in Populations, Population is a Group of Interbreeding Individuals who Share a Gene Pool, Some Genes are Monomorphic, and Others are Polymorphic, Population Genetics is Concerned with Allele and Genotype Frequencies, Hardy-Weinberg Equilibrium, Hardy—Weinberg Equation can be Used to Calculate Genotype Frequencies Based on Allele Frequencies, Nonrandom Mating may Occur in Natural and Human Populations, Factors that Change Allele Frequencies in Population, Mutations Provide the Source of Genetic Variation, In Small Populations, Allele Frequencies can be Altered Random Genetic Drift, Migrations between two Populations can Alter Allele Frequencies, Natural Selection Favours the Survival of the Fittest, Balanced Polymorphisms may Exist due to Heterozygote Superiority or

Heterogeneous Environments, Kettlewell Studied Industrial Melanism in the Moth Biston betularia as a Modern Example of Natural Selection, Genetic Load is the Negative Consequence of Genetic Variation.

12. **Ecological Genetics** **277—295**

Ecological Niches, Bottlenecks in Population Size and Founder Effects, Population Diversification and Speciation, Rapid Reproductive Isolation in Stephanomeria Derivatives, Clines, Microheterogeneity Due to Assortative Mating, Physical Isolation and Renewed Contact, Types of Reproductive Isolation Mechanisms, Premating Isolation, Postmating Isolation, Mechanisms for Enhancing Outbreeding, Stable Polymorphisms, Gel Electrophoresis, Euphausia Adaptive Strategies and Polymorphism, Coarse and Fine grained Environments and Polymorphism.

13. **Evolutionary Genetics** **296—323**

Origin of Species, Biological Species is a Group of Reproductively Isolated Individuals, Speciation Usually Occurs via a Branching Process called Cladogenesis, Divergent Evolution can be Allopatric, Parapatric, or Sympatric, Muntzing was Able to Re-create an Allopolyploid Species by Selective Breeding, Evolution can Proceed Gradually or be Punctuated by Periods of Rapid Change, Molecular Evolution, Homologous Genes are Derived from a Common Ancestral Gene, Variation in Gene Sequences can be Used to Construct Phylogenetic Trees, Scientists can Examine the Relationships between Living and Extinct Flightless Birds by Analyzing Ancient DNA and then Comparing DNA Sequences, Genetic Variation at the Molecular Level Commonly is Associated with Neutral Changes in Gene Sequences; Non-neutral Changes are Acted on by Natural Selection, Evolution is Associated with Changes in Chromosome Structure and Number.

Index **324—332**

1

Mendelian Principles

The gene is the focal point of the discipline of modern genetics. In all lines of genetic research, it is the gene that provides the common unifying thread to a great diversity of experimentation. Geneticists are concerned with the transmission of genes from generation to generation, with the nature of genes, with the variation in genes, and with the ways that genes function to dictate the features that constitute any given species.

In this chapter we trace the birth of the gene as a concept. We shall see that genetics is, in one sense, an abstract science: most of its entities began as hypothetical constructs in the minds of geneticists and were later identified in physical form if the reasoning was sound.

The concept of the gene (but not the word) was first set forth in 1865 by Gregor Mendel. Until then, little progress had been made in understanding heredity. The prevailing notion was that the spermatozoon and egg contain a sampling of essences from the various parts of the parental body; at conception, these essences are somehow blended to form the pattern for the new individual. This idea of *blending inheritance* evolved to account for the fact that offspring typically show some characteristics similar to those of both parents. However, attempts to expand and improve this theory never led to much progress.

As a result of his research with pea plants, Mendel proposed instead a theory of *particulate inheritance*. A genetic determinant of a specific character is passed on from one generation to the next as a unit, without any blending of the units. This model explained many observations that could not be explained by blending inheritance. It also proved a very fruitful framework for further progress in understanding the mechanism of heredity. For many reasons, the importance of Mendel's ideas was not recognized until about 1900 (after his death). His report was then rediscovered by three scientists after each independently had reached the same conclusions. Historically, then, Mendel's work was irrelevant to the development of ideas about heredity. However, his achievement is important, and his analysis provides a good example of basic genetic reasoning, so we now examine Mendel's work.

Mendel's Experiments

Mendel's studies provide an outstanding example of good scientific technique. He chose research material well suited to study of the problem at hand, designed his experiments carefully, collected large amounts of data, and used mathematical analysis to show that the results were consistent with his explanatory hypothesis. The predictions of the hypothesis were then tested in a new round of experimentation. (Some historians of science believe that Mendel may have "fudged" his data to fit his hypothesis, but we shall accept his reports at face value in our discussion here.)

Mendel studied the garden pea (*Pisum sativum*) for two main reasons. First, peas were available through a seed merchant in a wide array of distinct shapes and colours that are very easily identified and analyzed. Second, peas left to themselves will *self* (self-pollinate) because the male parts (anthers) and female parts (ovaries) of the flower—which produce the pollen and the eggs, respectively—are enclosed in a petal box, or keel. The gardener or experimenter can cross (cross-pollinate) any two plants at will. The anthers from one

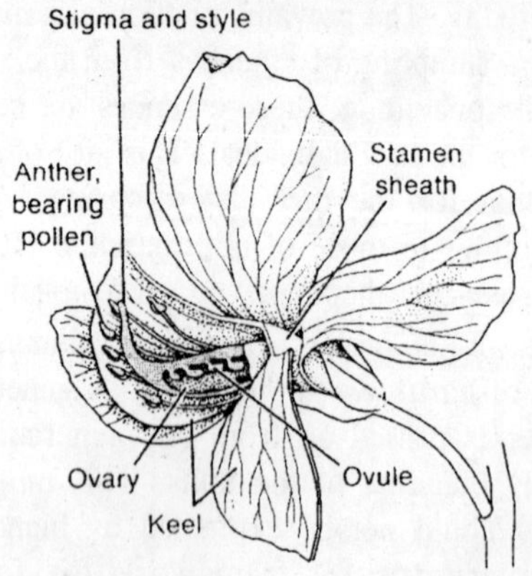

Fig. 1.1. Cutaway view of reproductive parts of a pea flower.

plant are clipped off to prevent selfing; pollen from the other plant is then transferred to the receptive area with a paintbrush or on transported anthers. Thus the experimenter can readily choose to self or to cross the pea plants.

Other practical reasons for Mendel's choice of peas were that they are cheap and easy to obtain, take up little space, have a relatively short generation time, and produce many offspring. These considerations are typical of those involved in the choice of organism for any piece of genetic research. This is a crucial decision and is nearly always based upon not only scientific criteria but also a good measure of expediency.

Plants Differing in One Character

First, Mendel chose several characters to study. It is important to clarify the meaning of character in this sense. Here, a *character* is a specific property of an organism; geneticists use this term as a synonym for characteristic or trait.

For each of the characters he chose, Mendel obtained lines of plants, which he grew for two years to make sure they were pure lines. A *pure line* is a population that breeds true for the particular character being studied; that is, all offspring produced by selfing or crossing within the population show the same form for this character. By ensuring pure lines in his research material, Mendel made a clever beginning: he had established a basis of identifiable constant behaviour, so that any changes observed following deliberate manipulation in his research would be scientifically meaningful; in effect, he had set up a control experiment.

Two of the lines Mendel grew proved to breed true for the character of flower colour. One line bred true for purple flowers, and the other for white flowers. Any plant in the purple-flowered line—when selfed, or when crossed with others from the same line—produced seeds that all grew into plants with purple flowers. When these plants in turn were selfed or crossed within the line, their progeny also had purple flowers, and so on. The white-flowered line similarly produced only white flowers through all generations. Mendel obtained seven pairs of pure lines for seven characters, with each pair differing in respect to only one character.

Each pair of Mendel's plant lines can be said to show a *character difference*, a contrasting difference between two lines of organisms (or between two organisms) with respect to one particular character. The differing lines (or individuals) represent different forms that the character may take: they can be called character forms, character variants, or phenotypes. The useful term phenotype (derived from Greek) literally means "the form that is shown." The term is extensively used in genetics, and we use it in this discussion, even though such words as gene and phenotype were not coined or used by Mendel. We shall describe Mendel's results and hypotheses in terms of more modern genetic language.

Each character represented by two contrasting phenotypes. Contrasting phenotypes for a particular character are the starting point for any genetic analysis, by Mendel or by the modern geneticist. Of courses, the definition of characters is somewhat arbitrary; there are many different ways to "split up" an organism into characters. For example, consider the following different ways of staling the same character and character difference.

Character	*Phenotypes*
Flower colour	Red versus white
Flower redness	Presence versus absence
Flower whiteness	Absence versus presence

In many cases, the description chosen is a matter of convenience (or chance). Fortunately, the choice of definition does not alter the final conclusions of the analysis, except, in the names used.

We turn now to some of Mendel's specific experimental results. In our discussion, we shall follow his analysis of the lines breeding true for flower colour.

In one of his early experiments, Mendel used pollen from a white-flowered plant to pollinate a purple-flowered plant. These plants from the pure lines are called the *parental generation* (P). All the plants resulting from this cross had purple flowers. This progeny generation is called the *first filial generation* (F_1). (The subsequent generations in such an experiment are called F_2, F_3, and so on.)

Mendel also made a *reciprocal cross*. In most plants, any cross can be made in two ways, depending on which phenotype is used as male (♂) or female (♀). For example, the two crosses

phenotype A♀ × phenotype B♂

phenotype B♀ × phenotype A♂

are reciprocal crosses. Mendel's reciprocal cross, in which a white flower was pollinated by a purple-flowered plant, produced the same result in the F_1. Mendel concluded that it makes no difference which way the cross is made. If one parent is purple-flowered and the other white-flowered, all plants in the F_1 are purple-flowered. The purple flower colour in the F_1 generation is identical to that in the purple-flowered parental plants. In this case, the inheritance obviously is not a simple blending of purple and white colours to produce some intermediate colour. To maintain a theory of blending inheritance, we would have to assume that the purple colour is somehow "stronger" than the white colour, completely overwhelming any trace of the white phenotype in the blend.

Next, Mendel selfed the F_1 plants, allowing the pollen of each flower to fall on the stigma within its petal box. He obtained 929 pea seeds from this selfing (the F_2 individuals) and planted them. Interestingly, some of the resulting plants were white-flowered; the white phenotype had reappeared! Mendel then did something that, more than anything else, marks the birth of modern genetics: he *counted* the numbers of plants with each phenotype. This procedure seems obvious to modern biologists after another century of quantitative scientific research, but it had seldom if ever been used in genetic studies before Mendel's work. There were 705 purple-flowered plants and 224 white-flowered plants. Mendel observed that the ratio of 705:224 is almost a 3:1 ratio (in fact, it is 3:1:1).

Mendel repeated this breeding procedure for six other pairs of pea character differences. He found the same 3:1 ratio in the F_2 generation for each pair. By this time, he was undoubtedly beginning to believe in the significance of the 3:1 ratio and seeking an explanation. The white phenotype is completely absent in the F_1 generation, but it reappears (in its full original form) in one-tourth of the F_2 plants. It is very difficult to devise an explanation of this result in terms of blending inheritance. Even though the F_1 flowers are purple, the plants still must carry the *potential* to produce progeny with white flowers.

Table 1.1. Results of all Mendel's crosses in which parents differed for one character

Parent phenotypes	F_1	F_2	F_2 *ratio*
1. Round × wrinkled seeds	All round	5474 round; 1850 wrinkled	2.96:1
2. Yellow × green seeds	All yellow	6022 yellow; 2001 green	3.01:1
3. Purple × white petals	All purple	705 purple; 224 white	3.15:1
4. Inflated × pinched pods	All inflated	882 inflated; 299 pinched	2.95:1
5. Green × yellow pods	All green	428 green; 152 yellow	2.82:1
6. Axial × terminal flowers	All axial	651 axial; 207 terminal	3.14:1
7. Long × short stems	All long	787 long; 277 short	2.84:1

Mendel inferred that the F_1 plants receive from their parents the ability to produce both the purple phenotype and the white phenotype, and that these abilities are retained and passed on to future generations rather than blended. Why is the white phenotype not expressed in the F_1 plants? Mendel invented the terms *dominant* and *recessive* to describe this phenomenon without explaining the mechanism. In modern terms, the purple phenotype is dominant to the white phenotype, and the white phenotype is recessive to the purple. Thus, the phenotype of the F_1 provides the operational definition of dominance.

Mendel made another important observation when he individually selfed the F_2 plants. In this case, he was working with the character of seed colour. Because this character can be observed without growing plants from the peas, much larger numbers of individuals can be counted. (In this species, the colour of the seed is characteristic of the offspring—the seed itself—rather than of the parent plant.) Mendel used two pure lines of plants with yellow and green seeds, respectively. In a cross between one plant from each line, he observed that all of the F_1 peas were yellow. Symbolically,

P yellow × green

↓

F_1 all yellow

Therefore, in this character, yellow is dominant and green is recessive.

Mendel grew F_1 plants from these yellow F_1 peas and selfed the plants. Of the resulting F_2 peas, 3/4 were yellow and 1/4 were green — the 3:1 ratio again. He then grew plants from 519 of the yellow F_2 peas and selfed each of these F_2 plants. When the peas appeared (the F_3 generation), he found that 166 of the plants had only yellow peas. The remaining 353 plants bore both yellow and green peas on the same plant. Counting all the peas from these plants, he again obtained a 3:1 ratio of yellow to green peas. The green F_2 peas all proved to be pure-breeding green peas; that is, selfing produced only green peas in the F_3 generation.

In summary, all of the F_2 green peas were pure-breeding greens like one of the parents. Of the F_2 yellows, about 2/3 were like the F_1 yellows (producing yellow and green seeds in a 3:1 ratio when selfed), and the remaining 1/3 were like the pure-breeding yellow parent. Thus, the study of the F_3 generation revealed that the apparent 3:1 ratio in the F_2 generation could be more accurately described as a 1:2:1 ratio.

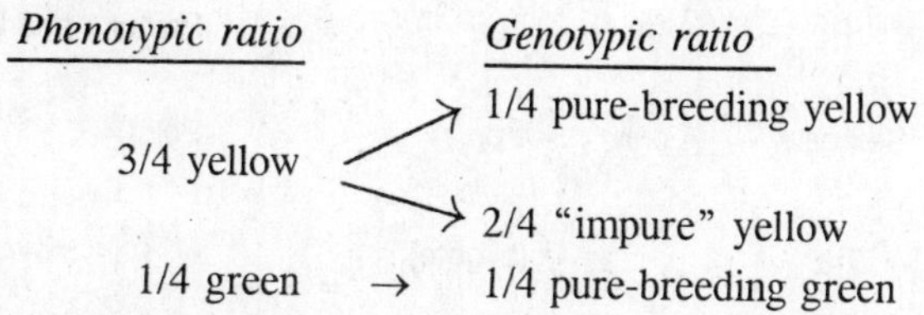

Further studies showed that these 1:2:1 ratios existed in all of the apparent 3:1 ratios that Mendel had observed. Thus the problem really was to explain the 1:2:1 ratio.

Mendel's explanation was a classic example of a model or hypothesis derived from observation, a model that was well suited for testing by further experimentation. Mendel deduced the following explanation.

1. There are hereditary determinants of a particulate nature. (He saw no blending of phenotypes, so he was forced to this particulate notion.) We now call these determinants genes.
2. Each adult pea plant has two genes—a *gene pair*—in each cell for each character studied. The reasoning here was obvious: the F_1 plants, for example, must have had one gene responsible for the dominant phenotype, called the *dominant gene*, and one gene for the recessive phenotype, which showed up only in later generations, called the *recessive gene*.
3. The members of each gene pair segregate (separate) equally into the gametes. In animals the gametes are readily identifiable as eggs and sperms. Plants produce eggs and sperms too, but these forms are less easily identified as such.
4. Consequently, each gamete carries only one member of each gene pair.
5. The union of gametes, to form the first cell (or zygote) of a new progeny individual, is random and occurs irrespective of which member of a gene pair is carried.

These points can be illustrated diagrammatically for a general case, using *A* to represent a dominant gene, and *a* the recessive gene (as Mendel did), much as a mathematician uses symbols to represent abstract entities of various kinds.

The whole model made beautiful sense of the data. However, many beautiful models have been knocked down under test: Mendel's next job was to test it. He did this by taking (for example) an F_1 yellow and crossing it with a green. A 1:1 ratio of yellow to green seeds could be predicted in the next generation. If we use *Y* to stand

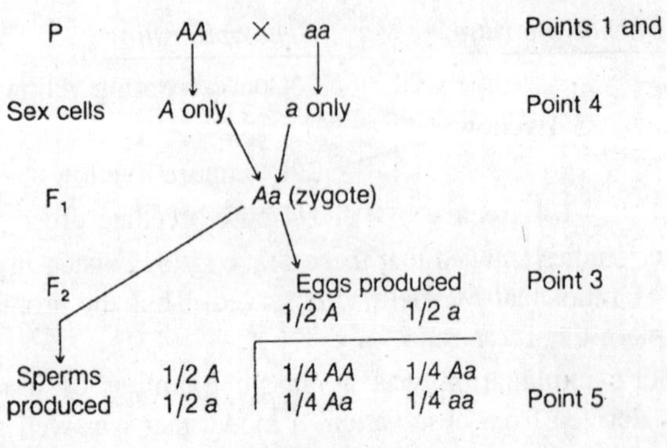

Fig. 1.2. Symbolic representation of the P, F_1, and F_2 generations in Mendel's system involving a character differene determined by one gene difference.

for the dominant gene causing yellow seeds and *y* to stand for the recessive gene causing green seeds. In this experiment, he obtained 58 yellow and 52 green seeds, a very close approximation to the predicted 1:1 ratio, and confirming the equal segregation of *Y* and *y* in the F_1 individual. This concept of *equal segregation* has been given formal recognition as Mendel's first law.

- *Mendel's First Law*: The two members of a gene pair segregate (separate) from each other into the gametes, so that one-half of the gametes carry one members of the pair and the other one-half of the gametes carry the other member of the gene pair.

Now we need to introduce some more terms. The individuals represented by *Aa* are called *heterozygotes*, or sometimes, *hybrids*, whereas those in pure lines are called *homozygotes*. Thus, an *AA* plant is said to be homozygous for the dominant gene, sometimes called *homozygous dominant*; an *aa* plant, is homozygous for the recessive

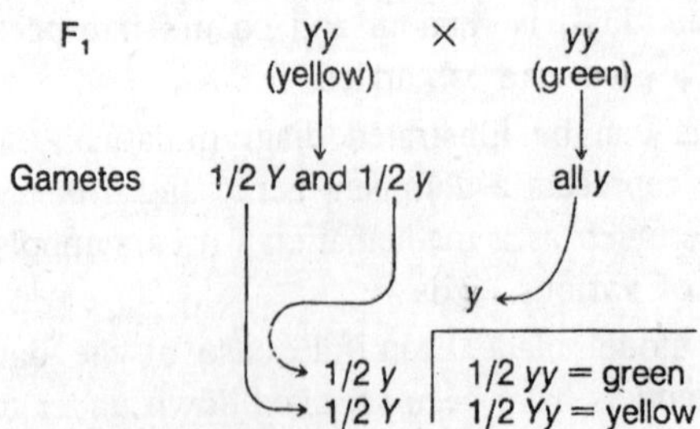

Fig. 1.3. Predicted consequences of crossing an F_1 yellow with any green.

gene, or *homozygous recessive*. The designated genetic constitution with respect to the character or characters under study is called the genotype. Thus, *YY* and *Yy,* for example, are different genotypes even though the seeds of both types are of the same phenotype (that is, yellow). You can see that in such a situation the phenotype can be thought of simply as the outward manifestation of the underlying genotype.

Note that the expressions dominant and recessive have been used in conjunction with both the phenotype *and* the gene. This is accepted usage. The dominant phenotype is established in analysis by the appearance of the F_1. Obviously, however, a phenotype (which is merely a description) cannot really exert dominance. Mendel showed that the dominance of one phenotype over another is in fact due to the dominance of one member of a gene pair over the other.

Let's pause to let the significance of this work sink in. What Mendel had done was to develop an analytical scheme for the identification of major genes regulating any biological character or function. Starting with two different phenotypes (purple and white) of one character (petal colour), he was able to show that the difference was caused by differences centering on one gene pair. Let's take the two forms of the petal-colour character as an example. Modern geneticists would say that Mendel's analysis had identified a major gene for petal colour. What does this mean? It means that in these organisms there is a *kind* of gene that has i) profound effect on the colour of the petals. This gene can exist in different forms: the dominant form of the gene (represented by *C)* causes purple petals, and the recessive form of the gene (represented by *c*) causes white petals. The forms *C* and *c* are called *alleles* (or alternative forms) of that gene for petal colour. They are given the same letter symbol to show that they are forms of the same kind of gene. One could express this another way by saying that there is a kind of gene, called phonetically a "see" gene, with alleles *C* and *c.* Any individual pea plant will always have two "see" genes, forming a gene pair, and the actual members of the gene pair can be either *CC*, *Cc,* or *cc*. Notice that although the members of a gene pair can produce different, effects, they obviously both affect the same character.

Students often find the term allele confusing, and the reason is probably that the words allele and gene are used interchangeably in some situations. For example, "dominant gene" and "dominant allele" both refer to the same thing in an interchangeable way. This stems from the fact that the forms (alleles) of any type of gene are of course genes themselves.

Plants Differing in Two Characters

The experiments described thus far dealt with a single gene pair—that is, a *monohybrid* system; we considered the allelic forms of a gene affecting one character. The next obvious question is what happens when a *dihybrid* cross is made, involving genes affecting two different characters. We can use the same symbolism that Mendel used to indicate the genotype of seed colour—*Y* and *y*—and seed shape—*R* and *r*.

A pure-breeding line of *RRyy* plants, on selling, produces seeds that are round and green. Another pure-breeding line is *rrYY*; on selfing, this line produces wrinkled yellow seeds *(r* is a recessive allele of the seed-shape gene arid produces a wrinkled seed). When Mendel crossed plants from these two lines, he obtained round yellow F_1 seeds, as expected. The results in the F_2 are complex. Mendel performed similar experiments using other pairs of characters in many other dihybrid crosses: in each case, he obtained 9:3:3:1 ratios. So, he had another phenomenon to explain, some more numbers to turn into an idea.

He first checked to see whether the ratio for each gene pair in the dihybrid cross is the same as that for a monohybrid cross. If you look at only the round and wrinkled phenotypes and add up all the seeds falling into these two classes, the totals are 315 + 108 = 423 round, and 101 + 32 = 133 wrinkled. Hence the monohybrid 3:1 ratio still prevails. Similarly, the ratio for yellow to green is (315 + 101):(108 + 32) = 416 : 140 ≅ 3: 1. From this clue, Mendel concluded that the two systems of heredity are independent. He was mathematically astute enough to realize that the 9:3:3:1 ratio is nothing more than a random combination of two independent 3:1 ratios.

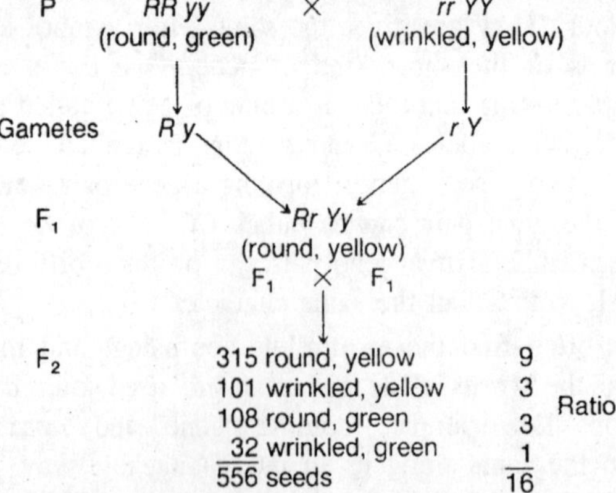

Fig. 1.4. The F_2 generation resulting from parents differing in two characters.

This is a convenient point to introduce some elementary rules of probability that we will often use throughout this book.

Rules of Probability

1. *Definition of probability.*

$$\text{Probability} = \frac{\text{The number of times an event is expected to happen}}{\text{The number of opportunities for it to happen}}$$

For example, the probability of rolling a four on a die in a single trial is written

$$p(\text{of a four}) = 1/6$$

because the dic has six sides. If each side is equally likely to turn up, then the average result should be one four for each six trials.

2. *The product rule.* The probability of two independent events occurring simultaneously is the product of each of their respective probabilities. For example, with two dice we have independent objects, and

$$p(\text{of two fours}) = 1/6 \times 1/6 = 1/36$$

3. *The sum rule.* The probability of *either one* of two mutually exclusive events occurring is the sum of their individual probabilities. For example, with two dice,

$$p(\text{of two fours } \textit{or} \text{ two fives}) = 1/36 + 1/36 = 1/18$$

In the pea example, the F_2 of a dihybrid cross can be predicted if the mechanism for putting *R* or *r* into a gamete is *independent* of the mechanism for putting *Y* or *y* into a gamete. The frequency of gamete types can be calculated by determining their probabilities according to the rules just given. That is, if you pick a gamete at random, the *probability* of picking a certain type of gamete is the same as the frequency of that type in the population.

We know from Mendel's first law that

$$Y \text{ gametes} = y \text{ gametes} = 1/2$$
$$R \text{ gametes} = r \text{ gametes} = 1/2$$

For an *RrYy* plant, the probability that a gamete will be *R* and *Y* is written $p(R\ Y)$; similarly, $p(Ry)$ denotes the probability that a gamete will be *R* and *y*. By the product rule, therefore,

$$p(RY) = 1/2 \times 1/2 = 1/4$$
$$p(Ry) = 1/2 \times 1/2 = 1/4$$
$$p(ry) = 1/2 \times 1/2 = 1/4$$
$$p(rY) = 1/2 \times 1/2 = 1/4$$

Thus we can represent the F_2 generation by a giant grid named (after its inventor) a Punnett square.

The probability of 1/16 shown for each box in the square is derived by further application of the product rule. The assortment of alleles into the male gametes and that into the female gametes are independent events. Thus, for example, the probability (or frequency) of *RRYY* zygotes (combining an *RY* male gamete with an *RY* female gamete) will be $1/4 \times 1/4 = 1/16$. Grouping all the types, we find the 9:3:3:1 ratio (now not so mysterious) in all its beauty.

round, yellow	9/16 or 9
round, green	3/16 or 3
wrinkled, yellow	3/16 or 3
wrinkled, green	1/16 or 1

The concept of independence of the two systems (round or wrinkled versus yellow or green) is important. This concept of *independent assortment* has been generalized to give the statement now known as Mendel's second law.

- *Mendel's Second Law*. During gamete formation the segregation of one gene pair is independent of other gene pairs.

A note of warning; we shall see later that the phenomenon of gene linkage is an important exception to Mendel's second law.

Note how Mendel's counting' led to the discovery of such unexpected regularities as the 9:3:3:1 ratio, and how a few simple assumptions (such as equal segregation and independent assortment) can explain this ratio that initially seems so baffling. Although it was unappreciated at the time, Mendel's approach was to provide the key to an understanding of genetic mechanisms.

Of course, Mendel went on to test this second law. For example, he crossed an F_1 dihybrid *RrYy* with a double-homozygous recessive strain *rryy*. Such a cross involving a homozygous recessive is now known as a *testcross*. For his testcross, Mendel predicted that the dihybrid *RrYy* should produce the gametic types *RY, rY, Ry,* and *ry* in equal frequency—that is, as shown along one edge of the Punnett square, in the frequencies 1/4, 1/4, 1/4, and 1/4. On the other hand, because it is homozygous, the *rryy* plant should produce only one gamete type (*ry*), regardless of equal segregation or independent assortment. Thus the progeny phenotypes should be a direct reflection of the gametic types from the *RrYy* parent (because the *ry* contribution from the *rryy* parent does not alter the phenotype indicated by the other gamete).

♂ / ♀	*R Y* 1/4	*R y* 1/4	*r y* 1/4	*r Y* 1/4
R Y 1/4	*RR YY* 1/16	*RR Yy* 1/16	*Rr Yy* 1/16	*Rr YY* 1/16
R y 1/4	*RR Yy* 1/16	*RR yy* 1/16	*Rr yy* 1/16	*Rr Yy* 1/16
r y 1/4	*Rr Yy* 1/16	*Rr yy* 1/16	*rr yy* 1/16	*rr Yy* 1/16
r Y 1/4	*Rr YY* 1/16	*Rr Yy* 1/16	*rr Yy* 1/16	*rr YY* 1/16

Key:

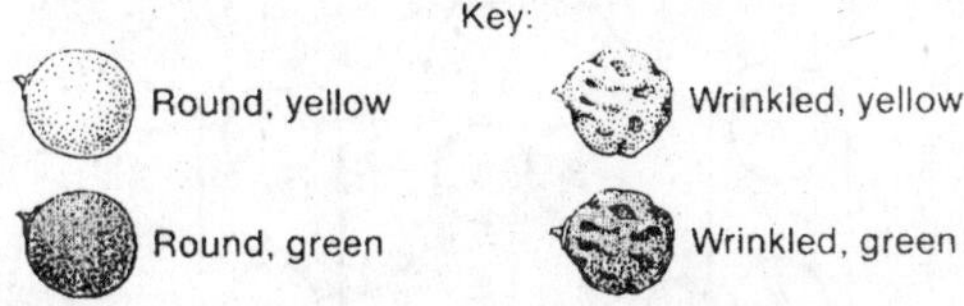

Fig. 1.5. Symbolic representation of the genetic and phenotypic constitution of the F_2 *generation resulting from parents differing in two characters.*

Mendel predicted a 1:1:1:1 ratio of *RrYy, rrYy, Rryy,* and *rryy* progeny from this testcross, and his prediction was confirmed. He tested the concept of independent assortment intensively on four different gene pairs and found that it applied to every combination.

Of course, the deduction of equal segregation and independent assortment as abstract concepts that explain the observed facts leads immediately to the question of what structures or forces are responsible for generating them. The idea of equal segregation seems to indicate that both alleles of a pair actually exist in some kind of orderly, paired configuration, from which they can separate cleanly during gamete formation. If any other gene pair behaves independently in the same way, then we have independent assortment. But this is all speculation at this stage of our discussion, as it was after the

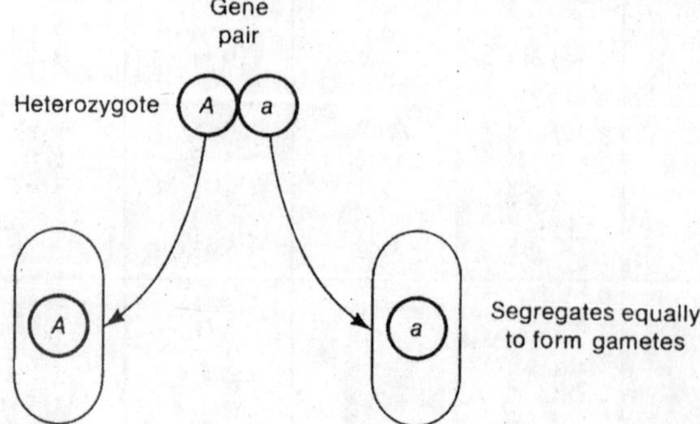

Fig. 1.6. Diagrammatic visualization of the equal segregation of one gene pair into gametes.

rediscovery of Mendel's work. The actual mechanisms are now known, and we shall discuss them later. (We shall see that it is the chromosomal location of genes that is responsible for then-equal segregation and independent assortment.) The key point to appreciate

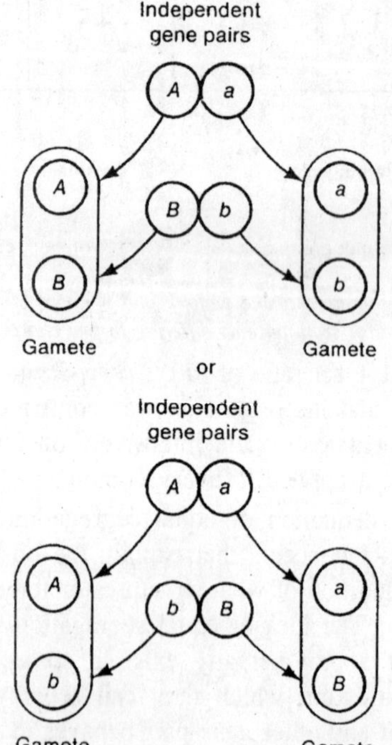

Fig. 1.7. Diagrammatic visualization of the segregation of two independent gene pairs into gametes.

at this stage is that the ground rules for genetic analysis were established by Mendel. His work made it possible to infer the existence and nature of hereditary particles and mechanisms without ever seeing them. All such theories were based on the analysis of phenotype frequencies in controlled crosses; this is the experimental approach still used in much of modern genetics.

Methods for Working Problems

We pause here for a few words on the working of problems. The Punnett square is sure and graphic, but it is unwieldy; it is suited only for illustration, not for efficient calculation.

A branch diagram is useful for solving some problems. For example, the 9:3:3:1 ratio can be derived by drawing a branch diagram and applying the product rule to determine the frequencies. (Note the use of the convention that *R*-represents both *RR* and *Rr*. That is, either allele can occupy the space indicated by the dash.)

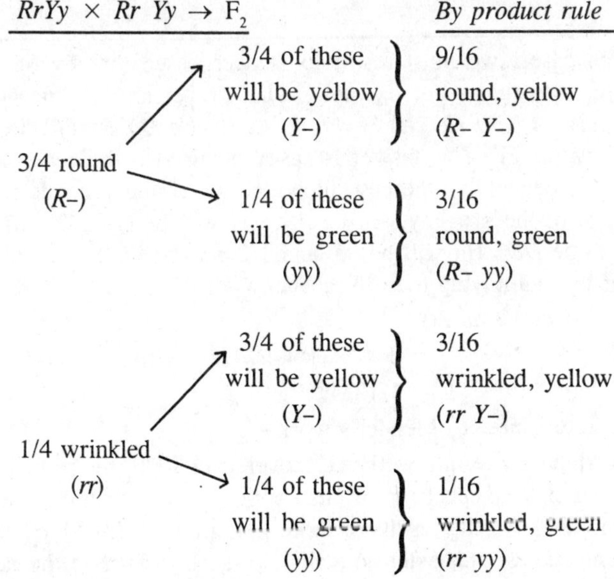

The branch diagram is, of course, a graphic expression of the product rule.

The diagram can be extended to a trihybrid ratio (such as *Aa Bb Cc* × *Aa Bb Cc)* by drawing another set of branches on the end. However, as the number of gene pairs increases, the number of identifiable phenotypes rises startlingly, and the number of genotypes climbs even more steeply. With such large class numbers, even the branch method becomes unwieldy.

Table 1.2. Rise in number of genotypic classes as the power of the number of segregating gene pairs

Number of segregating gene pairs	*Number of phenotypic classes*	*Number of genotypic classes*
1	2	3
2	4	9
3	8	27
4	16	81
.	.	.
.	.	.
.	.	.
n	2^n	3^n

In such cases, we must resort to devices based directly on the product and sum rules. For example, what proportion of progeny from the cross *Aa Bb Cc Dd Ee Ff* × *Aa Bb Cc Dd Ee Ff* will be *AA bb Cc DD ee Ff*? The answer is easily obtained if the gene pairs all assort independently, thereby allowing use of the product rule. That is, 1 /4 of the progeny will be *AA*, 1/4 will be *bb,* 1/2 will be *Cc,* 1/4 will be *DD,* 1/4 will be *ee,* and 1/2 will be *Ff,* so we obtain the answer by multiplying these frequencies:

$$p(AA\ bb\ Cc\ DD\ ee\ Ff) = 1/4 \times 1/4 \times 1/2 \times 1/4 \times 1/4 \times 1/2 = 1/1024$$

Let us return now to Mendel's work.

When Mendel's results were rediscovered in 1900, his principles were tested in a wide spectrum of eukaryotic organisms (those whose cells contain nuclei). The results of these tests showed that Mendelian genetics (sometimes with extensions that we shall discuss in the next three chapters) is universally applicable. Mendelian ratios (such as 3:1, 1:1, 9:3:3:1, and 1:1:1:1) were extensively reported, suggesting that equal segregation and independent assortment are fundamental hereditary processes found throughout nature. His laws are not merely laws about peas, but laws of the genetics of eukaryotic organisms in general. The experimental approach used by Mendel can be extensively applied in plants. However, in some plants, and in animals, the technique of selfing is impossible. This problem can be circumvented by intercrossing identical genotypes. For example, an F_1 animal resulting from the mating of parents from differing pure lines can be mated to its F_1 siblings (brothers or sisters) and an F_2 produced. The F_1 individuals are genetically identical, so the F_1 cross amounts to a selfing.

Simple Mendelian Genetics in Humans

Other systems present some special problems in the application of Mendelian methodology. One of the most difficult, yet most interesting, is the human species. Obviously, controlled crosses cannot be made, so human geneticists must resort to a scrutiny of established matings in the hope that informative matings have been made by chance. The scrutiny of established matings is called *pedigree analysis*. A member of a family who first comes to the attention of a geneticist is called the *propositus*. Usually the phenotype of the propositus is exceptional in some way—for example, a dwarf. The investigator then traces the history of the character shown to be interesting in the propositus back

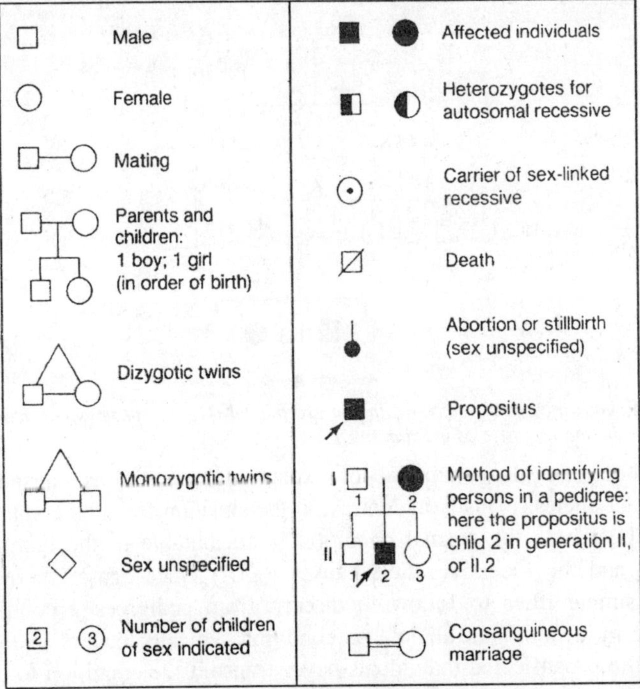

Fig. 1.8. Symbols used in human pedigree analysis.

through the history of the family, and a family tree or pedigree is drawn up using certain standard symbols.

Many human diseases and other exceptional conditions are determined by simple Mendelian recessive alleles. There are certain clues in the pedigree that must be sought. Characteristically the condition appears in progeny of unaffected parents. Furthermore, two affected individuals cannot have an unaffected child. Quite often such recessive alleles are revealed by consanguineous matings—for example, cousin marriages. This is particularly true of rare conditions where chance matings of heterozygotes are expected to be extremely rare. It has been estimated, for example, that first-cousin marriages account for about 18 to 24 percent of albino children and 27 to 53 percent of children with Tay-Sachs disease; both are rare recessive conditions. Some other examples of disease-causing recessive alleles in humans are those for cystic fibrosis and phenylketonuria (PKU). Of course, variants that are not regarded as diseases also may be caused by recessive alleles. These may be rare as in albinism or common as in light eye colour (blue or green) in North American populations.

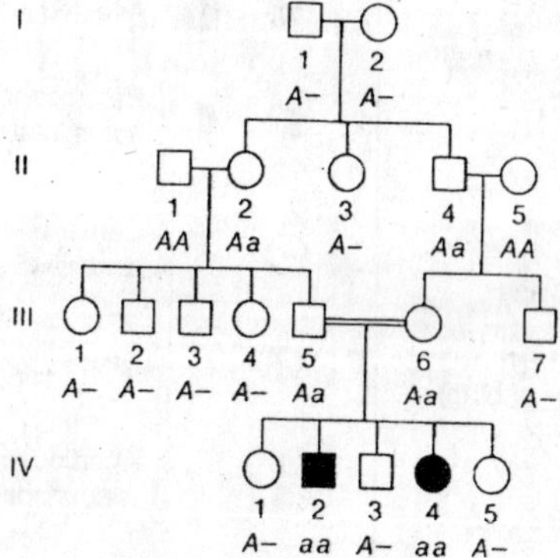

Fig. 1.9. Illustrative pedigree, involving an exceptional recessive phenotype determined by the recessive Mendelian allele a.

There are also examples of exceptional conditions caused by dominant alleles. (This is in contrast to the situation for such conditions as PKU, where the normal condition is attributable to the dominant allele, and the recessive allele causes PKU.) Once again, there are some simple rules to follow to discern from pedigrees a condition caused by a dominant allele: the condition typically occurs in every generation; unaffected individuals never transmit the condition to their offspring; two affected parents may have unaffected children; and the

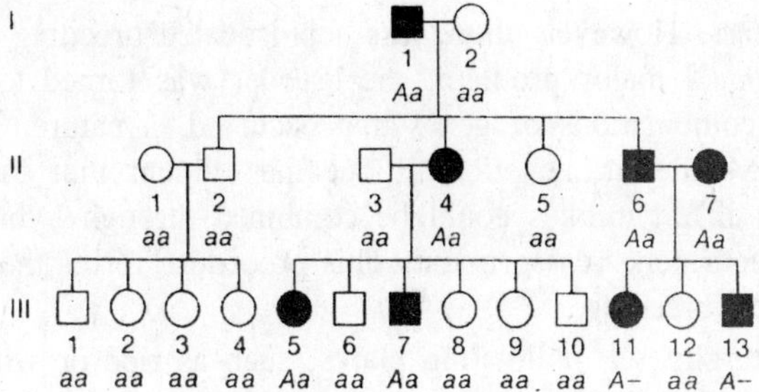

Fig. 1.10. Illustrative pedigree involving an exceptional dominant phenotype determined by the dominant Mendelian allele A.

condition is passed, on average, to one-half of the children of an affected individual. As with recessive alleles acting in a Mendelian manner, both sexes may be equally affected. Achondroplasia (a kind of dwarfism), Huntington's chorea, and brachydactyiy (very short fingers) are examples of exceptional conditions in humans caused by dominant alleles.

Notice that, in this kind of Mendelian analysis, there is again the notion of identification of genes affecting major biological function, this time in humans. Pedigrees for PKU, for example, demonstrate that there is a kind of gene controlling the character we might call "normal PKU function." The two alleles stand for presence and absence. This identification is an important step toward discovery of the precise way in which the abnormal allele is failing and its possible correction. The medical applications of such genetic analysis obviously are far-reaching. In fact, medical genetics is today a key part of medical training. Simple pedigree analyses have extensive use, not only in such medical research but also in the day-to-day counseling of prospective parents who fear genetic disease in their children.

Simple Mendelian Genetics in Agriculture

There has been an interest in plant breeding since prehistoric times. The methods used by Neolithic farmers were probably the same as those used until the discovery of Mendelian genetics. Basically the approach was to select superior phenotypes from seeds or plants derived from natural populations. Particularly desirable were pure lines of favourable phenotype, because these lines produced constant results over generations of planting. Without the knowledge of Mendelian genetics, how is it possible to develop pure lines? It so happens that self-pollinating plants, such as many crop plants, naturally lend lo be homozygous, and any heterozygous gene pairs become homozygous over generations of selfing. Thus pure lines have developed automatically over the years. However, these less sophisticated breeding practices suffered from a major problem: the breeder was forced to rely on favourable combinations of genes that occurred in nature. With the advent of Mendelian genetics, it became evident that favourable qualities in different lines could be combined through hybridization and subsequent gene reassortment. This procedure forms the basis of modern plant, breeding.

For naturally self-pollinating plants, such as rice or wheat, two pure lines (each of different favourable genotype) are hybridized by manual cross-pollination, and hence an F_1 is developed. The F_1 is then allowed to self, and its heterozygous gene pairs assort to produce many different genotypes, some of which represent desirable new combinations of the parental genes. A small proportion of these new genotypes will be pure-breeding already, but if not, several generations of selfing will produce homozygosity of the relevant genes.

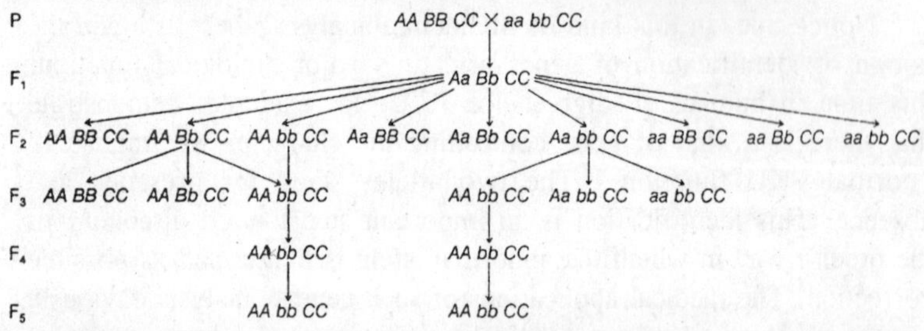

Fig. 1.11. The basic technique of plant breeding.

An example of genetic improvement in a species more familiar to most is the tomato. Anyone who has read a recent seed catalog will be familiar with the abbreviations V, F, and N next to a listed tomato variety. These represent, respectively, resistance to the pathogens *Verticillium*, *Fusarium*, and nematodes—resistance that has been crossed into the tomatoes, typically from wild forms of tomato, Another familiar phenomenon in tomatoes is determinate as opposed to indeterminate growth pattern. Determinate plants are bushier and more compact, and they do not need as much staking. Determinate growth is caused by a recessive allele *sp* (self-pruning), which has been crossed into modern varieties. Another useful allele is *u* (uniform ripening); this allele eliminates the green patch or shoulder around the stem on the ripe fruit.

Such examples could be listed for many pages. The point is that simple Mendelian genetics (as in this chapter) has provided agricultural

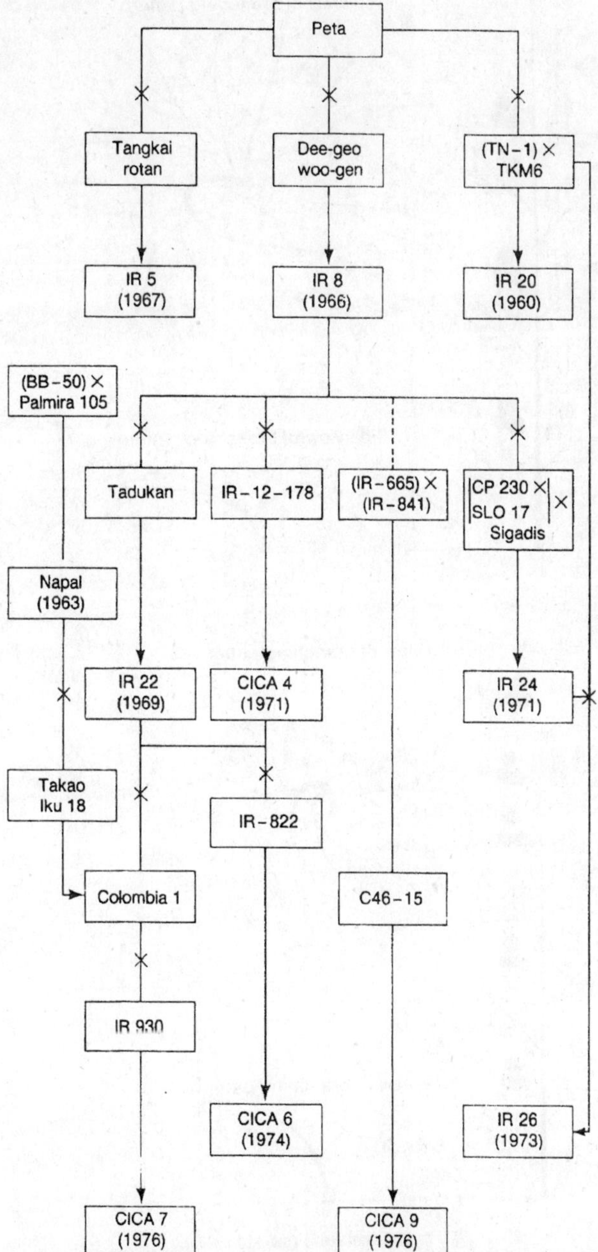

Fig. 1.12. The complex pedigree of modern rice varieties.

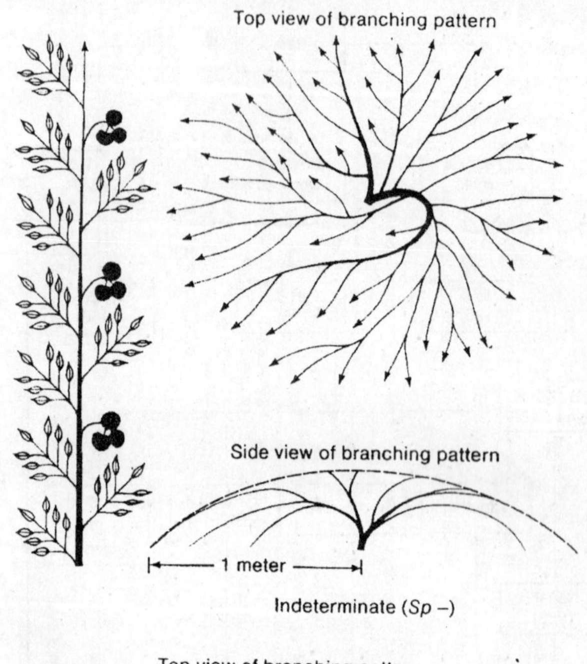

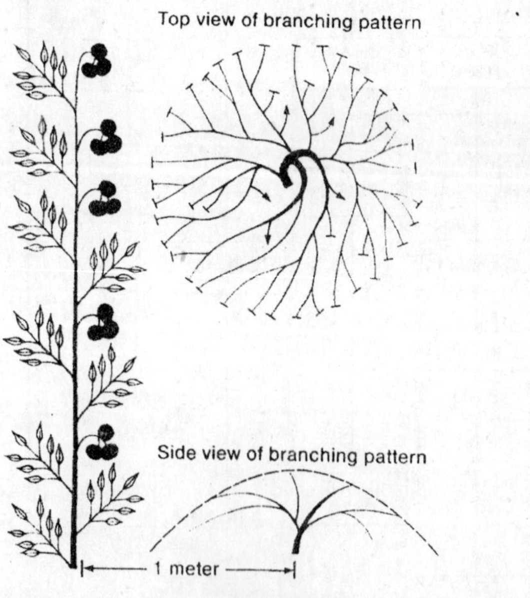

Fig. 1.13. Growth characteristics of indeterminate (Sp–) and determine (sp sp) tomatoes.

plant breeding with its rationale and its modern methods. Entire complex genotypes may be constructed from an array of ancestral lines, each showing some desirable feature. When we think of genetic engineering, we think of the genetic techniques of the 1970s and 1980s, but genetic engineering for plant improvement began long ago.

Mendelian genetics also has provided a formal theoretical basis for animal breeding, enabling a greater efficiency than under traditional practices. More recent techniques such as the use of frozen semen, artificial insemination, frozen embryos, and surrogate mothers in livestock species have enabled breeders to amplify the number of offspring of a specific genotype, a number normally limited by the lifespan of the maternal animal.

Origin of Variants

Genetic analysis, as we have seen, must start with parental differences. Without variants, no genetic analysis is possible. Where do these variants, this raw material for genetic analysis, come from? This is a question that can be answered in full only in later chapters. Briefly, most of the variants used by Mendel (and by ancestral and modern breeders of plants and animals) arise spontaneously, without the deliberate action of geneticists, in nature or in the breeders' populations. Undoubtedly, some of these variants would be "weeded out" by natural selection, but they may be kept alive through their deliberate nurture by geneticists, either for some specific breeding program or for the elucidation of the basic biological mechanisms of which the different forms are variants.

Genetic Dissection

We have seen that the genetic analysis of variants can identify a gene involved in an important biological process. This is a central aspect of modern genetics; this process is referred to as genetic dissection of biological systems. Mendel was the first genetic surgeon. Using genetic analysis, he was able to identify and distinguish among the several components of the hereditary process in a way as convincing as if he had microdissected those components. The fact that the genes he was using were for pea shape, pea colour, and so on was largely irrelevant. Those genes were being used simply as *genetic markers*, which enabled Mendel to trace the hereditary processes of segregation and assortment. A genetic marker is a variant allele that is used to label a biological structure or process throughout the course of an experiment. It is almost as though Mendel were able to "paint" two alleles different colours and send them through a cross to see how they behaved! Genetic markers are now routinely used in genetics and in all of biology to study all sorts of processes that the marker genes themselves do not directly affect.

Probably without realizing it, Mendel had also invented another aspect of genetic dissection in which the precise genes used *were* important. In this type of analysis, genetic variants are used in such a way that, by studying the variant gene function, we can make inferences about the "normal" operation of the gene and the process it controls. Let's consider the petal-colour gene in peas again. This is a major gene affecting an important biological function, which is the colour of the petals. Undoubtedly the success or failure of a plant in nature would largely hinge on its having the right kind of petal colour, presumably to attract the appropriate pollinator. So Mendel had pinpointed a gene for a process of central importance to the biology of peas, thus opening the way to extensive further study. Genetic dissection is a versatile tool of modern biological research.

Once a gene has been identified, affecting (say) petal colour in peas, we have called it a major gene, but what does this really mean? It is major in that it is obviously having a profound effect on the colour of the petals. But can we conclude that it is *the* single most important step in the determination of petal colour? The answer is no, and the reason may be seen in an analogy. If we were trying to discover how a car engine works, we might investigate this by pulling out various parts and observing the effect on the running of the engine. If a battery cable were disconnected, the engine would stop; we might erroneously conclude that this cable is the most important part of the running of the engine. Obviously, other parts are equally necessary, and their removal could also stop or seriously cripple the engine. In a similar way it can be shown that several genes can be identified, all of which have a major and similar effect on petal colouration.

Mendel's work has withstood the test of time and has provided us with the basic groundwork for all modern genetic study. Yet his work went unrecognized and neglected for 35 years following its publication. Why? There are many possible reasons, but here we shall consider just one. Perhaps it was because biological science at that time could not provide evidence for any real physical units within cells that might correspond to Mendel's genetic particles. Chromosomes had certainly not yet been studied, meiosis not yet described, and even the full details of plant life cycles had not been worked out. Without this basic knowledge, it may have seemed that Mendel's ideas were mere numerology.

2

Mendelian Analysis

We have seen that Mendel's laws seem to hold across the entire spectrum of eukaryotic organisms—that is, we can identify analogous phenomena that reveal segregation and independent assortment. These laws form a base for predicting the outcome of simple crosses. However, it is only a base; the real world of genes and chromosomes is more complex than is revealed by Mendel's laws, and exceptions and extensions abound. These situations do not invalidate Mendel's laws. Rather, they show that there are more situations than can be explained by segregation and independent assortment of gene pairs and that these situations must be accommodated into the fabric of genetic analysis. This is the challenge we now must meet. Of course, one extension has already been accommodated—sex linkage. This chapter presents a grab bag of other extensions, and the two following chapters discuss two major extensions. We shall see that, rather than creating a hopeless and bewildering situation, these complexities combine to form a precise and unifying set of principles for the genetic analyst. These principles interlock and support each other in a highly satisfying way that has provided great insight into the mechanics of inheritance.

Variations on Dominance Relations

Dominance is a good place to start. Mendel observed (or at least reported) full dominance (and recessiveness) for all the seven gene pairs he studied. He may have been selective in his choice of pea characters to study, because variations on the basic theme crop up quite often in analysis. The problems center on the phenotype of the heterozygote. Some examples will illustrate this. In four-o'clock plants, when a pure line with red petals is crossed to a pure line with white petals, the F_1 have *not* red petals but pink! If an F_2 is produced, the result is

1/4 red petals	1 C_1C_1
1/2 pink petals	2 C_1C_2
1/4 white petals	1 C_2C_2

The occurrence of an intermediate phenotype in the heterozygote introduces the possibility of *incomplete dominance*. The precise position of the heterozygote on the phenotypic "scale" defines several possibilities. In practice it is often difficult to determine exactly where on the scale the heterozygote is, however.

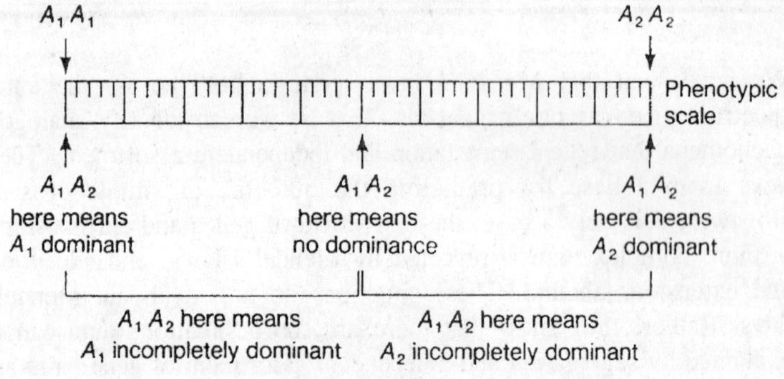

Fig. 2.1. Summary of dominance relationships.

The phenotype of the heterozygote is also the key in the phenomenon of *codominance*, in which the heterozygote shows the phenotypes of *both* the homozygotes. In a sense, then, codominance is no dominance at all! A good example is found in the M-N blood-group gene pair in humans. Three blood groups are possible—M, N, and MN—and these are determined by the genotypes L^ML^M, L^NL^N, and L^ML^N, respectively. Blood groups actually represent the presence of an immunological antigen on the surface of red blood cells. People of genotype L^ML^N have both antigens. This example shows how the heterozygote can show both phenotypes. Another interesting example of dominance is found in the human disease sickle-cell anemia. The gene pair concerned affects the oxygen transport molecule hemoglobin, the major constituent of red blood cells. The three genotypes have different phenotypes, as follows:

Hb^AHb^A: Normal. Red blood cells never sickled.

Hb^SHb^S: Severe, often fatal anemia. Red blood cells sickle-shaped.

Hb^AHb^S: No anemia. Red blood cells sickle only under abnormally low oxygen concentrations.

In regard to anemia the Hb^A allele is dominant. In regard to blood cell shape there is incomplete dominance. Penally, as we shall now see, in regard to hemoglobin there is codominance! Luckily, the two types of hemoglobin concerned, dictated by the two alleles, have different chemical charges, and this property can be used in a technique called electrophoresis to make a more refined distinction.

In electrophoresis, mixtures of proteins can be separated on the basis of their charges. A small sample of protein (here, hemoglobin from red blood cells) is placed in a small well, cut in a slab of gel. A powerful electric field is applied across this gel, and the hemoglobin moves according to the degree of electrostatic charge. The gel is actually a supporting web containing electrolyte solution in its spaces: hence the hemoglobin can move easily through the field. After an appropriate time, the gel can be stained for protein. The blotches are the stained hemoglobin. We see that homozygous normal people have one type of hemoglobin (A), and anemics have another slow-moving type (S). The heterozygotes have both types, A and S.

Sickle-cell anemia illustrates the somewhat arbitrary nature of the terms incomplete dominance and codominance. The type of dominance depends on the phenotypic level at which the observations are being made—organismal, cellular, or molecular.

Multiple Alleles

Early in the history of genetics, it became clear that it is possible to have more than two forms of one kind of gene. Although only two actual alleles of a gene can exist in a diploid cell (and only one in a haploid cell), the total number of possible different allelic forms that might exist in a population of individuals is often quite large. This situation is called *multiple allelism*, and the set of alleles itself is called an *allelic series*. The concept of allelism is a crucial one in genetics, so we consider several examples. The examples themselves serve also to introduce important areas of genetic investigation.

ABO Blood Group in Humans

The human ABO blood-group alleles afford a modest example of multiple allelism. There are four blood types (or phenotypes) in the ABO system. The allelic series includes three major alleles, which can be present in any pairwise combination in one individual; thus only two of the three alleles can be present in any one individual. Note that the series include cases of both complete dominance and codominance. In this allelic series, the alleles I^A and I^B each determine a unique form of one type of antigen; the allele i determines a failure to produce either form of that type of antigen.

Table 2.1. ABO blood groups in human

Blood phenotype	*Genotype*
O	ii
A	I^AI^A or I^Ai
B	I^BI^B or I^Bi
AB	I^AI^B

C Gene in Rabbits

Another example of multiple allelism involves a larger allelic series determining coat colour in rabbits. The alleles in this series are C (full colour), c^{ch} (chinchilla, a light grayish colour), c^h (Himalayan, albino with black extremities), and c (albino). You will have noted the use of a superscript to indicate an allele of a type of gene; this symbolism often is necessary because more than the two symbols C and c are needed to describe multiple alleles. In this series, dominance is in the order of the alleles.

Table 2.2. C gene in rabbits

Coat colour phenotype	*Genotype*
Full colour	CC or Cc^{ch} or Cc^h or Cc
Chinchilla	$c^{ch}c^{ch}$ or $c^{ch}c^h$ or $c^{ch}c$
Himalayan	c^hc^h or c^hc
Albino	cc

Operational Test of Allelism

Now that we have seen two examples of allelic series, it is a good time to pause and consider a question. How do we know that a set of phenotypes is determined by alleles of the same type of gene? In other words, what is the operational test for allelism? For now, the answer is simply the observation of Mendelian single-gene-pair ratios in all combinations crossed. For example, consider three pure-line phenotypes in a hypothetical plant species. Line 1 has round spots on the petals; line 2 has oval spots on the petals; and line 3 has no spots on the petals. Suppose that crosses of the three lines yield the following results:

These results prove that we are dealing with three alleles of a single gene that affects petal spotting. We can choose any symbols we

Cross	F_1	F_2
1 × 2	all round-spotted	3/4 round, 1/4 oval
1 × 3	all round-spotted	3/4 round, 1/4 unspotted
2 × 3	all oval-spotted	3/4 oval, 1/4 unspotted

wish. Because we don't know which phenotype is the wild-type, we could follow the rabbit system and use *S* for the round-spotted allele, s^{o} for the oval-spotted allele, and *s* for the unspotted allele. Alternatively, we could use S^{r} for round, S^{o} for oval, and *s* for unspotted. The convention provides no firm rules about whether to use capital or small letters, particularly for the alleles in the middle of the series that are dominant to some of their alleles but recessive to others.

Clover Chevrons

Clover is the common name for plants of the genus *Trifolium*. There are many species—some native to North America and some that grow here as introduced weeds. The red-flowered and white-flowered clovers are familiar to most people. These are *different* species, but each shows considerable variation among individuals in the curious V or "chevron" pattern on the leaves. Much genetic research has been done with white clover; the different chevron forms (and the absence of chevrons) constitute an allelic series in this species. Furthermore, several types of dominance relations may be seen in the allele combinations.

Incompatibility Alleles in Plants

It has been known for millennia that some plants just will not self-pollinate. A single plant may produce both male and female gametes, but no seeds will ever be produced. The same plants, however, will cross with certain other plants, so obviously they are not sterile. This phenomenon is called *self-incompatibility*. Of course, the pea plants used by Mendel were not self-incompatible, for he was able to self his plants with ease. We now know that incompatibility has a genetic basis and that there are several different genetic systems acting in different self-incompatible species. These systems form very nice examples of multiple allelism. One of the most common systems is found in many dicot and monocot plants, including sweet cherries, tobacco, petunias, and evening primroses. In each of these species, one gene determines compatibility/incompatibility relations, with many different allelic forms of the gene possible in different plants of any one species.

If a pollen grain bears an S allele that is also present in the maternal parent, then it will not grow; however, if that allele is not in the maternal tissue, then the pollen grain produces a pollen tube containing the male nucleus, and this tube effects fertilization. The number of *S* alleles in a series in one species can be very large (over 50 in the evening primrose and clover), and cases of more than 100 alleles have been reported in some species.

Tissue Incompatibility in Humans

When an organ or a tissue graft is medically necessary in a human, the success or failure of the graft depends on the genotypes of the host and the donor. If the two are mismatched, rejection of the graft eventually occurs, often accompanied by death of the host.

The rejection system is based on two important genes called *HLA-A* and *HLA-B,* which determine the immunological acceptability of the introduced tissue. Each gene has a series of allelic forms. There are eight alleles for *HLA-A,* designated *A1, A2, A3, A9, A10, A11, A28,* and *A29*. By coincidence, there are also eight alleles for *HLA-B,* designated *B5, B7, B8, B12, B13, B14, B18,* and *B27.* (The two genes do not show independent assortment, but this need not concern us now.

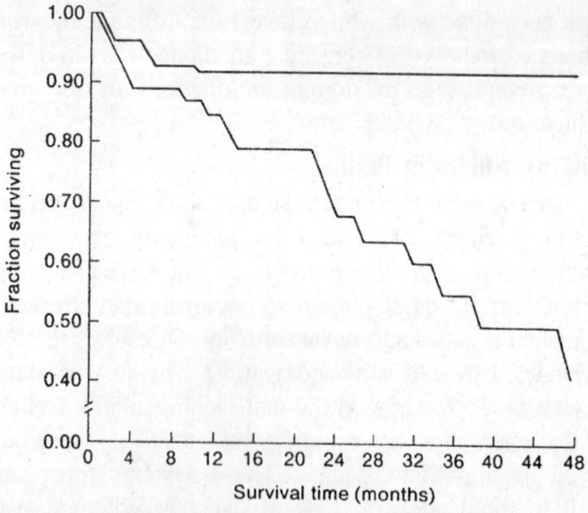

Fig. 2.2. The success of kidney transplants in HLA-matched and HLA-mismatched sibling pairs.

For the moment we shall simply use these allelic series as excellent examples of multiple allelism in humans.)

The HLA type of an individual is determined by testing his or her lymphocytes (white blood cells) with a battery of standard antibodies from donors of known HLA type. In a transplant, the recipient's immune system will recognize the graft as "foreign" and reject it if an allele is present in the donor tissue that is not present in the recipient. In the typing test, this rejection reaction is observed as lymphocyte death. The immune system will not reject tissue that lacks some alleles present in the recipient—it only rejects "strange" alleles.

Table 2.3. Success or failure of transplants determined by HLA types

Transplant	*Recipient genotype*	*Donor genotype*	*Result*
1	*A1 A2, B5 B5*	*A1 A1, B5 B7*	Rejected (because of *B7*)
2	*A2 A3, B7 B12*	*A1 A2, B7 B7*	Rejected (because of *A1*)
3	*A1 A2, B7 B5*	*A1 A2, B7 B7*	Accepted
4	*A2 A3, B7 B5*	*A3 A3, B5 B5*	Accepted

We might wonder what function these HLA genes normally serve—obviously, tissue transplants have not been a normal aspect of human evolution! One possible answer is that some HLA alleles seem to confer resistance or susceptibility to certain specific diseases. Thus these alleles may have played some role in our evolutionary history, in response to some environmental selection pressure. This supposition is supported by the finding that certain ethnic groups and races show a preponderance of specific alleles in their populations. The study of such alleles is important both in deciphering the history of human evolution and in developing modern medical techniques—not only for transplants, but also for the general study of resistance to disease.

Another very interesting possibility emerging from current research is that the HLA genes may be intimately involved in cancer. Presumably, one of the normal functions of the HLA genes is to recognize cancer cells as a kind of "foreign" agent within the body. The development of cancer cells may be a fairly common event in healthy individuals, with these cells being recognized and destroyed before they cause noticeable damage. A cancerous tumour- may be the result of a failure of this detection-and-destruction system at some stage. The HLA genes also play an important role in the normal immune response.

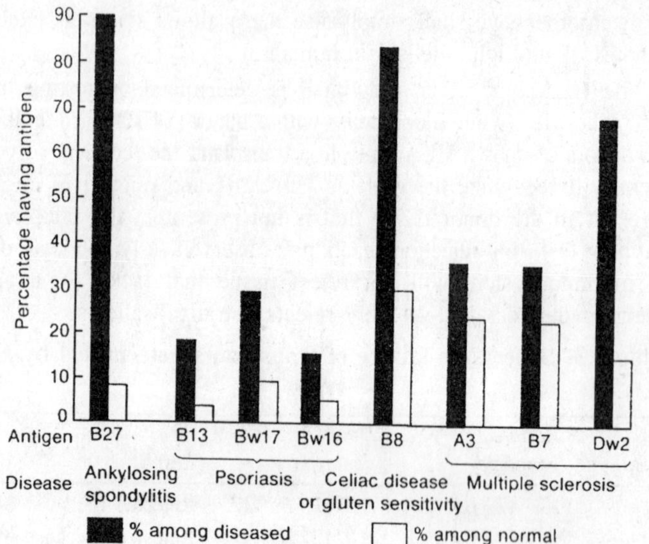

Fig. 2.3. Histograms showing effects of specific HLA alleles on predisposition to certain diseases in humans.

As a postscript to this section, it is worth noting that several different "kinds" or "species" of cell-surface anti-genes have been discovered. Each of these kinds is, of course, specified by one kind of gene. The different forms one kind of antigen can take are determined by the multiple alleles of the gene concerned.

Lethal Genes

Normal wild-type mice have coats with a rather dark overall pigmentation. In 1904, Lucien Cuenot studied mice having a lighter coat colour called yellow. Mating a yellow mouse to a normal mouse from a pure line, Cuenot observed a 1:1 ratio of yellow to normal mice in the progeny. This observation suggests that a single gene determines this aspect of coat colour and that an allele for yellow is dominant to an allele for normal colour. However, the situation became more confusing when he crossed yellow mice with one another. The result was always the same, no matter which yellow mice were used:

$$\text{yellow} \times \text{yellow} \begin{array}{l} \nearrow \text{2/3 yellow} \\ \searrow \text{1/3 normal colour} \end{array}$$

Two things are of interest in these results. First, the 2:1 ratio is a departure from Mendelian expectations. Second, Cuenot failed to detect (in any generation of breeding) a homozygous yellow mouse.

Cuenot suggested the following explanation for these results. A cross between two heterozygotes would be expected to yield a Mendelian genotype ratio of 1:2:1. If one of the homozygous classes died before birth, the live births would then show a 2:1 ratio of heterozygotes to the surviving homozygotes. In other words, the allele A^Y for yellow is dominant to the normal allele A with respect to its effect on colour, and it also acts as a *recessive lethal* allele with respect to a character we could call "survival." Thus a mouse with the homozygous genotype A^YA^Y dies before birth and is not observed among the progeny. All surviving yellow mice must be heterozygous A^YA, so a cross between yellow mice will always yield the following results:

$$A^YA \times A^YA \begin{array}{l} \nearrow \ 1/4\ AA \quad \text{normal} \\ \rightarrow \ 2/4\ A^YA \quad \text{yellow} \\ \searrow \ 1/4\ A^YA^Y \quad \text{die before birth} \end{array}$$

The expected Mendelian ratio of 1:2:1 would be observed among the zygotes, but it is altered to a 2:1 ratio of viable progeny because of the lethality of the A^YA^Y genotype. This hypothesis was confirmed by the removal of uteri from pregnant females of the yellow × yellow cross; one-fourth of the embryos were found to be dead.

The A^Y allele has effects on two characters: coat colour and survival. Such genes known to have more than one distinct phenotypic effect are called *pleiotropic* genes. It is entirely possible that both effects of the A^Y pleiotropic allele are the result of the same basic cause, which promotes yellowness of coat in a single dose and death in a double dose.

Lethal genes are quite common, even in humans. Some lethal effects occur in utero, others much later—in infancy, in childhood, or even in adulthood. Only rarely is a lethal gene associated with a distinguishable heterozygous phenotype, as in yellow mice. One other example you may be familiar with is the tail-less condition in Manx cats. The determining gene is homozygous lethal. Typically, the only detectable effect of a lethal gene is on survival of the individual, In some cases, a specific abnormality can be pin-pointed as the cause of death. For example, a recessive lethal gene in rats may have its pleiotropic effects traced to a basic problem in the nature of the rat's cartilage.

The lethality of an allele of a gene is often dependent on the environment in which the organism develops. Whereas certain alleles would be lethal in virtually any environment, others are viable in one

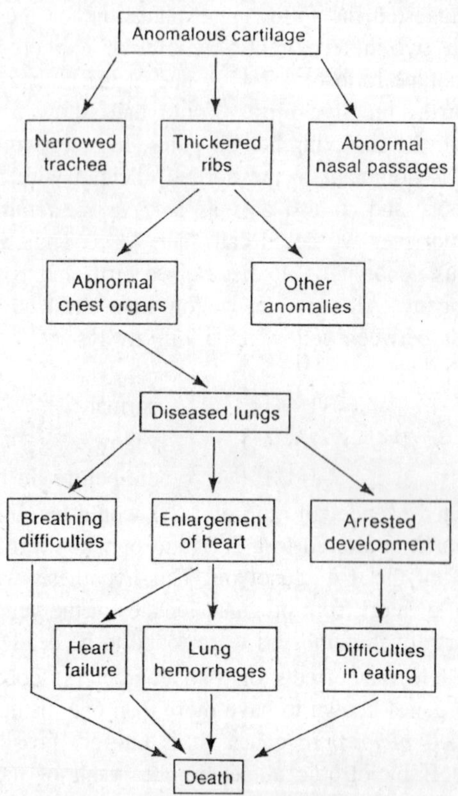

Fig. 2.4. Diagram showing how one specific lethal gene causes death in rats.

environment but lethal in another. For example, remember that many of the phenotypes favoured and selected by agricultural breeders would almost certainly be wiped out in nature in competition with the normal members of their population. Modern grain varieties provide good examples; it is only the careful nurturing by the farmer that has maintained such phenotypes for our benefit.

In practical genetics, we very commonly encounter situations where perfect expected Mendelian ratios are consistently skewed in one direction by one allele. For example, in the cross *Aa* × *aa,* we predict a progeny phenotypic ratio of 50 percent *A*– and 50 percent *aa,* but we might consistently observe a ratio such as 55%:45% or 60%:40%. In such a case, the *a* allele is said to be *subvital*—a kind of partial lethality, expressed in only some individuals. The partial lethality may range from 0 to 100 percent depending on the rest of the genome and on the environment.

SEVERAL GENES AFFECTING THE SAME CHARACTER

We saw earlier that, in a genetic dissection, the identification of a major gene affecting a character does not mean that this is the *only* gene affecting that character. An organism is a highly complex machine in which all functions interact to a greater or lesser degree. At the level of genetic determination, the genes likewise can be regarded as interacting. A gene does not act in isolation; its effects depend not only on its own functions but also on the functions of other genes (and on the environment). In many cases, the complex interactions of major genes are detectable through genetic analysis, and we now look at some examples. Typically, the mode of interaction is revealed by *modified Mendelian ratios*.

Coat Colour in Mammals

A character that has been extensively studied at the genetic level is coat colour in mammals. These studies have revealed a beautiful set of examples of the interplay between different genes in the determination of one character. The best studied mammal in this regard is the mouse, because of its small size and its short reproductive cycle. However, the genetic determination of coat colour in mice has direct parallels in other mammals, and we shall look at some of these as our discussion proceeds. There are at least five interacting major genes involved in determining the coat colour of mice: *A, B, C, D,* and S.

A gene

The wild-type allele *A* produces a phenotype called agouti. Agouti is an overall grayish colour with a "brindled" or "mousy" appearance. It is common in many mammals in nature. The effect is produced by a band of yellow on the hair shaft. In the nonagouti phenotype (determined by the allele *a*), the yellow band is absent, so the coat colour appears solid.

The lethal yellow A^Y allele is another form of this gene. Another form is a^t, which gives a "black-and-tan" effect involving a cream-coloured belly with dark pigmentation elsewhere. We shall not include these two alleles in the following discussion.

B gene

There are two major alleles of the *B* gene. The allele *B* gives the normal agouti colour in combination with *A*, but with *aa* it gives solid black. The genotype *A-bb* gives a colour called cinnamon ("mousy" brown), and *aa bb* gives solid brown.

The following sample cross illustrates the inheritance pattern of the *A* and *B* genes.

AA bb (cinnamon) × *aa BB* (black)
or *AA BB* (agouti) × *aa bb* (brown)
↓
F_1 all *Aa Bb* (agouti)
↓
F_2 9 *A–B–*(agouti)
3 *A–bb* (cinnamon)
3 *aa B* – (black)
1 *aa bb* (brown)

In horses, no *A* agouti gene seems to have survived the generations of breeding, although such a gene does exist in certain wild relatives of the horse. The colour we have called brown in mice is called chestnut in horses, and this phenotype also is recessive to black.

C gene

The wild-type allele *C* permits colour expression, and the allele *c* prevents colour expression. The *cc* constitution is said to be *epistatic* to the other colour genes. The word epistatic literally means "standing upon"; the *c* allele in homozygous condition "stands on" (blots out) the expression of other genes concerned with coat colour. The *cc* animals, lacking coat colour, are called albino. Albinos are common in many mammalian species, but albinos have also been occasionally reported among birds, snakes, and fish. Epistatic genes produce interesting modified ratios, as seen in the following sample cross (where we assume that both parents are *aa):*

BB cc (albino) × *bb CC* (brown)
or *BB CC* (black) × *bb cc* (albino)
↓
F_1 all *Bb Cc* (black)
↓
F_2 9 *B-C-*(black) 9
3 *bb C-*(brown) 3
3 *B-cc* (albino) } 4
1 *bb cc* (albino) }

A phenotypic ratio of 9:3 :4 is observed. This ratio is the signal for inferring gene interaction of the type called recessive epistasis. In some other organisms, dominant epistasis is observed. (What ratio is produced in dominant epistasis?)

We have already encountered the c^h (Himalayan) allele in rabbits. It exists also in other mammals, including mice (also called Himalayan) and cats (called Siamese).

It should be pointed out here that the term epistasis is often used in a different way (mainly in population genetics) to describe *any* kind of gene interaction.

***D* gene**

The *D* gene controls the intensity of pigment specified by the other coat-colour genes. The genotypes *DD* and *Dd* permit full expression of colour in mice, but *dd* "dilutes" the pigment to a milky appearance. Dilute agouti, dilute cinnamon, dilute brown, and dilute black coats all are possible. A gene of this nature is called a *modifier gene*. In the following sample cross, we assume that both parents are *aa CC*.

BBdd (dilute black) × *bbDD* (brown)
or *BB DD* (black) × *bbdd* (dilute brown)
↓
F_1 all *Bb Dd* (black)
↓
F_2 9 *B-D-*(black)
3 *B-dd* (dilute black)
3 *bb D-*(brown)
1 *bb dd* (dilute brown)

In horses, the *D* allele shows incomplete dominance.

***S* gene**

The *S* gene controls the presence or absence of spots. The genotype S-results in no spots, and *ss* produces a spotting pattern called piebald in both mice and horses. This pattern can be superimposed on any of the coat colours discussed earlier—with the exception of albino, of course.

By this time, the point of the discussion should be obvious. Normal coat appearance in wild mice is produced by a complex set of interacting genes determining pigment type, pigment distribution in the individual hairs, pigment distribution on the animal's body, and the presence or absence of pigment. Similar situations exist for any character in any organism.

Examples of Gene Interaction in Other Organisms

Some other kinds of gene interaction are best illustrated in other organisms. Peas provide a good example of one important situation. Two different, independently obtained pure lines of pea plants are both white-petaled. When these lines are crossed, all of the F_1 have purple flowers. The F_2 shows both purple and white plants in a ratio of 9:7. How can this result be explained? By now, you should immediately suspect that the 9:7 ratio is a modification of the Mendelian 9:3:3:1 ratio. The explanation is that two different genes in the pea have similar effects on petal colour. Let us represent the alleles of these genes by *A, a, B,* and *b.*

white strain 1 × white strain 2

AA bb × *aa BB*

↓

F_1 all *Aa Bb* (purple)

↓

F_2	9 *A-B-*(purple)	9
	3 *A-bb* (white)	7
	3 *aa B-*(white)	
	1 *aa bb* (white)	

Both gene pairs affect petal colour. Whiteness can be produced by a recessive allele of either gene pair, but purpleness is a phenotype produced by a combination of the dominant alleles of *both* gene pairs. This phenomenon is called *complementary gene action*, a term that satisfactorily describes how the two dominant alleles are uniting to produce a specific phenotype—in this example, purple pigment. (Note that we might have mistakenly inferred purpleness to be a specific phenotype of one gene pair if we had only one white strain available for our crosses.)

Another important kind of interaction is *suppression* of one gene by another. This interaction is illustrated by the inheritance of the production of a chemical called malvidin in the plant genus *Primula*. Malvidin production is determined by a single dominant gene *K*. However, the action of this dominant gene may be suppressed by a nonallelic dominant suppressor *D*. The following pedigree is informative:

Recessive suppression of both dominant and recessive genes also is known. The suppressor gene may have its own associated phenotype or (as in the malvidin example) have no known phenotypic effect other than the suppression.

KK dd (malvidin) × *kk DD* (no malvidin)

↓

F_1 all *Kk Dd* (no malvidin)

↓

F_2 9 K-D-(no malvidin)
3 *kk D*-(no malvidin) } 13
1 *kk dd* (no malvidin)
3 *K-dd* (malvidin) 3

Our final example of gene interaction introduces a concept that we shall develop in a later chapter. It concerns the genes that control fruit shape in the plant called shepherd's purse. Two different lines have fruits of different shapes: one is "round," the other "narrow." Are these two phenotypes determined by two alleles of a single gene? A cross between the two lines reveals an F_1 with round fruit; this result is consistent with the hypothesis of a single gene pair. However, the F_2 shows a 15:1 ratio of round to narrow. Again, this ratio immediately suggests a modification of the 9:3:3:1 Mendelian ratio, and it can be explained in terms of two *duplicate genes*. Apparently, round fruits are produced as a result of the presence of at least one dominant allele of either gene. The two genes appear to be identical

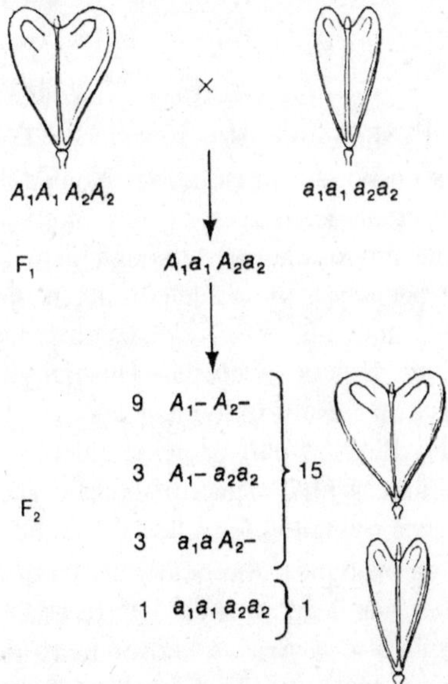

Fig. 2.5. Inheritance pattern of duplicate genes controlling fruit shape in shepherd's purse.

in function. (Contrast this 15:1 ratio with the 9:7 ratio obtained from complementary genes, where *both* dominant genes are necessary to produce a specific phenotype.)

Gene interactions also can be detected in haploid organisms. We have already discussed a gene in the fungus *Neurospora* where one allele causes albino asexual spores, as opposed to the normal pinkish-orange colour produced by the wild-type allele of the same gene. A cross *al* × *al*$^+$ gives 1/2 *al* and 1/2 *al*$^+$ progeny. Another interesting gene, *ylo,* gives yellow asexual spores, and the cross *ylo* × *ylo*$^+$ gives 1/2 *ylo* and 1/2 *ylo*$^+$ progeny. When an *al* culture is crossed with a *ylo* culture, the resulting progeny are 1/4 yellow, 1/4 wild-type, and 1/2 albino. How can we explain this result? The answer is a kind of epistasis: *al and ylo* are forms of separate genes, each of which can affect the normal production of pink pigment. The genotypes in the cross are the following:

Haploid parental culture	*al ylo*$^+$ (albino) × *al*$^+$ *ylo* (yellow) ↓	
Transient diploid	*al*$^+$/*al*, *ylo*$^+$/*ylo* ↓ meiosis	
Progeny (culture from sexual spores)	1/4 *al ylo* (albino) 1/4 *al ylo*$^+$ (albino)	1/4
	1/4 *al*$^+$ *ylo* (yellow)	1/4
	1/4 *al*$^+$ *ylo*$^+$ (normal)	1/4

Penetrance and Expressivity

Clearly, genes do not act in isolation. A gene does *not* determine a phenotype by acting alone; it does so only in conjunction with other genes and with the environment. Although geneticists do routinely ascribe a particular phenotype to an allele of a gene they have identified, we must remember that this is merely a convenient kind of jargon designed to facilitate genetic analysis. This jargon arises from the ability of geneticists to isolate individual components of a biological process and to study them as part of genetic dissection. Although this logical isolation is an essential aspect of genetics, the message of this chapter is that a gene cannot act by itself.

In the preceding examples, the genetic basis of the dependence of one gene on another has been worked out. In other situations, where the phenotype ascribed to a gene is known to be dependent on other factors but the precise nature of those factors has not been established, the terms *penetrance* or *expressivity* may be very useful in describing the situation. We have already encountered penetrance in the discussion of lethal alleles. *Penetrance* is defined as the percentage of individuals with a given genotype who exhibit the phenotype associated with that genotype. For example, an organism may be of genotype *aa* or A– but may not express the phenotype normally associated with its genotype—because of the presence of modifiers, epistatic genes, or suppressors in the rest of the genome or because of a modifying effect of the environment. Penetrance can be used to describe such an effect when the exact cause is not known.

Expressivity, on the other hand, describes the degree or extent to which a given genotype is expressed phenotypically in an individual. Again, the lack of full expression may be due to the rest of the genome or to environmental factors. Obviously, both penetrance and expressivity variation are integral components of the concept of norm of reaction.

Human pedigree analysis and predictions in genetic counseling can often be thwarted by the phenomena of penetrance and expressivity. For example, if a disease-causing allele is not fully penetrant (as usually is the case), it is difficult to give a clean genetic bill of health to any individual involved as part of a disease pedigree—for example, individual R. On the other hand, pedigree analysis can sometimes identify individuals who do not express but almost certainly have a disease genotype—for example, individual Q. The exceptions to Mendelian analysis covered in this chapter and following chapters are

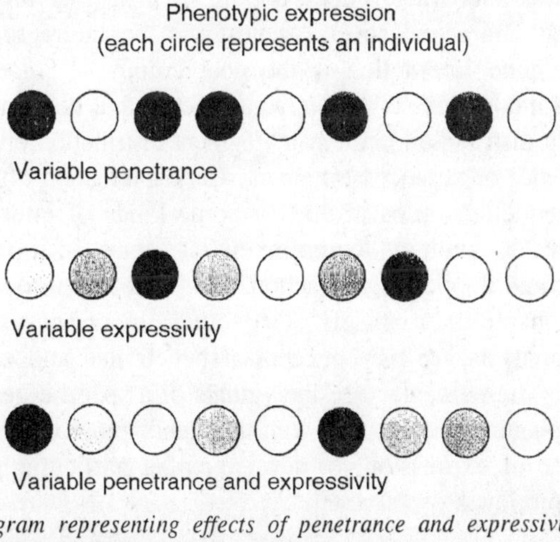

Fig. 2.6. Diagram representing effects of penetrance and expressivity through a hypothetical character "pigment intensity."

not the only ones encountered in routine genetic analysis; they are simply among the most common. Each exception—whether, for example, multiple allelism, incomplete dominance, epistasis, or variable expressivity—has its key recognition features at the experimental level. The analyst, must be constantly on the lockout for these and any other results that might indicate the uniqueness of any given situation. Such signals often lead to the discovery of new phenomena and to the opening up of new research areas. Although it has been shown that Mendel's laws apply to all eukaryotic organisms, these laws are only a base for understanding heredity. The real world of genes and chromosomes is much more complex. In addition to the full dominance that Mendel observed in his experiments, incomplete dominance and codominance may exist. In incomplete dominance, the phenotype of a heterozygote is intermediate between those of the homozygotes. In codominance, the heterozygote shows the phenotypes of both homozygotes.

In his experiments, Mendel reported genes with two forms. It was later discovered that in fact a gene may have more than two forms. This situation is known as multiple allelism. The members of an allelic series may exhibit any type of dominance relationship with the other members. The genes controlling the rejection of incompatible tissues in humans are an example of multiple allelism.

One gene may affect more than one character. Such genes are known as pleiotropic genes. An example is the A^Y allele mice, which affects both coat colour and survival. The identification of a major gene affecting a character does not mean that it is the only gene affecting that character: several genes may be interacting. A good example of gene interaction is the coat colour of mice, which is produced by a complex set of interacting genes that determine pigment type, pigment distribution in the hair, pigment distribution on the animal, and the presence or absence of pigment. Gene interaction often produces modified Mendelian ratios in the F_2. Some kinds of interaction have specific names, such as complementary gene action, epistasis, suppression, and duplicate gene action. Gene interaction occurs in both diploid and haploid organisms. Two other important extensions to Mendelian analysis are the concepts of penetrance and expressivity. Penetrance is the percentage of individuals of a specific genotype who express the phenotype associated with that genotype. Expressivity refers to the degree of expression, or severity, of a particular genotype at the phenotypic level.

3

Genetic Material

Earth originated about 5 billion years ago and life originated about 3 billion years ago and that too in water on this earth. All life that has appeared on earth is descended from organisms that were originally very simple. The physical and chemical evolution on earth that preceded the appearance of organisms (prebiotic evolution) was followed by an organismal evolution over the last 4 billion years which culminated in organisms of extraordinary complexity, such as ourselves. Note that present-day organisms—those in existence the past few thousand years—would occupy much less horizontal space in figure than the vertical line marking the interval at 0, for even if 1000 vertical lines could be drawn between 0 and -1, each thin line would represent 1000 thousand (that is, 1 million) years!

It has been reported in this latest thin line of time a great variety of organisms—plant and animal, microscopic to mammoth, incredibly diverse—making up some 2 million kinds, or species. These organisms range from viruses and bacteria through protozoans, sponges, corals and jellyfish, flat, round, and segmented worms, shellfish and starfish, spiders and insects, finned fishes, amphibians, reptiles, birds, mammals, algae and fungi, mosses, ferns, and seed plants in a bewildering tangle of forms of life. By collcting information about these various organisms and by organizing such information—that is, by developing the science of biology—we can group facts about organisms, and establish principles and generalities that apply to many species.

Such common threads among species are revealed by studying their structue, form, function and composition. All organisms have a common origin, and all of them require one or more *cells* and many cell products for their structure and function. Cells vary in size and complexity from the tiny and relatively simple cell of a bacterium, about 100 of which can fit across the dot of an *i*, to the giant and relatively complex yolk, which is the single cell of a chicken or an ostrich egg.

Features of a typical cell can be suggested by imagining a plastic bag containing a very porous sponge which is saturated with a thick vegetable soap. This sponge, in turn, surrounds a smaller plastic bag containing noodle soup. The outer plastic bag represents the *cell membrane*, or the outside limit of the cell. The inner plastic bag represents the *nuclear membrane*, the outside limit o the *nucleus*. The sponge represents the *endoplasmic reticulum*, which is a network of membranes (forming channels often interconnected) that also connect the other two membranes. The soups and membranes make up *protoplasm*, which is called *cytoplasm* outside the nucleus. The vegetables in the cytoplasm include various types of bodies (membrane-bound bodies are called *organelles*), such as *ribosomes*, *Mitochondria* and, in green cells, *chloroplasts*. The noodles of the nucleus represent the *chromosomes*. One or a few chromosomes are also present in each mitochondrion and chloroplast. Nucleus-containing cells are *eukaryotic* cells, and organisms composed of such cells are *eukaryotes*. The cells of bacteria and blue-green algae, on the other hand, can be likened to a single plastic bag, the size of a mitochondrion or smaller, containing soup with relatively few vegetables and only a noodle or two. Such cells consists largely of a small mass of protoplasm bounded by a cell membrane, containing ribosomes and one or a few chromosomes. Lacking a nucleus the cells are said to be *prokaryotic*; the cells also lack an endoplasmic reticulum, mitochondria, and chloroplasts. Although all *prokaryotes* are single-celled organisms, eukaryotes may also be composed of single cells, as are protozoans such as amebae or of as many as trillions of cells, as in each human being.

The common features of cell structure and form are accompanied by common features of cell function. The thousands of chemical reactions and physical changes that occur in protoplasm comprise the cell's *metabolism*. Metabolism occurs primarily at sites and surfaces provided by membranes, organelles, and large molecules (including chromosomes). The reactions that synthesize more energy-containing substances or protoplasm are *anabolic*; the reactions which degrade energy-providing substances or protoplasm are *catabolic*. By drawing raw materials and energy from the environment and processing them through metabolism, cells are able to maintain themselves, grow, and divide to produce more cells. All of today's organisms are characterized functionally, therefore, by their capacity for two basic functions; (1) *self-maintenance*, which involves growth, replacement, and/or repair of parts of an organism; and (2) *self-reproduction*, which involves making more of the same kind of organisms.

Requirements for the Genetic Material

The genetic material is of central importance to cell function and therefore must fulfil a number of basic requirements:

1. It must contain the information for cell structure, function, and reproduction in a stable form. This information is encoded in the sequence of basic building blocks of the genetic material.
2. It must be possible to replicate the genetic material accurately such that the same genetic information is present in descendant cells and in successive generations.
3. The information coded in the genetic material must be able to be decoded to produce the molecules essential for the structure and function of cells.
4. The genetic material must be capable of (infrequent) variation. Specifically, mutation and recombination of the genetic material are the foundations for the evolutionary process.

The nucleic acids, *deoxyribonucleic acid* (DNA) and ribonucleic acid (RNA), meet all these requirements.

Presence of Genetic Material

Organisms have unique requirements at the structural level, at the functional level and at the macromolecular level. Each organism must possess a facility or factor that (1) persists during the entire existence of the organism, (2) is repeated in each of its progeny, (3) is different in different organisms, and (4) contains the information needed for synthesizing characteristic proteins and nucleic acids. For purposes of scientific investigation, we assume that we are dealing with a material factor rather than a spiritual one. Since this material factor must contain the instructions for the creation, or genesis, of an organism, we can call it *genetic material*. All the different organisms that have appeared on earth are likely to have had a single ancestor, (1) there is only one basic tye of genetic material, and (2) the genetic material can be changed yet still be reproduced. The single most important feature of all organism is, with this view, genetic material. This material is characterized by being preserved, replicated, capable of replicating its modifications, and contains the information for unique protein and nucleic acid. The study of the properties and functions of genetic material thus comprises the core of the study of all organisms, that is, of the science of biology.

Some mature viruses, or *virions*, are composed only of nucleic acid and protein in combination. Clearly, their genetic material must be one or the other or some combination of these two types of macromolecule.

A Genetic View

The concept of the gene has been the focus of some hundred years of work to establish the basis of heredity. A gene is a sequence of DNA that carries the information representing a protein. Until very recently this would have been an adequate (if incomplete) bio-chemical description. The sequence of DNA could be identified as a continuous stretch of nucleotides, related to the protein sequence by a colinear read-out. But now it is clear that the sequence representing protein is not always continuous; it may be interrupted by sequences not concerned with specifying the protein. So genes may be in pieces that are put together during the process of gene expression. The large number or genes that make up the genome of any species are organizd into a comparatively small number of chromosomes. The genetic material of each chromosome consists of an extremely long stretch of DNA, containing many genes in a linear order. How many genes are present in total has been a puzzle for a long time.

Recently the view that each gene may reside by itself as a unique entity has been superseded by the realization that, in many cases, there may be clusters of related genes that constitute small families. The genome generally has been viewed as rather stable, subject to changes in overall constitution and organization only on an extra-ordinarily slow evolutionary time scale. This contrasts with recent evidence that in some instances there may be rearrangements that occur regularly; and there may be components of the genome that are relatively mobile. The concept of the gene has therefore undergone an evolution in which, although many of its traditional properties remain, exceptions have been found to show that none constitutes an absolute rule. Starting from the discovery of the gene as a fixed unit of inheritance, its properties were defined in terms of its residence at a definite position on the chromosome, this in turn leading to the view that the genetic material of the chromosome is a continuous length of DNA representing many genes.

A gene may exist in alternative forms that result in the expression of a different characteristic (such as red versus white flower colour). These forms are called *alleles*. The law of independent segregation states that these alleles do not affect each other when present in the same plant, but segregate unchanged by passing into different gametes when the next generation forms. In a true-breeding organism, a *homozygote*, both alleles are the same. But a mating between two parents each of which is homozygous for a different allele generates a hybrid or *heterozygote*. If one allele is *dominant* and the other is *recessive*, the organism will have the appearance or *phenotype* only of the dominant type (so the heterozygote is indistinguishable from the true-breeding dominant parent). But Mendel's first law recognizes that the genetic constitution or *genotype* of the hybrid comprises the presence of both alleles. This is revealed, when the hybrid is crossed with another hybrid to form the second generation. The critical point is that the alleles do not mix, but are physical entities whose interaction is at the level of expression.

Alleles may exhibit *incomplete* (*partial*) *dominance* or no dominance (sometimes known as *codominance*). In the latter case the heterozygote is distinguished from the homozygotes by properties that are intermediate between them. In the snap-dragon, for example, a cross between red and white generates hybrids with pink flowers. Although there is no dominance, the same rule is observed that the first hybrid (F_1) generation is uniform; and the same ratios are generated in a further hybrid cross, although three phenotypes can be distinguished instead of two.

The Independence of Different Genes

Mendel's second law which is the independent assortment summarizes of different genes. When a plant that is dominant for two different characters is crossed with a parent that is recessive for both, as before the F_1 consists of plants uniformly of the dominant type. But in the next hybrid cross, two classes of plants are seen. One consists of the two *parental types*. The other consists of new phenotypes, representing plants with the dominant feature of one parent and the recessive feature of the other. These are called *recombinant types*; and they occur in both possible (*reciprocal*) combinations. The ratios in which these occur can be explained by supposing that gamete formation involves an entirely random association of one of the alleles for the first character with one of the alleles for the other. All four possible types of gamete are formed in equal proportion, and associate at random in forming the zygotes of the next generation. Once again, the characteristic ratio of phenotypes conceals a greater variety of genotypes, which can be confirmed by the *backcross* to the recessive parent. Essentially the backcross provides a method for looking directly at the genotype of the organism that is being examined.

The law of independent assortment therefore says that the behaviour of any pair (or greater number) of genes can be predicted overall by the rules of mathematical combination. Thus the assortment of one gene does not influence the assortment of the other. Implicit in this concept is the view that assortment is a matter of statistical probability and not an exact result. The ratios of progeny types will approximate increasingly closely to the predicted numbers as a greater number of crosses are performed.

Chromosomes in Heredity

When the chromosomal theory of inheritance was simultaneously proposed by Sutton and by Boveri, it resolved the discussion that had been continuing for some time on the role of chromosomes. They had already been implicated in heredity, although in a somewhat hazy manner, in 1903 it was realized that their properties corresponded exactly with those ascribed to Mendel's particulates units of inheritance. The cell theory established in the middle of the nineteenth century proposed that all organisms are composed of cells, and that these can arise only from preexisting cells.

Early cytology showed that a "typical" cell consists of a dense nucleus separated by a membrane from the less-dense surrounding cytoplasm. Within the nucleus, the granular region of *chromatin* could be recognized by its reaction with certain stains. Not long after Mendel's work, it was found that the chromatin consists of a discrete number of thread-like particles, the *chromosomes*. Chromosomes can be visualized in most cells only during the process of cell division.

The two types of division in sexually reproducing organisms explain both the perpetuation of the genetic material and the process of inheritance as predicted by Mendel's laws. During growth of the organism, the *cell cycle* falls into two parts. The long period of *interphase* represents the time during which the cell engages in its synthetic activities and reproduces its components. Then the short period of *mitosis* is an interlude during which the actual process of division into two daughter cells is accomplished. The products of the series of mitotic divisions that generate the entire organism are called the *somatic cells*. At the end of mitosis, each daughter cell can be seen to start its life with two copies of each chromosome. These are called *homologous*. The total number of chromosomes in what is known as the *diploid set* can therefore be described as *2n*. The typical somatic cell exists in the diploid state. During interphase, a growing cell duplicates its chromosomal material. This is not evident at the time and becomes apparent only during the subsequent mitosis.

During mitosis, each chromosome appears to split longitudinally to generate two copies. These are called *sister chromatids*. At this point, the cell contains *4n* chromosomes, organized as *2n* pairs of sister chromatids. In other words, there are two (homologous) copies of each sister chromatid pair. The series of events, by which mitotic division is accomplished. Essentially the sister chromatids are pulled toward opposite poles of the cell, so that each daughter cell receives one member of each sister chromatid pair. Now these are chromosoems in their own right. The *4n* chromosomes present at the start of division have been divided into two sets of *2n* chromosomes. This process is repeated in the next cell cycle. Thus mitotic division ensures the constancy of the chromosomal complement in the somatic cells.

Meiosis generates cells that contain the *haploid* chromosome number, *n*. This involves two, successive divisions. Once again, the chromosomes have been duplicated previously, so that the cell enters division with *4n*. At the time of first division, the homologous pairs of sister chromatids *synapse* or *pair* to form *bivalents*. Each bivalent contains all four of the cell's copies of one homologue. The first division causes each bivalent to segregate into its two component sister chromatid pairs. This generates two sets of *2n* chromosomes, each set consisting of *n* sister chromatid pairs.

Now the second meiotic division follows, in which both of the sets of *2n* divide again. This division resembles the mitotic division, since one member of each sister chromatid pair segregates to a different daughter cell. The overall result of meiosis is to divide the starting number of *4n* chromosomes into four haploid cells. These may then give rise to mature eggs or sperm. In forming these gametes, homologous of paternal and maternal origin are separated, so that each gametes gains only one of the two homologues of its parent.

A critical feature in relating this process to the predictions of Mendel's laws was the realization that nonhomologous chromosomes undergo segregation independently, so that either member of one homologous pair enters the gamete at random with either member of a different homologous pair.

Genes occur in allelic pairs, one member of each pair having been contributed by each parent; the diploid set of chromosomes results from the contribution of a haploid set by each parent. The assortment of nonallelic genes into gametes should be independent of origin, nonhomologous chromosomes undergo independent segregation. The critical provision is that each gamete obtains a complete haploid set, and this is fulfilled whether viewed in terms of factors or chromosomes.

Arrangement of Genes

Linear

The proof that genes reside on chromosomes required a demonstration that a particular gene is always present on a particular chromosome. This was provided by the properties displayed by a variant of the fruit fly *Drosophila melanogaster* obtained by Morgan in 1910. This white-eyed male appeared spontaneously in a line of flies of the usual red eye colour.

The red eye colour is the *wild type* of character. The white eye is the *mutant* phenotype. The event responsible for generating the mutant phenotype is a genetic change called a *mutation*. Most mutations prevent the gene in which they occur from functioning properly or at all; and so the study of mutation is for the most part the study of inactive alleles. Sometimes it is possible for a mutation to be reversed by a genetic change that restores the original state of the genetic material. This is called a *reversion* to the wild type. The white mutation could be located on a particular chromosome because of its association with sexual type.

In many sexually reproducing organisms there is an exception to the rule that chromosomes occur in homologous pairs whose separation (disjunction) at meiosis produces identical haploid sets. Male and female sets may differ visibly in chromosome constitution, the most common form of difference being the replacement of one member of a homologue pair with a different chromosome in one of the sexes. This pair is referred to as the *sex chromosomes*, and the remaining homologous pairs are called the *autosomes*.

The chromosome complements of the two sexes can be described as 2A + XX and 2A + XY, where the haploid set of autosomes is denoted A and the two sex chromosomes are X and Y. The sex with the complement of 2A + XX is called *homogametic*; if forms gametes only of the type A + X. The sex with the complement of 2A + XY is called *heterogametic*; it forms equal proportions of gametes of the types A + X and A + Y. The random union of gametes from one sex with gametes from the other sex perpetuates the equal sex ratio at zygote formation. In *Drosophila* the females are homogametic.

A critical prediction of Mendel's laws is that the results of a genetic cross should be the same regardless of orientation—that is, irrespective of which parent introduces which allele. But the reciprocal crosses with white eye in *Drosophila* give different results.

The cross of white male × red female gives the entirely red F_1 expected if red is dominant and white is recessive. But in the F_2 all the white-eye flies that reappear are males. In the reciprocal cross of red male × white female, all the F_1 males are white-eyed and all the females are red eyed. Crossing these gives an F_2 with equal proportions of white and red eyes in each sex. This pattern of inheritance exactly follows that of the sex chromosomes. If the alleles for red and white eyes are carried on the X chromosome, with no locus for eye colour present on the Y chromosome, the phenotype of a male will be determined by the single allele present on its X chromosome. This allele will be transmitted to all of its daughters and none of its sons. This is the typical pattern of *sex linkage*.

The separation of chromosomes seen at meiosis explains the independent assortment of genes that are carried on different chromosomes. But the number of genetic factors is much greater than the number of chromosomes. Each chromosome may bear as many as 1000 to 10000 genes. The quickening pace of genetics after the turn of the century was shown by the passage of only three years between the report of the first eye-colour mutant and Stretevant's study in 1913 of the pattern of inheritance of six sex-linked mutations. Morgan had shown in 1911 that each of these factors shows the same sex-linked pattern as white/red eye colour. By the same logic each must therefore be carried by the X chromosome.

Morgan proposed that the cause of genetic linkage is the simple mechanical result of the location of the factors in the chromosomes. He suggested that the production of genetic recombinant classes can be equated with the process of *crossing over* that is visible during meiosis. Early in meiosis, at a stage when there are four copies of each chromosome organized in a bivalent, pairwise exchanges of material occur between the closely associated (synapsed) homologue pairs. This exchange is called a *chiasma*.

In maize and in *Drosophila*, suitable *translocations* had occurred in which a part of one chromosome had broken off and become attached to another. This allows the translocation chromosome to be distinguished by its appearance from the normal chromosome. In suitable crosses, it is then possible to show that the formation of genetic recombinants occurs only when there has been a physical crossing-over between the appropriate chromosome regions. As the distance between the genes increases, the probability of crossing-over between them will increase. Thus if crossing-over is responsible for recombination, genes lying near each other will be tightly linked, and genetic linkage will decrease with physical distance apart. Reversing the argument, genetic linkage can be taken to be measure of physical distance.

The concept that genes on the same chromosome are linked was extended by Sturtevant with the proposal that the extent of recombination between them can be used as a *map distance* to measure their relative locations. This is expressed as the percent recombination; that is,

$$\text{Map distance} = \frac{\text{Number of recombinants} \times 100}{\text{Total number of progeny}}$$

Distance is given in terms of *map units*, defined by 1 map unit (or centiMorgan) equal 1% recombination.

Thus if two genes A and B are 10 units apart, and it is 5 units from B to a further gene C, the direct measure of distance between A and C will be close to 15. The genes can therefore be placed in *linear order*. A crucial point in the construction of maps is that the distance between genes does not depend on the *alleles* that are used, but only on the gene *loci*. The locus defines the *position* on the chromosome at which the gene for a particular trait resides; the various alternative forms of the gene—that is, the alleles used in mapping—all reside at the *same* location.

The genetic mapping is concerned with identifying the locations of gene loci, which are fixed and in a linear order. In a mapping experiment the same result is obtained irrespective of the particular combination of alleles. Sturtevant concluded that his results "form a new argument in favour of the chromosome view of inheritance, since they strongly indicate that the factors investigated are arranged in a linear series, at least mathematically."

This last qualification is interesting. Although a genetic map can be constructed that represents a chromosome as a linear array of genes, this does not prove that the genetic content of the chromosome is *physically* a continuous array of genes. Many other models were considered before it became clear that a chromosome contains a single thread of genetic material. Linkage is not displayed between all pairs of genes located on a single chromosome.

The maximum recombination that occurs in the 50% predicted by Mendel's second law. (Although there is a high probability that recombination will occur between two genes lying far apart on a chromosome, each individual recombination event involves only two of the four associated chromosomes, thus generating 50% crossover between the genetic factors).

The factors that are well separated may show no direct linkage, in spite of their presence on the same chromosome. However, each can be linked via intermediate factors, and so a genetic map can be extended beyond the limit of 50 map units directly measurable between any pair of genes.

In fact, because of the distortion of linkage measured directly between markers that are not closely linked, map distances usually are based on measurements between fairly close factors, and may be subject to corrections from the simple percent recombination. The *linkage group* includes all those genes that can be connected either directly or indirectly by linkage relationships. Those close together show direct linkage; those more than 50 map units apart in practice assort independently.

As linkage relationships are extended, each gene identified in an organism can be placed into linkage with a group of previously mapped genes. Thus the genes fall into a discrete number of linkage groups. Genes in one linkage group always show independent assortment with regard to genes located in other linkage groups. The number of linkage groups is the same as the number of chromosomes.

The relative lengths of the linkage groups are similar to the actual relative sizes of the chromosomes. The example of *Drosophila* (where it happens to be particularly easy to measure the chromosome lengths). Mendel's concept of the gene as a discrete particulate factor can therefore be extended into the concept that the chromosome constitutes a linear unit divided into many genes, whose physical arrangement may underlie their genetic behaviour.

Identification of DNA as the Genetic Material

Two classic experiments led to the identification of DNA as the genetic material and, in so doing, laid the foundation for molecular genetics. These experiments are described in the following section.

Transformation Experiments

The development of the current idea that DNA is the genetic material began with an observation in 1928 by Fred Griffith, who was studying the bacterium responsible for human pneumonia—i.e., *Streptococcus pneumoniae* or *Pneumonoccus*. The virulence of this bacterium was known to be dependent on a surrounding polysaccharide capsule that protects it from the dense systems of the body. This capsule also causes the bacterium to produce smooth-edged (S) colonies on an agar surface. It was known that mice were normally killed by S bacteria. Griffith then isolated a rough-edged (R) colony mutant, which proved to be both nonencapsulated and nonlethal. He subsequently made a significant observation—namely, whereas both R and heat-killed S were nonlethal, a mixture of live R and heat-killed S was lethal. Furthermore, the bacteria isolated from a mouse that had died from such a mixed infection were only S—i.e., the live R had somehow been replaced by or *transformed* to S bacteria. Several years later it was shown that the mouse itself was not needed to mediate this

(a) S strain kills mouse

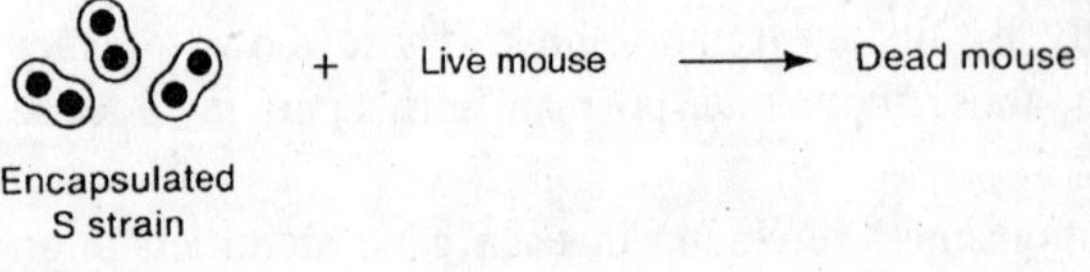

(b) R strain does not kill mouse

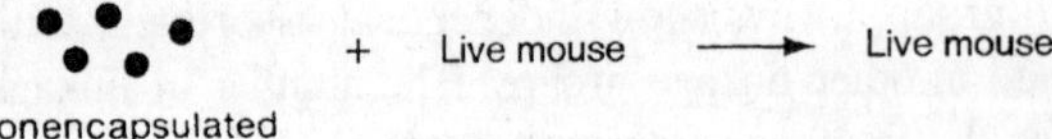

(c) Heat-killed S strain does not kill mouse

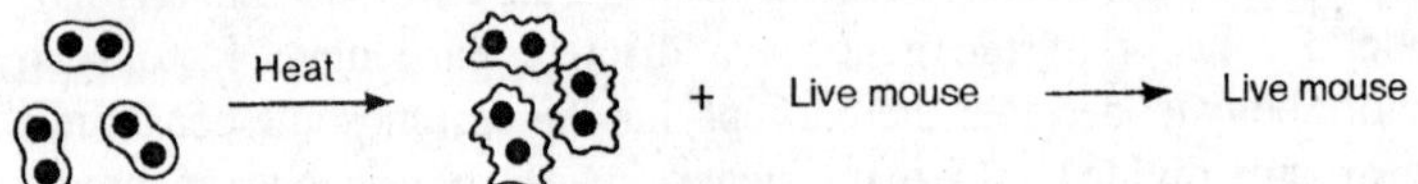

(d) R strain plus heat-killed S strain, both of which are separately nonlethal, kill mouse

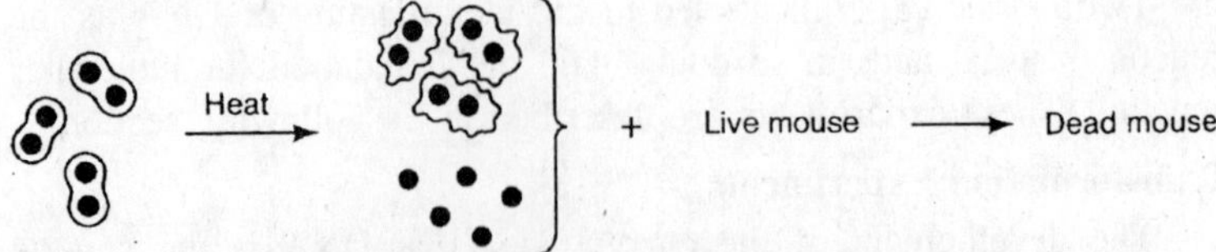

Fig. 3.1. The Griffith experiment showing conversion of a nonlethal bacterial strain to a lethal form by a cell extract.

transformation because when a mixture of R and heat-killed S was grown in a culture fluid, living S cells were produced. A possible explanation for this surprising phenomenon was that the R cells restored the viability of the dead S cells; but this idea was eliminated by the observation that living S cells grew even when the heat-killed S culture in the mixture was replaced by a cell extract prepared from broken S cells, which had been freed from both intact cells and the capsular polysaccharide by centrifugation. Hence, it was concluded that the cell extract contained a *transforming principle*, the nature of which was unknown.

15 years later when Oswald Avery, Colin MacLeod, and Maclyn McCarty partially purified the transforming principle from the cell extract and demonstrated that it was DNA. These workers modified known schemes for isolating DNA and prepared samples of DNA from S bacteria. They added this DNA to a live R bacterial culture; after a period of time they placed a sample of the S-containing R bacterial culture on an agar surface and allowed it to grow to form colonies. Some of the colonies (about 1 in 10^4) that grew were S type.

To show that this was a permanent genetic change, they dispersed many of the newly formed S colonies and placed them on a second agar surface. The resulting colonies were again S type. If an R colony

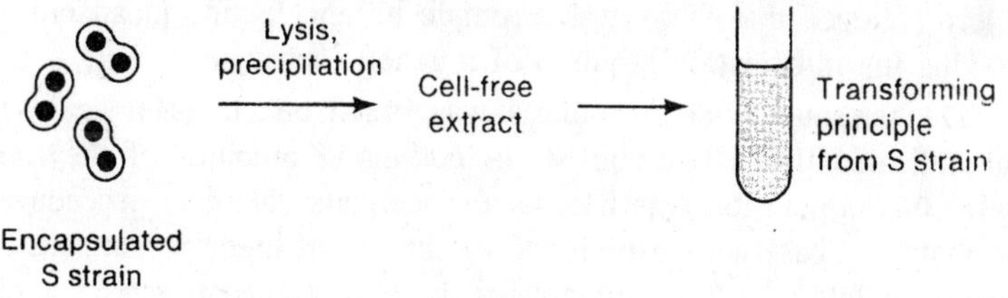

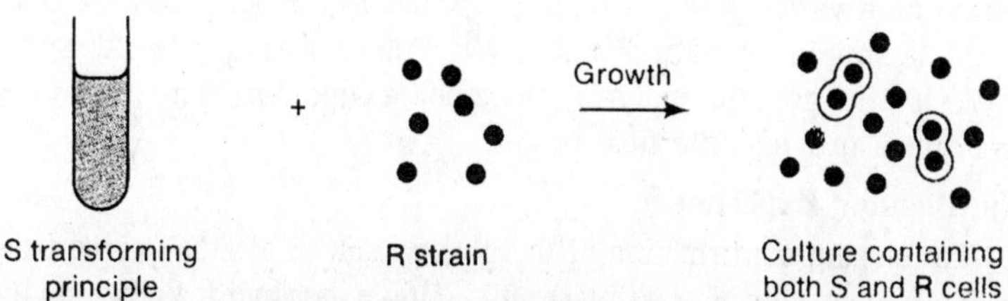

Fig. 3.2. The transformation experiment.

arising from the original mixture was dispersed, only R bacteria grew in subsequent generations. Hence the R colonies retained the R character, whereas the transformed S colonies bred true as S. Because S and R colonies differed by a polysaccharide coat around each S bacterium, the ability of purified polysaccharide to transform was also tested, but no transformation was observed. Since the procedures for isolating DNA then in use produced DNA containing many impurities, it was necessary to provide evidence that the transformation was actually caused by the DNA alone.

This evidence was provided by the following four procedures.

1. Chemical analysis showed that the major component was a deoxyribose-containing nucleic acid.
2. Physical measurements showed that the sample contained a highly viscous substance having the properties of DNA.
3. Experiments demonstrated that transforming activity is not lost by reaction with either (a) purified proteolytic (protein-hydrolyzing) enzymes--trypsin, chymotrypsin, or a mixture of both--or (b) ribonuclease (an enzyme that depolymerizes RNA).
4. It was demonstrated that treatment with materials known to contain DNA-depolymerizing activity (DNase) inactivated the trans-forming principle.

The transformation experiment was not accepted by the scientific community as proof that DNA is the genetic material, because it was widely believed that DNA was a simple tetranucleotide incapable of carrying the information required of a genetic substance.

The tetranucleotide hypothesis was based on chemical analyses that indicated that DNA consists of equimolar amounts of the four bases; this conclusion was based on inadequate chemical procedures for analyzing base composition and on the use of higher organisms as sources of DNA; in these organisms the base composition is not far from being equimolar. However, as techniques improved and a greater range of organisms was examined, it was found that the base composition varies widely from one species to the next and that DNA is not a simple small molecule. With these results, the idea that DNA is the genetic material became acceptable. The following experiment provided the final proof.

The Blendor Experiment

An elegant confirmation of the genetic nature of DNA came from an experiment with *E. coli* phage T2. This experiment, known as the *blendor experiment* because a kitchen blendor was used as a major piece of apparatus, was performed by Alfred Hershey and Martha Chase, who demonstrated that the DNA injected by a phage particle into a bacterium contains all of the information required to synthesize progeny phage particles. A single particle of phage T2 consists of DNA (now known to be a single molecule) encased in a protein shell.

The DNA is the only phosphorus-containing substance in the phage particle; the proteins of the shell, which contain the amino acids methionine and cysteine, have the only sulfur atoms. Thus, by growing phage in a nutrient medium in which radioactive phosphate ($^{32}PO_4^{3-}$) is the sole source of phosphorus, phage containing radioactive DNA can be prepared. If instead the growth medium contains radioactive sulfur as $^{35}SO_4^{3-}$, phage containing radioactive proteins are obtained. If these two kinds of labeled phage are used in an infection of a bacterial host, the phage DNA and the protein molecules can always be located by their radioactivity. Hershey and Chase used these phage to show that ^{32}P but not ^{35}S is injected into the bacterium. Each phage T2 particle has a long tail by which it attaches to sensitive bacteria.

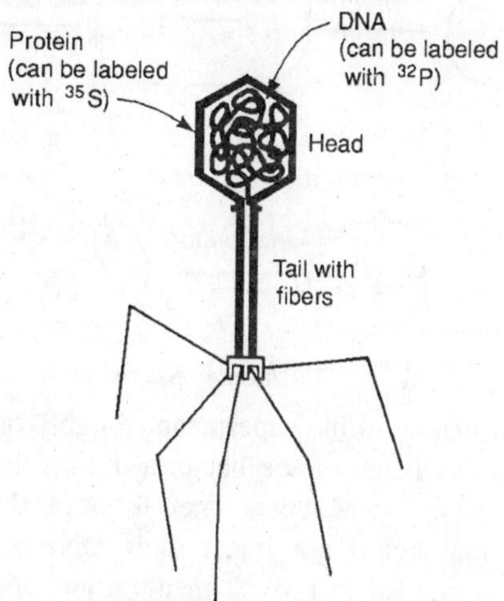

Fig. 3.3. Diagram of T2 phage.

Hershey and Chase showed that an attached phage can be torn from a bacterial cell wall by violent agitation of the infected cells in a kitchen blendor. Thus, it was possible to separate an adsorbed phage from a bacterium and determine the component(s) of the phage that could not be shaken free by agitation—presumably, those components had been injected into the bacterium.

In the first experiment ^{35}S-labeled phage particles were adsorbed to bacteria for a few minutes. The bacteria were separated from unadsorbed phage and phage fragments by centrifuging the mixture and collecting the sediment (the pellet), which consisted of the phage-bacterium complexes. These complexes were resuspended in liquid and blended. The suspension was again centrifuged, and the pellet, which now consisted almost entirely of bacteria, and the supernatant were collected. It was found that 80 percent of the ^{35}S label was in the supernatant and 20 percent was in the pellet. The 20 percent of the ^{35}S that remains associated with the bacteria was shown some years later to consist mostly of phage tail fragments that adhered too tightly to the bacterial surface to be removed by the blending. A very different result was observed when the phage population was labeled with ^{32}P. In this case 70 percent of the ^{32}P remained associated with the bacteria in the pellet after blending and only 30 percent was in the supernatant.

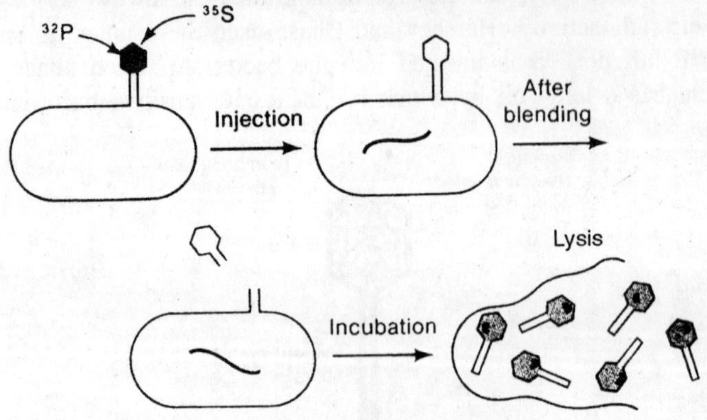

Fig. 3.4. The blendor experiment.

Of the radioactivity in the supernatant roughly one-third could be accounted for by breakage of the bacteria during the blending. (The remainder was shown some years later to be a result of defective phage particles that could not inject their DNA.) when the pellet material was resuspended in growth medium and reincubated, it was found to be capable of phage production. Thus, the ability of a bacterium to synthesize progeny phage is associated with transfer of ^{32}P, and hence of DNA, from parental phage to the bacteria.

Transfer Experiments

Another series of experiments, known as *transfer experiments*, supported the interpretation that genetic material contains ^{32}P but not

^{35}S. In these experiments progeny phage were isolated, after blending, from cells that had been infected with either ^{35}S- or ^{35}P-containing phage and the progeny were then assayed for radioactivity, the idea being that some parental genetic material should be found in the progeny. It was found that no ^{35}S but about half of the injected ^{32}P was transferred to the progeny. This result indicated that though ^{35}S might be residually associated with the phage-infected bacteria, it was not part of the phage genetic material. The interpretation (now known to be correct) of the transfer of only half of the ^{32}P was that progeny DNA is selected at random for packaging into protein coats and that all progeny DNA is not successfully packaged.

Properties of a Genetic Material

To comprise the genes, DNA must carry the information to control the synthesis of the enzymes and proteins within a cell or organism, self-replicate with high fidelity yet show a low level of mutation, and be located in the chromosomes.

Control of the Enzymes

The growth, development, and functioning of a cell are controlled by the proteins within it, primarily its enzymes. Thus the nature of a cell's phenotype is controlled by the protein synthesis within that cell. The genetic material must determine the presence and effective amounts of the enzymes in a cell. For example, given inorganic salts and glucose, an *E. coli* cell can synthesize, through its enzyme-controlled biochemical pathways, all of the compounds it needs for growth, survival, and reproduction.

An enzyme is a protein that acts as a catalyst of a specific metabolic process without itself being markedly altered by the reaction. Most reactions that are catalyzed by enzymes could occur anyway, but only under conditions too extreme for them to take place within living systems. For example, many oxidations occur naturally at high temperatures. Enzymes allow these reactions to occur within the cell by lowering what is called the *activation energy* (ΔG) of a particular reaction. Most metabolic processes, such as the biosynthesis or degradation of molecules, occur in pathways in which an enzyme facilitates each step in the pathway.

Each reaction product in the pathway is altered by an enzyme that converts it to the next product. The enzyme threonine dehydratase, for example, converts threonine into α-ketobutyric acid. Enzymes are composed of folded polymers of amino acids; the sequence of amino acids determines the final structure of an enzyme. The genetic material determines the sequence of the amino acids.

The three-dimensional structure of enzymes permits them to perform their function. An enzyme combines with its substrate or substrates (the molecules it works on) at a part of the enzyme called the *active site*. The substrates "fit" into the active site, which has a shape that allows only the specific substrates to enter. This view of the way an enzyme interacts with its substrates is called the *lock-and-key model* of enzyme functioning. When the substrates are in their proper position in the active site of the enzyme, the particular reaction that the enzyme catalyzes takes place. The reaction products then separate from the enzyme and leave it free to repeat the process.

Enzymes can work at phenomenal speeds. Some can catalyze as many as a million reactions per minute. Not all of the cell's proteins function as catalysts. Some are structural proteins, such as keratin, the main component of hair. Other proteins are regulatory—they control the rate of production of other enzymes. Still others are involved in different functions; albumins, for example, help regulate the osmotic pressure of blood.

Replication

The genetic material must replicate itself precisely so that every daughter cell receives an exact copy. Some *mutability*, or the ability to change, is also required because we know that the genetic material has changed, or evolved, in the history of life on earth. In their 1953 paper, Watson and Crick had already worked out the replication process based on the structure of DNA. Mutability is also a natural derivative of this process.

Location

It has been known since the turn of the century that genes, the discrete functional units of genetic material, are located in chromosomes within the nuclei of eukaryotic cells: The behaviour of chromosomes during the cellular division stages of mitosis and meiosis mimics the behaviour of genes. Thus the genetic material in eukaryotes must be a part of the chromosomes.

For a long time, proteins were considered the most probable cell components to be genetic material because they have the necessary molecular complexity. Amino acids can be combined in an almost unlimited variety, creating thousands and thousands of different proteins. The first proof that the genetic material is deoxyribonucleic acid (DNA) was provided in 1944 by Oswald Avery and his colleagues. The Watson and Crick model in 1953 ended a period when DNA was thought to be the genetic material, but its structure was unknown.

Storage and Transmission of Genetic Information by DNA

The information possessed and conveyed by the DNA of a cell is of several types:

1. The sequence of amino acids in every protein synthesized by the cell.
2. A start and a stop signal for the synthesis of each protein.
3. A set of signals that determine whether a particular protein is to be made and how many molecules are to be made per unit time.

This information is contained in the sequence of DNA bases. The amino acid sequence and the start and stop signals are not obtained directly from the DNA base sequence but via RNA intermediates. That is, DNA serves as a template for synthesis of specific RNA molecules called messenger RNA. The base sequence of the mRNA is complementary to that of one of the DNA strands and it is form the mRNA sequence that the amino acid sequence is translated. In particular, the base sequence is "read" in order in groups of three bases called *triplets* or *codons* and each group corresponds to a particular amino acid or to a start or stop codon.

The two-stage process has the advantage that the DNA molecule neither has to be used very often nor has to enter the protein-synthetic mechanism. Since protein synthesis occurs continually, one can appreciate why DNA evolved with bases having groups capable of forming hydrogen bonds. First, by specific patterns of hydrogen-bonding—that is, base-pairing—the base sequence of DNA is easily transcribed into an RNA molecule by means of an enzymatic system that adds a base to the polymerizing end of an RNA molecule only if that base can hydrogen-bond with the base being copied. The use of hydrogen-bonding enables the cell to use less genetic material to hold information than if van der Waals forces had been selected in the course of evolution; this is because van der Waals forces are so weak that the error frequency in transmitting information would be much greater than with hydrogen bonds unless more bases (or other molecules) were used for complementary binding.

Transmission from Parent to Progeny

When a cell divides, each daughter cell must contain identical genetic information; that is, each DNA molecule must become two identical molecules, each carrying the information that was contained in the parent molecule. This duplication process is called *replication*. Once again, the value of bases capable of hydrogen-bonding is apparent. That is, to make a replica, the replication system need only require that the base being added to the growing end of a new chain be capable of hydrogen-bonding to the base being copied.

Chemical Stability of DNA

In long-lived organisms a single DNA molecule may have to last one hundred years or more. Furthermore, the information contained in the molecule is passed on to successive generations for millions of years with only small changes. Thus, DNA molecules must have great stability. The sugar-phosphate backbone of DNA is extremely stable. The C—C bonds in the sugar are resistant to chemical attack under all conditions other than strong acid at very high temperatures.

The phosphodiester bond is a little less stable and can be hydrolyzed at room temperature at pH 2, but this is not a physiological condition. In considering the stability of the phosphodiester bond, one can see immediately why 2′-deoxyribose rather than ribose is the constituent sugar in DNA. The phosphodiester bond in RNA is rapidly broken in alkali. The chemical mechanism for this breakage requires the OH group on the 2′ carbon of the sugar. In DNA, because deoxyribose is used, there is no 2′-OH group, so the molecule is exceedingly resistant to alkaline hydrolysis. The N-glycosylic bond is also very stable, though it would not be so if the bases were not rings. An alteration in the chemical structure of a base definitely means loss of genetic information.

In a cell there are certainly a large number of chemical compounds that can attack a free base. In considering this problem one can see immediately the value of the double helix. One aspect of this is that the molecule isredundant in the sense that identical information is contained in both strands; that is, the base sequence of one strand is complementary to that of the other strand. In fact, there exist in cells elegant repair systems that can remove an altered base and then by reading the sequence on the complementary strand, can replace the correct base. The more important aspect of this double-helical structure is that the nature of the bases and the duplex structure of DNA provides extreme protection against chemical attack.

The bases are hydrophobic rings having charged groups that contain the genetic information. It is therefore the charged groups that need protection. The hydrophobic nature of the base causes the bases to stack so extensively that water is almost completely excluded from the stacked array. This has the effect that water-soluble compounds are often unable to come into close contact with the "dry" stack of bases. The bases themselves, with the exception of cytosine, are very stable. At a very low rate, however, cytosine is deaminated to form uracil.

$$\text{Cytosine} + H_2O \longrightarrow \text{Uracil} + NH_3$$

Deamination is a disastrous change because the deamination product, uracil, pairs with adenine rather than with guanine. This has two effects: (1) an incorrect base will appear in mRNA and (2) an adenine instead of a guanine will occur in newly replicated DNA strands. There exists an intracellular system, for removing uracil from DNA and replacing it with thymine. The necessity for eliminating a uracil formed by deamination explains why DNA utilizes thymine and not uracil as the base complementary to adenine. If uracil were the normal DNA base, there would be no way to distinguish a correct uracil from an incorrect uracil produced by deamination of cytosine. By using thymine, the cell follows the rule: Always remove a uracil from DNA because it is unwanted. It is not obvious why RNA uses

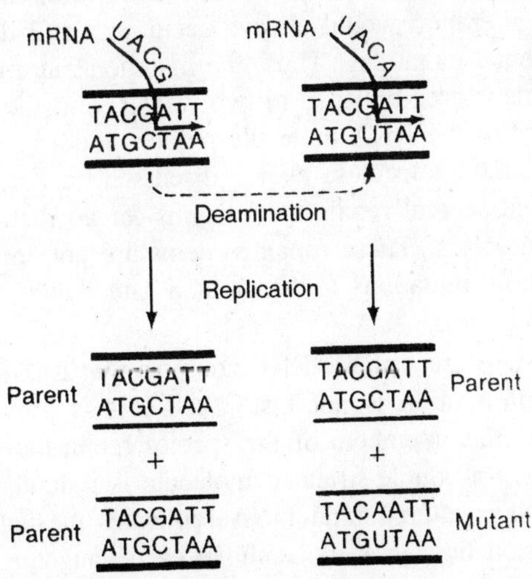

Fig. 3.5. The effect of deamination of cytosine to form uracil in the base sequences of mRNA and of two daughter DNA molecules.

uracil and not thymine nor why DNA evolved with cytosine rather than with some base that would not deaminate. This may have been an evolutionary accident. It is possible, though, that the original DNA and RNA molecules both contained cytosine and uracil because these were the only pyrimidines available in the primordial sea. Cells then gained the ability to methylate uracil to form thymine because this provides a cell with a criterion for eliminating the result of a cytosine→uracil conversion. It is significant that the final step in the synthesis of thymidylic acid is the methylation of deoxyuridylic acid. The thymidylic acid is then converted to the triphosphate needed for incorporation into DNA.

Mutation

The process by which a base sequence changes is called *mutation*. There are two main mechanisms of mutation:

1. A *chemical alteration* of the base that gives it new hydrogen-bonding properties and causes a new base to be present in a newly replicated daughter molecule.
2. A *replication error* by which an incorrect base or an extra base is accidentally inserted in the daughter molecules.

On the average, mutational changes are deleterious and lead to cell death. Therefore, it is important that not too many mutations occur in a single DNA molecule because otherwise the rare advantageous alteration would always occur in a cell destined to die by virtue of lethal mutations. Thus, the mutation rate must be low or controlled. This is accomplished in two ways. First, the hydrophobic, water-free core of the DNA molecule reduces the accessibility of the DNA to attacking molecules, as we discussed earlier. Second, the cell has evolved several repair mechanisms for correcting alterations and replication errors. These repair systems are not entirely efficient though and allow mutations to occur at a rate that is very low but useful in the long run.

The mutations are usually deleterious, so that it is important that the parental information is not lost. Such loss is prevented in two ways: (1) The other members of the species retain the parental base sequence and (2) a double-stranded molecule is redundant. Normally only one strand is altered, and DNA replicates in such a way that after cell division the DNA molecule in each daughter cell receives only one of the parental single strands. Thus, it is possible for one of the daughter DNA molecules to be normal and the other to be mutant; the mutant may not be able to survive but the cell with the parental base sequence will. If the mutant is better equipped to survive and multiply than the parent organisms or any other member of the species, then after a great many generations, Darwin's principle of survival of the fitest will lead to ultimate replacement of the parental genotype by the mutant phenotype in nature.

RNA as Genetic Material in Small Viruses

As more and more viruses were identified and studied, it became clear that many of them contain RNA and proteins, but no DNA. In all cases so far studied, it is clear that these "RNA viruses" store their genetic information in nucleic acids rather than in proteins just like all other organisms, although in these viruses the nucleic acid in RNA. One of the first experiments that established RNA as the genetic material in RNA viruses was the so-called reconstitution experiment of H. Fraenkel-Conrat and B. Singer, published in 1957. Fraenkel-Conrat and Singer's simple, but definitive, experiment was done with tobacco mosaic virus (TMV), a small virus composed of a single molecule of RNA encapsulated in a protein coat. Different strains of TMV can be identified on the basis of differences in the chemical composition of their protein coats.

By using the appropriate chemical treatments, one can separate the protein coats of TMV from the RNA. Moreover, this process is reversible; by mixing the proteins and the RNA under appropriate conditions, "reconstitution" wall occur, yielding complete, infective TMV particles. Raenkel-Contrat and Singer took two different strains of TMV, separated the RNAs from the protein coats, and reconstituted "mixed" viruses by mixing the proteins of one strain with the RNA of the second strain, and vice versa. When these mixed viruses were used to infect tobacco leaves, the progeny viruses produced were always found to be phenotypically and genotypically identical to the parent strain from which the RNA had been obtained. Thus, the genetic information of TMV is stored in RNA, not protein.

4

Chromosome

We have previously discussed the control of gene expression in prokaryotes and bacteriophages. Compared to eukaryotes, bacteriophages and prokaryotes are relatively simple. Of fundamental importance is that, in these lower forms, the operon model of induction and repression of transcription is a unifying theme for control of gene expression. Despite nuances such as catabolite repression and attenuator control, the operon model provides a relatively clear picture of how genes are turned on and off in phages and prokaryotes.This model does not exist for eukaryotes. In attempting to elucidate models for control of gene expression in eukaryotes, we must take one very important factor into account: the complexity of the structure of the eukaryotic chromosome. In this chapter, we cover the current understanding of how these very large structures are organized.

Eukaryotic Cell

Eukaryotes and prokaryotes are the two superkingdoms of organisms. The following comparisons, using *E. coli* as a general model for prokaryotes, show how much more complex eukaryotes are:

1. An *E. coli* chromosome contains approximately 4.2×10^6 base pairs of DNA. The haploid human genome contains nearly one thousand times as much DNA.
2. An *E. coli* cell has very little internal structure. Eukaryotes have a number of internal organelles and an extensive lipid membrane system, including the nuclear envelope itself.
3. Eukaryotic DNA is in the form of nucleoprotein, a DNA-histone protein complex. Although a few histonelike proteins have been

found in *E. coli*, its chromosomal DNA is not complexed with protein to anywhere near the same extent.

4. An *E. coli* cell is small (0.5 to 5.0 μm in length for bacteria). Eukaryotic cells are generally larger than prokaryotes (10 to 50 μm in length for animal tissue cells).
5. The messenger RNA of *E. coli* is translated while it is being transcribed. Eukaryotic messenger RNA is modified within the nucleus before it is transported out for translation in the cytoplasm.
6. Almost no messenger RNA isolated from eukaryotic cells, including the messenger RNA of animal viruses, has been found to be polycistronic (containing many genes). Most prokaryotic messenger RNAs are polycistronic.
7. Most *E. coli* genes are parts of inducible or repressible operons; there are almost no operons in eukaryotes.
8. *E. coli* exists as a simple, single cell. Although some prokaryotes do aggregate, sporulate, and show a few other limited forms of differentiation, they are primarily one-celled organisms. And, although some eukaryotes are single-celled (e.g., yeast), the essence of eukaryotes is differentiation. In human beings, a zygote gives rise to every other cell type in the body in a relatively predictable manner.

Eukaryotic Chromosome

DNA Arrangement

Evidence that the eukaryotic chromosome is *uninemic*—that is, contains one double helix of DNA—comes from several sources.The best data are provided by radioactive-labeling studies, first done by J. Taylor and his colleagues in 1957. If a eukaryote is allowed to undergo one DNA replication in the presence of tritiated (^{3}H-) thymidine, each of the daughter chromatids would be expected to contain a double helix with one unlabeled DNA template strand and one labeled strand of newly synthesized bases. This configuration is expected on the basis of semiconservative replication, with each chromatid containing one double helix. A second round of DNA replication, in the absence of ^{3}H-thymidine, should produce chromosomes in which one chromatid would have unlabeled DNA and one would have labeled DNA. As expected, one chromatid of every pair is labeled and one is not.

In another kind of experiment, R. Kavenoff, L. Klotz, and B. Zimm demonstrated that *Drosophila* nuclei contained pieces of DNA of the size predicted from their DNA content, based on the premise

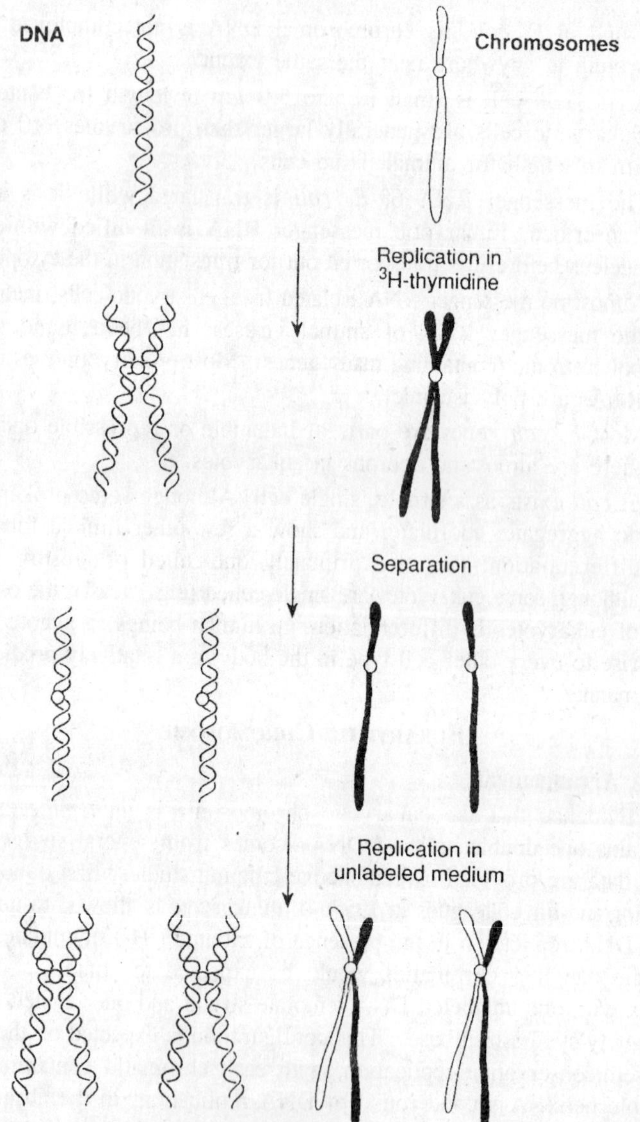

Fig. 4.1. Radioactive labeling of a uninemic eukaryotic chromosome following semiconservative replication.

that each chromosome contains one DNA molecule. They isolated the DNA and measured the size of the largest DNA molecules using the *viscoelastic* property of DNA, the rate at which stretched molecules relax. From other sources, primarily UV absorbance studies, it was estimated that the largest *Drosophila* chromosome had about 43×10^9 daltons of DNA. Results from the viscoelastic measurements indicated the presence of DNA molecules of between 38 and 44×10^9 daltons. Viscoelastic measurements of inversions, which changed the ratio of the arms but not the overall size of the chromosome, yielded similar results. However, a translocation that radically changed the size of the chromosome to 59×10^9 daltons resulted in an equivalent change in the viscoelastic estimates to between 52 and 64×10^9 daltons.

The conclusion from these studies is that the largest *Drosophila* chromosome, and by extension every eukaryotic chromosome, contains a single DNA molecule running from end to end, encompassing both arms. The viscoelastic values were corroborated by carefully isolating and measuring the lengths of long DNA molecules, an especially difficult task given DNA's propensity to break. The longest molecule that the investigators found was 1.2 cm long, equivalent to between 24 and 32 $\times 10^9$ daltons, close to the predicted size. Thus, the evidence is in complete concordance with the simple uninemic model of eukaryotic chromosomal structure.

Nucleoprotein Composition

Nucleosome structure

Since each eukaryotic chromosome consists of a single, relatively long piece of duplex DNA, the average diploid cell contains many of these long pieces of DNA. For chromosomes to be properly distributed to each daughter cell during mitosis and meiosis, they must be condensed into structures that are more easily managed. Wrapping the DNA around "spools" of protein constitutes the first step in a series of coiling and folding processes that eventually result in the fully compacted chromosome we see at metaphase.

Interphase nuclei can be disrupted by placing them in a hypotonic liquid such as water. When this happens, chromatin material is released. When this material is observed under the electron microscope, small particles called *nucleosomes* can be seen. These are the spools that the DNA is wrapped around. They are made of *histone* proteins and associated DNA. The histones, a group of arginine- and lysine-rich basic proteins, have been well characterized. They are especially well suited to bind to the negatively charged DNA.

When chromatin is treated with micrococcal nuclease, individual nucleosomes can be isolated, indicating that the DNA between nucleosomes is accessible to digestion. The results of these studies

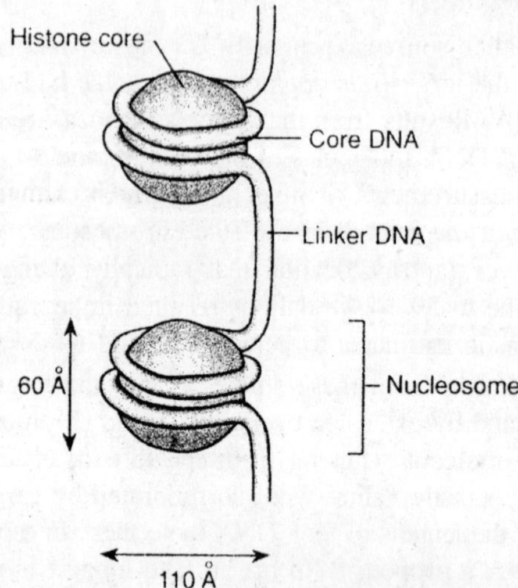

Fig. 4.2. The eukaryotic chromosome is associated with histone proteins to form nucleosomes.

indicate that a length of 168 base pairs (bp) of DNA, the core DNA, is intimately associated with the nucleosome, and another 50 to 75 base pairs, depending on species, connects the nucleosomes (linker DNA). When the quantities of the various histones were measured, there were two each of histones H2A, H2B, H3, and H4 per nucleosome and only one molecule of histone H1. Reconstitution and degradation studies have indicated that histone H1 is not a necessary component in the formation of nucleosomes. We believe that histone H1 is associated with the linker DNA as it enters and emerges from the nucleosome, although its exact position is not known with certainty. Histone H1 may be more off center and internally located than illustrated. The term *chromatosome* has been suggested for the core nucleosome plus the H1 protein, a unit that includes approximately 168 base pairs of DNA. Nucleosomes, then, are a first-order packaging of DNA; they reduce its length and undoubtedly make the coiling and contraction required during mitosis and meiosis more efficient.

When DNA is replicated, twice as many nucleosomes are needed since one double helix becomes two. Recent studies indicate that a parental nucleosome is partly disassembled during DNA replication and reassembled on one or the other daughter strand, apparently randomly. The other DNA strand has a new nucleosome constructed of histones from the cellular pool with the help of proteins called *chromatin assembly factors*; at least three of these factors are known.

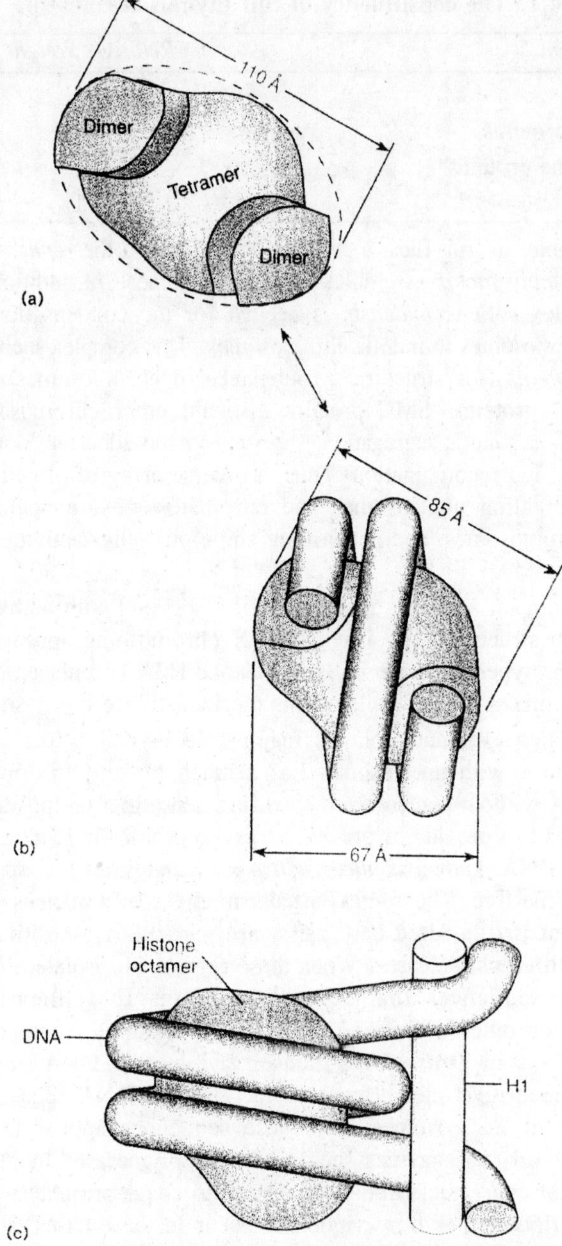

Fig. 4.3. Nucleosome structure.

Table 4.1. The constituency of calf thymus chromatin

Constituent	*Relative Weight*
DNA	100
Histone proteins	114
Nonhistone proteins	33
RNA	1

For example, in fruit flies, a protein complex called the *replication-coupling assembly factor* assembles new nucleosomes. In addition, a protein complex called *condensin* is needed for the condensation of interphasechromosomes to mitotic chromosomes. This complex includes two *SMC proteins* (for structural maintenance of chromosomes) and two non-SMC proteins. SMC proteins also aid other chromosomal activities, such as mitotic segregation, sister-chromatid adhesion, dosage compensation, and recombination. Thus, a diverse array of proteins is involved in creating nucleosomes and chromatosomes, condensing interphase chromosomes, and performing numerous other activities of chromosomes.

Dosage compensation has recently been associated with a change in nucleosome structure. The inactivated X chromosome appears to have a different type of histone present. Histone H2A is replaced by a variant called mH2A. The details of this mechanism are under study.

Nucleosomes apparently play a major role in controlling gene expression; DNA with nucleosomes has a much lower transcription rate than DNA without nucleosomes. It makes sense that the positions of nucleosomes can provide or prevent access to promoters. There are regions of the DNA, known as *nuclease-hypersensitive sites*, that appear to be nucleosome free. These sites, usually mutiples of a nucleosomal region of about two hundred base pairs, are particularly sensitive to digestion by different nucleases. When these regions are isolated, they usually have sequences that control functions in replication, transcription, or other activities of DNA. For example, numerous promoter regions in *Drosophila*, mouse, and human DNA are in nuclease-hypersensitive sites. Hence, some specific DNA sequences are kept free of nucleosomes, and these sequences appear to be recognized by various enzymes such as RNA polymerase. In many other cases, however, nucleosomes do appear to cover promoters and repress transcription. For transcription to occur in these cases, some form of *chromatin remodeling* must take place.

Two general classes of proteins are involved in chromatin remodeling. First are proteins that acetylate the N-terminal tails of the histones, a process that may cause the nucleosomes to bind the DNA less tightly and thus make it available for attachment of transcription factors. These enzymes are called *histone acetyl transferases* (HATs). Deacetylating enzymes have the reverse effect: They act to repress transcription. Second, a class of ATP-dependent proteins such as the SWI/SNF complex in yeast also affect chromatin remodeling. (Some workers called the proteins SWI because they were involved in mating type switching, and others called them SNF for sucrose nonfermenting.) The SWI/SNF complex is a group of eleven proteins involved in transcription activation in many genes, presumably allowing transcription factors to access promoters by remodeling chromatin. These proteins are able to reposition a nucleosome on DNA by sliding the nucleosome down the DNA.

We thus conclude that although nucleosomes serve as a general, first-order packing mechanism in eukaryotic DNA, they can be positioned precisely and can attenuate transcription. It is interesting to note that once transcription begins, RNA polymerase apparently moves along nucleosomed DNA by translocation of the histones by 75 to 80 base pairs without disrupting the nucleosome itself. This seems to be accomplished by the RNA polymerase moving the DNA and then re-forming the nucleosome in its wake.

Higher-order structure of chromatin

Since the nucleosome has a width of only 110 Å, and metaphase chromosomes appear to be constructed of a fiber having a diameter of about 2,400 Å, several additional levels of chromatin compaction lead to the metaphase chromosome. Various experiments, which change the ionic strength the chromatin is subjected to, indicate that the 110 Å DNA spontaneously forms a 300 Å, solenoidlike fiber with increased ionic strength. It seems that this fiber results from the coiling of the nucleosomal DNA. This 300 Å fiber is not, however, the final form of the DNA. We can account for the contraction of the 300 Å fiber to the 2,400 Å fiber found in metaphase chromosomes by the formation of a second solenoidlike structure from the winding of the 300 Å fiber.

If the histones are removed from a chromosome, the DNA billows out, leaving a proteinaceous structure termed a *scaffold*. This scaffold structure is formed from *nonhistone proteins*; two of them predominate, namely SC1 and SC2. SC1 has been identified as topoisomerase II. It

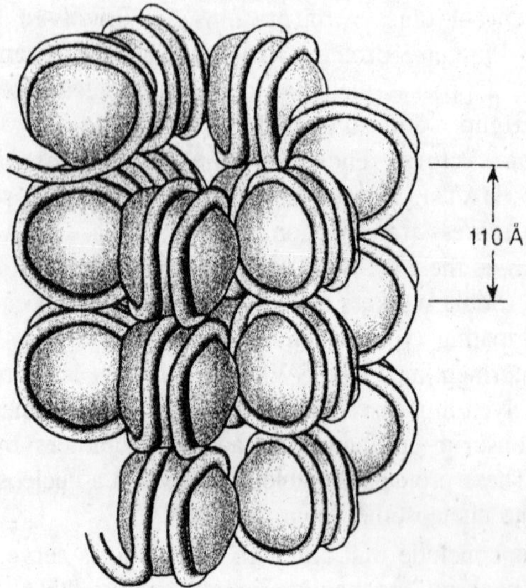

Fig. 4.4. Solenoid model for the formation of the 300-Å chromatin fiber.

would not be unreasonable to expect several hundred different proteins, many in minute quantities, to be associated with the chromosome and involved in replication, repair, and transcription.

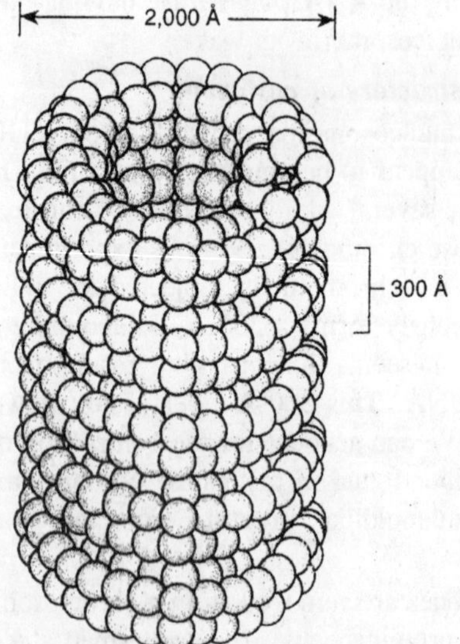

Fig. 4.5. The 2,000-Å fiber of the eukaryotic chromosome is a solenoidlike structure formed by the coiling of the 300-Å fiber, which itself is a solenoid.

Polyteny, puffs, and balbiani rings

Drosophila's salivary glands, as well as some other tissues of *Drosophila* and other diptera, contain giant banded chromosomes that result from the replication of the chromosomes and the synapsis of homologues without cell division (endomitosis). These chromosomes

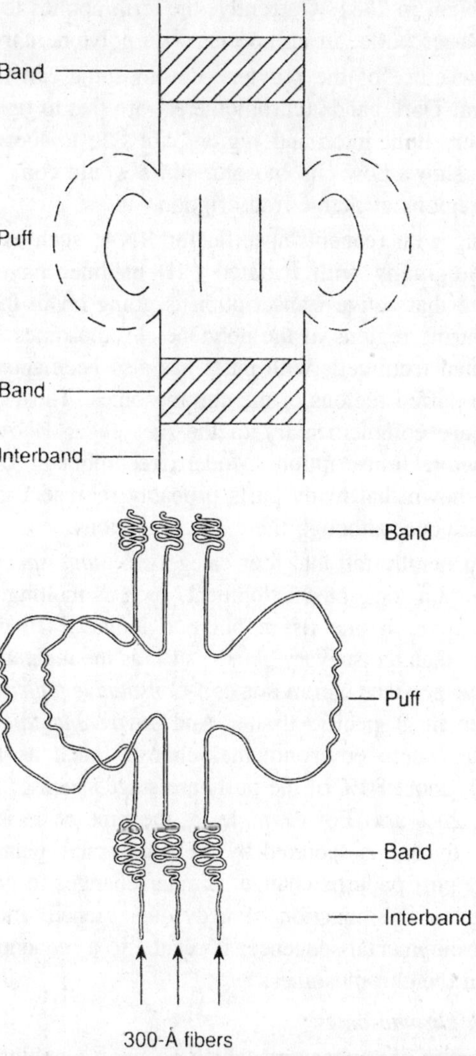

Fig. 4.6. Polytene chromosome with bands and a puff. Three of the approximately one thousand synapsed chromatids are shown diagrammatically on the below.

consist of more than one thousand copies of the same chromatid and appear as alternating dark bands and lighter interband regions. The dark bands are referred to as chromomeres. Also seen are diffuse areas called *chromosome puffs*. Chromosome puffs are also referred to as *Balbiani rings*. These rings were originally defined as puffs in the midge, *Chironomus*, whose polytene chromosomes were discovered by E. G. Balbiani in 1881. Currently, the term applies to all puffs, or at least the larger puffs, in all species with polytene chromosomes.

The structure of the polytene chromosome can be explained by the diagram. Dark bands (chromomeres) are due to tight coiling of the 300 Å fiber; light interband regions are due to looser coiling. The figure also shows how chromosome puffs would come about as fibers unfold in regions of active transcription.

Staining with reagents specific for RNA, such as toluidine blue, or autoradiography with tritiated (^{3}H) uridine, have been used to demonstrate that active transcription is going on in the puffs but not in neighbouring regions of the polytene chromosomes. The messenger RNA isolated from cells with puffs has also been shown to hybridize only to the puffed regions of the chromosomes. Thus, these regions of the DNA are complementary to the messenger RNA and represent areas of active transcription. Modern recombinant DNA techniques have also shown that many puffs probably represent the transcription of only one gene, although there are exceptions.

Puffs generally fall into four categories. *Stage-specific puffs* appear during a certain stage of development, such as molting. *Tissue-specific puffs* are active in one tissue but not another. (In dipteran larvae, tissues other than the salivary glands, such as the midgut and Malpighian tubules, have polytene chromosomes.) *Constitutive puffs* are active almost all the time in a specific tissue. And *environmentally induced puffs* appear after some environmental change, such as heat shock. In *Drosophila*, about 80% of the puffs are stage specific; in *Chironomus*, only about 20% are. For example, at the time of molt in insects, the hormone ecdysone is secreted by the prothoracic gland. At the same time, many puff patterns change. Similar changes in puff patterns can be induced by the injection of ecdysone. Hence, molting, a stage-specific developmental sequence, is related to a sequential transcription sequence in the chromosomes.

Lampbrush chromosomes

Lampbrush chromosomes, which occur in amphibian oocytes, are so named because their looped-out configuration has the appearance of a brush for cleaning lamps, now a relatively uncommon household item. The loops of the lampbrush chromosomes are covered by an RNA matrix and are the sites of active transcription. Presumably, the loops are unwindings of the single chromosome, similar to the unwindings in the polytene chromosome. Thus, under certain circumstances, such as in polytene chromosomal puffs and in lampbrush chromosomes, active transcription can be seen in the light microscope. Since only certain bands puff at any one moment in polytene chromosomes, and since the loops of lampbrush chromosomes are of various sizes (with some regions not looped at all), we have evidence of specific transcription. However, we have no indication, so far, of the nature of the control of that transcription.

Chromosomal Banding

Several chromosomal staining techniques reveal consistent banding patterns. By means of these patterns, all of the human chromosomes can be differentiated. Of possibly greater importance is the fact that these staining techniques have provided some insight into the structure of the chromosome. The techniques for staining the C, G, and R chromosomal bands will serve as an illustration.

G-bands are obtained with *Giemsa stain*, a complex of stains specific for the phosphate groups of DNA. Treatment of fixed chromatin with trypsin or hot salts brings out the G-bands. Giemsa stain enhances banding that is already visible in mitotic chromosomes. The banding pattern is caused by the arrangement of chromomeres. Under careful observation, the major G-bands prove to consist of many smaller chromomeres. This banding appearance has led D.

C-bands are Giemsa-stained bands after the chromosomes are treated with NaOH. The *C* is for "centromere," because these bands represent constitutive heterochromatin surrounding the centromeres. The DNA is also usually satellite rich. *Satellite DNA* differs in *buoyant density* from the major portion of cellular DNA. When eukaryotic DNA is isolated and centrifuged in CsCl, forming a density gradient, the majority of the DNA forms one band in the gradient at a single buoyant density. The buoyancy is determined by the G-C content of the DNA. However, smaller secondary bands are also usually present, indicating regions of DNA having sequences different from the majority of the cell's DNA. DNA isolated this way is referred to as *satellite DNA* because of the secondary, or satellite, bands formed in the density gradient. As we will see, this DNA is found primarily around centromeres and consists of numerous repetitions of a short sequence.

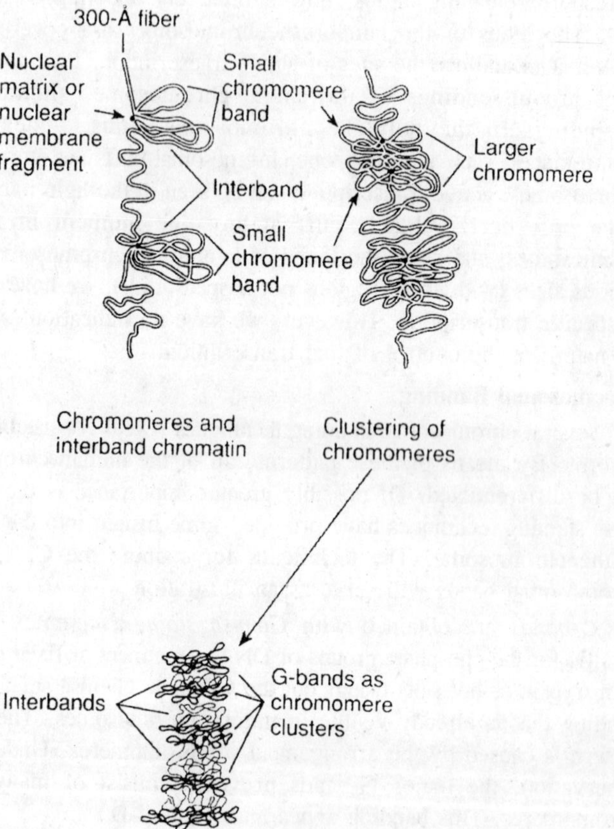

Fig. 4.7. Model of eukaryotic (mammalian) chromosomal banding.

R-bands are visible with a technique that stains the regions between G-bands. The chromosomes are fixed, stained with Giemsa, and then viewed with a phase contrast microscope. Since the dark-light pattern is the opposite of the G-band pattern, these bands are called *reverse bands*.

From the information gleaned from these staining techniques, D. Comings distinguished between three basic chromatin types: euchromatin, constitutive heterochromatin, and intercalary heterochromatin. Presumably, the only chromatin involved in transcription is *euchromatin*. *Constitutive heterochromatin* surrounds the centromere and is rich in satellite DNA. *Intercalary heterochromatin* is found throughout the chromosome. Thus, it becomes apparent that the eukaryotic chromosome is a relatively complex structure.

Centromeres and Telomeres

Centromeres

Two regions of the eukaryotic chromosome have specific functions—the centromere and the telomeres. The centromere is involved in chromosomal movement during mitosis and meiosis, whereas the telomeres terminate the chromosomes. As we pointed out, the terms *centromere* and *kinetochore*, while occasionally used interchangeably, are distinct. The kinetochore is the interface between the visible constriction in the chromosome (the centromere) and the microtubules of the spindle. The kinetochore of higher organisms (e.g., mammals) contains proteins and some RNA. Microscopically, it is a trilaminar structure, attached to chromatin at the inner layer and to microtubules at the outer layer.

Most of our knowledge of the genetics of centromeres has come from work in yeast (*Saccharomyces cerevisiae*). Cells did not maintain most artificially created yeast plasmids because they were lost during mitosis. However, plasmids were isolated that did replicate normally during cell division. Presumably, they contained centromeres, allowing them to replicate and move in synchrony with the host's chromosomes. Further genetic engineering made it possible to isolate smaller and smaller regions that could serve as centromeres. After sequencing the centromeres of fifteen of the sixteen yeast chromosomes, it was possible to conclude that the centromere from yeast is about 250 base pairs long with three consensus regions; we are defining a centromere as a sequence of DNA called the *CEN* locus or CEN region. Recent data indicate that this region may contain a single, modified nucleosome associated with region II. The 250 base-pair length of the CEN regions of yeast chromosomes is about 200 Å, the same as the diameter of a microtubule, indicating that only one micro-tubule attaches to each centromere during mitosis or meiosis in a yeast cell. This region is called a *point centromere*. Higher eukaryotes have larger centromeric regions that attach more microtubules. These regions are referred to as *regional centromeres*. Regional centromeres range from nineteen to one hundred kilobases (kb; 19,000–100,000 bases) with unique and satellite (repeated sequence) DNA that is heterochromatic and may include expressed genes. We know much less about regional centromeres than we do about point centromeres.

Telomeres

Since eukaryotic chromosomes are linear, each has two ends, referred to as *telomeres*, that not only mark the termination of the linear chromosome but also have several specific functions. Telomeres must prevent the chromosomal ends from acting in a "sticky" fashion, the way that broken chromosomal ends act. In other words, chromosomal ends must not elicit a DNA repair response. Telomeres must also prevent the ends of chromosomes from being degraded by exonucleases and must allow chromosomal ends to be properly replicated.

Most telomeres isolated so far are repetitions of sequences of five to eight bases. In human beings, the telomeric sequence is TTAGGG, repeated 300 to 5,000 times at the end of each chromosome. The human telomere was discovered by R. Moyzis and his colleagues when they probed the highly repetitive segment of human DNA. (Highly repetitive DNA, as its name implies, consists of numerous copies of a single sequence and usually comprises the satellite components of the cell's DNA; see next section.) When a probe for this sequence was applied to human chromosomes, the sequence was found at the tip of each chromosome in roughly the same quantity.This is a highly conserved sequence, found in all vertebrates studied as well as in unicellular trypanosomes. Similar sequences are found in various other eukaryotes; the first sequence was isolated by E. Blackburn and J. Gall in 1978.

When a linear DNA molecule is replicated, the 3' → 5' strand can be replicated to the end. The 5' → 3' strand, however, is replicated with RNA primers that are then degraded, leaving a short gap on the progeny strand. It is always the G-rich strand of telomeric DNA that ends up single-stranded, forming a 3' overhang of twelve to sixteen nucleotides. Thus, the normal replication process of a linear DNA molecule leaves an incomplete terminus. Hence, scientists suspected that there would be a unique mechanism for the replication of telomeres.

Telomeric sequences appear to be added de novo without, DNA template assistance by an enzyme called *telomerase*, discovered by E. Blackburn and her colleagues. This was seen when telomeres from another species were engineered into yeast cells. After a cell cycle, the yeast telomeric sequence had been added on at the ends of the foreign chromosome, the result, presumably, of the telomerase enzyme.

When Blackburn and her colleagues isolated telomerase, they discovered that a segment of RNA, about 160 base pairs, is an integral part of the enzyme. That RNA has a region that is complementary to the G-rich repeat of the telomeric DNA sequence of the species. After careful experimentation, including modifying the gene for the

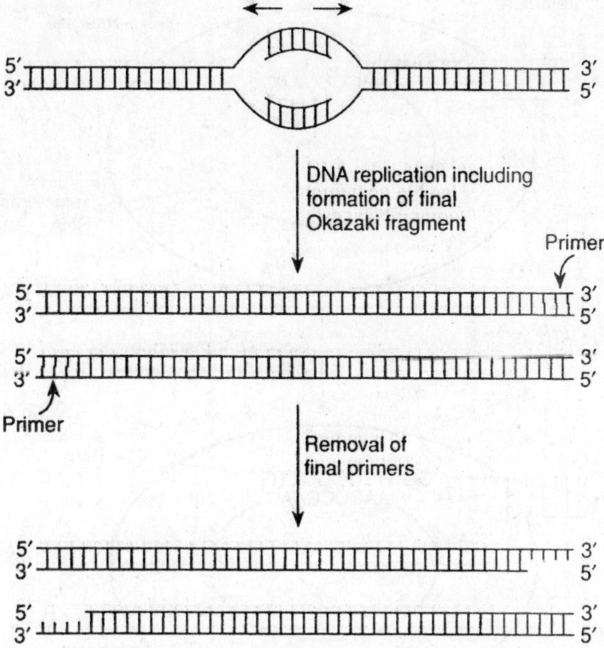

Fig. 4.8. Removal of final primers after the replication of linear DNA creates single-stranded ends.

telomerase RNA, Blackburn and her colleagues concluded that telomerase uses its RNA as a template for adding telomeric repeats to the ends of chromosomes. Telomerase is thus a reverse transcriptase, using RNA nucleotides as a template to polymerize DNA nucleotides.

Blackburn and her colleagues proposed that the first step in telomere extension is hybridization of the 3' end of the telomere with the RNA component of telomerase. Then, with the telomerase RNA as a template, the 3' end of the telomere is extended. Finally, a translocation step takes place that displaces the telomere in respect to the RNA, returning to the configuration at the beginning of the process. The single-stranded C-rich strand is then synthesized with DNA polymerase and DNA ligase.

Once telomeres have been added to the ends of eukaryotic chromosomes, different organisms use any of three different methods known to protect the ends of the chromosomes. First, the guanine-rich DNA can form complex structures. Biochemists have discovered that four guanines can form a planar *G-tetraplex*, with the four bases hydrogen bonded to each other. Several structures have been hypothesized

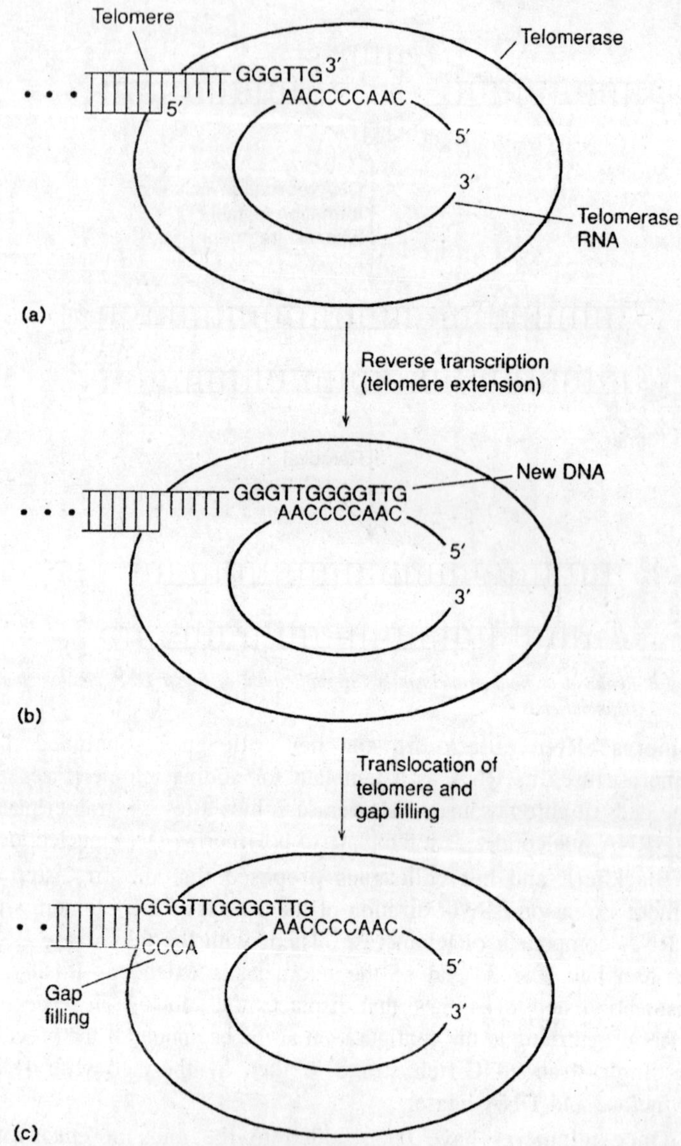

Fig. 4.9. Telomerase extends telomeres using telomerase RNA as a template Gap filling by DNA polymerase I and ligase complete the double helix.

to explain the novel ends of these chromosomes. Second, proteins have been discovered that bind to the 3' ends of telomeres. In the ciliate *Oxytricha nova*, a protein called the telomere end-binding protein (TEBP) attaches to the 3' ends of telomeres and protects them. Finally, a novel structure called the *t-loop* has been discovered at the end of mammalian telomeres. This loop forms at the ends of chromosomes under the direction of a protein called TRF2 (telomere repeat-binding factor), which causes the 3' end of the chromosome to loop around and interdigitate into the double helix, forming the loop.

How do cells keep track of the number of their telomeric repeats? Proteins have been isolated that bind to telomeres (Rap1 in *Saccharomyces cerevisiae*, TRF1 in human beings). By mutating these proteins or the telomeric sequences, scientists have changed the equilibrium number of telomeric repeats.This led to the current model that the cell counts the number of these proteins bound to the telomeres, not the number of telomeres directly, to know whether telomeres should be added. This is a very active area of research.

In yeast, protozoa, and other single-celled organisms, telomerase is active, keeping the ends of the chromo somes at the appropriate lengths. These cells can divide potentially forever. However, in most cells of higher organisms, telomerase is not active, and the ends of the chromosomes get shorter with each cell division. At a certain telomeric length, the cells no longer divide. However, if telomerase becomes active, and the ends of the chromosomes lengthen, a signal is conveyed to keep cells dividing, which can lead to cancerous growth. In fact, human telomerase was isolated from an immortal cell line (HeLa) derived from cervical cancer cells. Thus, attention is now turning to the possible clinical application of this knowledge: If telomerase can be deactivated in tumor cells, the cells may stop dividing or die, thereby eliminating the cancer. Further, studying normal telomere shortening, which appears to act as a biological clock, may help us understand the aging process and senescence.

C-Value Paradox

Why do eukaryotes have so much DNA, and why is there huge variation in the DNA content between species of comparable complexity? These questions define the *C-value paradox*, in which *C* refers to the quantity of DNA in a cell. For an example of the paradox, although human beings have 3.3 billion base pairs in the haploid genome, an amoeba has more than 200 billion base pairs. And although an average bony fish has over 300 billion base pairs of DNA in its haploid genome, the Japanese puffer fish has less than half a billion base pairs. If the basic bony fish pattern can be created with less than half a billion base pairs, why does the average bony fish have over 600 times that much DNA? What is this excess DNA doing? To explain the C-value paradox, researchers examined the repetitiveness of DNA, and more recently, probed and sequenced DNA to understand its properties.

DNA-DNA Hybridization

R. Britten and his colleagues, using the technique of *DNA-DNA hybridization*, first systematically analyzed the repetitiveness of the DNA within eukaryotes. When DNA is heated, it denatures or unwinds into single strands; when it cools, it renatures. The rate of renaturation depends on the DNA sequences. If the sample contains DNA with repeated sequences, it will hybridize faster than DNA that does not have repeated sequences. From these studies, Britten and his colleagues found that eukaryotic chromosomes contain regions of unique, moderately repetitive, and highly repetitive DNA. *Unique DNA* is, as its name implies, DNA with unrepeated sequences. *Repetitive DNA* is DNA whose sequences are repeated in the genome.

Satellite DNA, found around centromeres, is highly repetitive DNA with a unique repeat length of about two hundred base pairs. Given the quantity of satellite DNA per cell, there must be more than one million repetitions of this two-hundrednucleotide sequence in higher eukaryotes. At the other end of the spectrum is unique DNA, which makes up most of the transcribed genes of an organism. The rest of the DNA is repetitive DNA in a few to several hundred thousand copies. This repetitive DNA comprises at least three categories. One is "junk" DNA, DNA that is not useful to the organism, made up of untranscribed and parasitic sequences (selfish DNA). Another category is transcribed genes in many copies that have diverged from each other, such as antibody, collagen, and globin genes. We use the term *gene family* to refer to genes that have arisen by duplication, with or without divergence, from an ancestral gene. And finally, transcribed genes in many copies that are virtually identical, such as ribosomal RNA and histone genes, make up a third category of repetitive DNA.

Junk DNA

We saw that transposons in prokaryotes are generally viewed as selfish or parasitic: They serve no purpose to the cell. The transposons replicate on their own, increasing in number. Eukaryotic transposons are mostly *retrotransposons*, transposable elements that move by way of an RNA intermediate. That is, the retrotransposon is transcribed into RNA and then, by reverse transcription, converted to a cDNA that is then inserted into the genome. These elements can make up

50% of the eukaryotic genome, existing in hundreds of thousands of copies. They generally fall into two categories: LINES and SINES. *Long interspersed elements* (LINES), are up to seven thousand base pairs each and contain genes for reverse transcription, RNA binding, and endonuclease activity. They thus have the ability to jump by way of an RNA intermediate. Human DNA is believed to be composed of about 15% LINES.

Short interspersed elements (SINES) are generally derivatives of transfer RNA genes and do not have the ability to retrotranspose on their own. That is, in the past, their transcripts were modified, converted to cDNA by reverse transcription, and then reinserted into the host's genome. They rely on the reverse transcriptase provided by the genes of LINES or retroviruses. One group of SINEs not derived from transfer RNA is derived from the RNA of the signal recognition particle; members of this group occur in human beings in about five-hundred thousand copies of a three-hundred-basepair sequence. Because these sequences are cleaved by the restriction endonuclease *Alu*I, they are called the *Alu family*. The human genome is also permeated by remnants of at least a dozen distinct families of ancient retroviruses scattered throughout our chromosomes.

At this point, we can see some potential explanations for the C-value paradox. Much eukaryotic DNA is junk, apparently doing no harm. In some cases, 97% of the host genome is composed of junk DNA. Recent work seems to indicate that gross differences in DNA content between higher organisms may be due to the differing abilities of different species to rid themselves of this parasitic DNA. If it builds up without being removed, the DNA content of the species can soar. Thus, the wide differences in DNA content among higher eukaryotes mentioned at the beginning of this section have little to do with the complexity of the organism, but rather with the ability of the organism to remove junk DNA as it forms.

Expressed genes in many copies

Several types of genes create a product that is needed in such large quantity that one copy of the gene could not fulfill the cell's needs. We are familiar with the nucleolus, the site of the ribosomal RNA genes. Human beings have about two hundred copies of the major ribosomal RNA gene and about two thousand copies of the 5S ribosomal RNA gene. Fruit flies have about two hundred and one hundred copies, respectively, of the two genes. In some cases, the normal number of multiple copies of a gene is still not enough. The cell must then resort to *gene amplification*, a process whereby the cell increases the number of copies of the gene. For example, during oogenesis, ribosomal RNA genes (rDNA) are often amplified. In *Xenopus*, rDNA is amplified about one thousand times, which allows an oocyte to accumulate about 10^{12} ribosomes. The amplified DNA is in the form of small, circular, extrachromosomal molecules of DNA. Several models have been proposed as to how cells actually amplify their DNA. One model relies on unequal crossing over (as in *Bar* eye in *Drosophila*), whereas another model is based on unscheduled extra DNA replication in a region, followed by recombinational events that generate linear and circular forms of the excess DNA. It is not presently clear which model is correct.

In addition to ribosomal RNA genes, other genes are repeated, ensuring adequate gene products. The number and location of repeated genes are usually discovered by hybridization studies using probes, similar to the way that telomeric DNA was shown to be at the tips of the chromosomes. Repeated genes include the genes for transfer RNAs and histones. The average transfer RNA is repeated about a dozen times in *Drosophila*. Human beings have over thirteen hundred copies of transfer RNA genes in the haploid genome. In many species, the five histone genes form a repeated cluster, although each gene is transcribed independently, while prokaryotic operons are transcribed as a unit. The arrangement of histone genes may be more complex in higher forms. There are indications that in mammals, histone genes may lie in small groups or even as individual genes.

Several types of genes occur in similar but not identical forms—that is, an original gene was duplicated but, unlike histone or ribosomal RNA genes, the copies diverged in function. These gene families include globin genes, immunoglobulin genes, chorion protein (insect eggshell) genes, and *Drosophila* heat shock genes.

Globin gene family

Globins are oxygen-transporting and storage molecules found in animals, some plants, and microorganisms. In higher vertebrates, there are two types of globins: myoglobin, which stores oxygen in muscles, and hemoglobin, found in red blood cells. Myoglobins function as single molecules, whereas hemoglobins occur as tetramers, two each of two protein chains. Evolution in the globin gene family can be traced by comparative studies of globins in different species as well as molecular studies of globins within a species. Studying hemoglobins has provided a great deal of information on gene expression and evolution. We turn our attention to the globin gene family in human beings. During human development, four major hemoglobins appear: embryonic hemoglobin, Hb F, Hb A, and Hb A_2. Structurally, the ζ (Greek, zeta) subunit (a component of embryonic hemoglobin) is α-like, whereas the rest are β-like. Fetal hemoglobin has a higher affinity for oxygen than does adult hemoglobin, thus allowing fetuses to draw oxygen from their mother's blood. From a comparative study of the DNA sequences, the evolution of the various hemoglobin genes has been inferred.

The α genes are located in a cluster on chromosome 16; the β genes are located in a cluster on chromosome 11. These two clusters provide a clear case history of gene duplication, presumably by unequal crossing over, followed by divergence. Having a second or third copy of a gene allows one of the duplicates to diverge (and perhaps to become nonfunctional in the process), whereas the original still performs the required function.

Many diseases of genetic interest involve the hemoglobins. In fact, hemoglobinopathies, including sickle-cell anemia and the thalassemias, are the most common genetic disorders in the world population. The best-known mutation of a hemoglobin gene itself is the one that causes sickle-cell anemia, a mutation of the sixth amino acid of the β chain. In the homozygous state, the disease is usually fatal. However, heterozygotes show an increased resistance to malaria. One of the ramifications is that the sickle-cell allele is maintained at relatively high frequencies in malarial regions.

The *thalassemias* are a group of diseases that affect the regulation of the α and β hemoglobin genes. (*Thalassemia* comes from the Greek for "sea blood," because the disease is best known in individuals living around the Mediterranean Sea.) In α and β thalassemias, the α or β subunit, respectively, is present in very low quantities or entirely absent. Many of the genetic defects are deletions, possibly due to unequal crossing over within the globin gene complexes. T. Maniatis showed that β thalassemia is caused by a mutation in the β-globin gene that disrupts RNA splicing. The body compensates by forming γ_4 or β_4 hemoglobin in a thalassemias, or $\alpha_2\gamma_2$ or $\alpha_2\delta_2$ in β thalassemias. These are relatively unsuitable or inefficient responses; the diseases range from very mild to very severe and frequently fatal. More information is needed regarding the control of hemoglobin production in the thalassemias.

5

Chromosome in Inheritance

The term *genetic variation* refers to genetic differences between members of the same species, or those between different species. Variation in specific gene is called *allelic variation*. Allelic differences are due to small changes (i.e., mutations) that occur within a particular gene. In this chapter, our emphasis will shift to larger types of genetic changes that affect the chromosomal composition of eukaryotic organisms. As we will see, these changes may affect the expression of many genes and lead to interesting changes in phenotypes.

In the first section of this chapter, we will examine how the structure of a eukaryotic chromosome can be modified, either by altering the total amount of the genetic material or by rearranging the order of genes along a chromosome. A substantial change in chromosome structure, possibly affecting more than a single gene, is referred to as a *chromosome mutation*. In most cases, a mutant chromo some can be detected microscopically; its structure will differ visually from a normal chromosome. In this chapter, we will examine how chromosome mutations occur, how they are transmitted from parent to offspring, and how they affect the phenotype of the individual who inherits them.

The rest of the chapter will be concerned with changes in the total number of chromosomes. A change in chromosome number is called a *genome mutation*. This type of mutation comprises two subtypes: changes in the number of sets of chromosomes, and changes in the numbers of individual chromosomes with in a set. Natural variation in the number of sets of chromosomes is relatively common among different species, particularly in the plant kingdom. Within the same species, changes in chromosome number within a set can also occur, but such alterations usually are detrimental. In this chapter, we will explore how variation in chromosome number occurs, and describe examples where it has significant phenotypic consequences.

VARIATION IN CHROMOSOME STRUCTURE

We will begin our discussion of chromosome variation by considering several ways in which the structures of eukaryotic chromosomes can be altered. The chromosomes found in the nuclei of eukaryotic cells are long, linear molecules that carry hundreds or even thousands of genes. In this section, we will examine how the composition of a chromosome can be changed. As you will see, segments of a chromosome can be lost, duplicated, or rearranged in a new way.

The study of chromosomal variation is important for several reasons. First, geneticists have discovered that variations in chromosome structure can have major effects on the phenotype of an organism. For example, we now know that several human genetic diseases are caused by changes in chromosome structure. In other cases, an individual possessing a chromosomal rearrangement may have a normal phenotype yet have a high probability of producing offspring with genetic abnormalities. Later in this section, we will examine irregularities in gamete formation that account for this phenomenon. Finally, changes in chromosome structure have been an important force in the evolution of new species. In this section, we will examine the cellular mechanisms that underlie changes in chromosome structure. We will explore unusual events during meiosis that affect how altered chromosomes are transmitted from parents to offspring. Also, we will consider many examples where chromosomal alterations affect an organism's phenotypic characteristics.

Natural Variation Exists in Chromosome Structure

Before we begin to discuss how chromosome structure can be altered, we need to have a reference point for a normal set of chromosomes. To determine what the normal chromosomes of a species look like, a *cytogeneticist* microscopically examines the chromosomes from several members of the species. Inmost cases, two phenotypically normal individuals of the same species will have the same number and types of chromosomes.

To determine the chromosomal composition of a species, the chromosomes in actively dividing cells are karyotyped. A *karyotype* is a micrograph in which all the chromosomes within a single cell have been arranged in a standard fashion. As seen here, humans contain 46 chromosomes (23 pairs), fruit flies have 8 chromosomes (4 pairs), and corn has 20 chromosomes (10 pairs). Except for the sex chromosomes, which differ between males and females, most members of the same species have very similar karyotypes. By comparison, distantly related species (e.g., humans versus fruit flies) have very different karyotypes. Nevertheless, a karyotype of 46 chromosomes is normal for humans, as is a karyotype of 8 chromosomes for fruit flies.

The chromosomes from any given species can vary considerably in size and shape. Cytogeneticists have various ways to classify and identify chromosomes. The three most commonly used features are size, location of the centromere, and banding patterns that are revealed when the chromosomes are treated with stains. By convention, the chromosomes are numbered according to their size, with the largest chromosomes having the smallest numbers. For example, human chromosomes 1, 2, and 3 are relatively large, whereas 21 and 22 are the smallest. An exception to the numbering system are the sex chromosomes, which are designated with letters (for humans, X and Y).

Another distinguishing feature of eukaryotic chromosomes is the location of the centromere. Chromosomes are classified as *metacentric*, *submetacentric*, *acrocentric*, and *telocentric*. Because the centromere is not exactly in the center of a chromosome, each chromosome has a long arm (designated with the letter *q*) and a short arm (designated with the letter *p*). In the case of telocentric chromosomes, the short arm may be almost nonexistent. When preparing a karyotype, the chromosomes are aligned with the short arms on top and the long arms on the bottom. Since different chromosomes often have similar sizes and centromeric locations (e.g., compare human chromosomes 8, 9, and 10), geneticists must use additional methods to accurately identify each type of chromosome. For detailed identification, chromosomes are treated with stains to produce characteristic banding patterns. A human karyotype in which the chromosomes have been treated with a chemical dye called *Giemsa stain*. In some chromosomal regions, the stain binds heavily and produces a dark band; in other regions, the stain hardly binds at all and produces a light band. As shown here, the alternating banding pattern, known as *G bands*, is a unique feature for each chromosome.

The banding pattern of eukaryotic chromosomes is useful in several ways. First, individual chromosomes can be distinguished from each other, even if they have similar sizes and centromeric locations. For example, compare the differences in banding patterns between human chromosomes 8 and 9. These differences permit us to distinguish these two chromosomes even though their sizes and centromeric locations are very similar. Banding patterns are also used to detect changes in chromosome structure. As discussed next, chromosomal re arrangements or changes iii the total amount of genetic material are more easily detected when viewing banded chromosomes. Also, chromosome banding may reveal evolutionary relationships among the chromosomes of closely related species.

Mutations can Alter Chromosome Structure

Now that we understand that normal chromosomes can come in a variety of shapes and sizes, we can consider how the structures of normal chromosomes can be modified. As mentioned earlier, a chromosome mutation is a substantial change in the structure of a chromosome. There are two primary ways that the structure of chromosomes can be altered. First, the total amount of genetic material within a single chromosome can be increased or decreased significantly. Second, the genetic material in one or more chromosomes may be rearranged without affecting the total amount of material. These alterations are categorized as deficiencies, duplications, inversions, and translocations.

Deficiencies and duplications are changes in the total amount of genetic material within a single chromosome. When a *deficiency* occurs, a segment of chromosomal material is missing. In other words, the affected chromosome is deficient in a significant amount of genetic material. Deficiencies are also referred to as *deletions*. In contrast, a *duplication* is the opposite situation—an amount of genetic material is repeated compared with the normal parent chromosome.

Inversions and translocations are chromosomal rearrangements. An *inversion* involves a change in the direction of the genetic material along a single chromosome. For example, a segment of one chromosome has been inverted, so that the order of four C-bands is opposite to that of the parent chromosomes. A *translocation* occurs when one segment of a chromosome breaks off and becomes attached to a different chromosome. When a single piece of chromosome is attached to another chromosome, this is called a *simple translocation*. In other cases, two different types of chromosomes can exchange pieces, thereby producing two abnormal chromosomes carrying translocations. This situation is referred to as a *reciprocal translocation*.

Loss of Genetic Material in a Deficiency Tends to be Detrimental to an Organism

A chromosomal deficiency occurs when a chromosome breaks in one or more places and a fragment of the chromosome is lost. A normal chromosome has broken into two separate pieces. The piece without the centromere will be lost and degraded, Therefore, this event produces a chromosome with a *terminal deficiency*. A chromosome has broken in two places to produce three chromosomal fragments. The central fragment has become lost (and degraded), and the two outer pieces have reattached to each other. This process has created a chromosome with an *interstitial deficiency*. Deficiencies can also be created when recombination takes place at incorrect locations between two homologous chromosomes. The products of this type of aberrant recombination event are one chromosome with a deficiency and another chromosome with a duplication.

The phenotypic consequences of a chromosomal deficiency depend on the size of the deletion and whether it includes genes (or portions of genes) vital to the development of the organism. When deletions have a phenotypic effect, they are usually detrimental. Larger deletions tend to be more harmful, because more genes are missing. Many examples are known in which deficiencies have significant phenotypic influences. In humans, for example, a genetic disease known as *cri-du-chat syndrome* involves a deficiency in a segment of the short arm of human chromosome 5. Individuals who carry a single copy of this abnormal chromosome (along with a normal chromosome 5) display an array of abnormalities including severe mental retardation, facial anomalies, and an unusual catlike cry (the meaning of the French name for the syndrome). Some other human genetic diseases, such as Angelman syndrome and Prader Willi syndrome are due to a deficiency in chromosome 15. Many types of chromosomal deletions have also been associated with human cancers.

Deficiencies can be Detected Using Cytological, Genetic, and Molecular Techniques

A variety of experimental techniques are used to detect a deletion within a chromosome. These include microscopic, genetic, and molecular methods. Microscopic techniques, which are relatively quick and easy, can detect chromosomal deficiencies large enough to be seen with a light microscope. Cri-du chat and Prader Willi syndromes are chromosomal deletions easily detectable via light microscopy. Fairly small deletions, which may involve only one or a few genes, are often difficult to detect with simple microscopy. However, more complicated microscopy techniques, such as in situ hybridization, have enhanced our ability to detect small deletions. When a mutation causes a phenotypic effect, genetic analysis can sometimes show that it is due to a deletion. Deletions in particular genes can be identified by their inability to revert back to the wild-type allele. In addition, deletions can sometimes be revealed by a phenomenon known as *pseudodominance*. This occurs when one copy of a gene has been deleted from a chromosome and the remaining copy of a recessive allele on the homologous chromosome is phenotypically expressed. Under these conditions, the individual is hemizygous for the recessive allele.

Duplications Tend to be Less Harmful than Deletions

Duplications result in extra genetic material. They are usually caused by abnormal events during recombination. Under normal circumstances, crossing over occurs at analogous sites between homologous chromosomes. On rare occasions, the crossover may occur at two different sites on the homologues. This results in one chromatid with an internal duplication and another chromatid with a deletion. The chromosome with the extra genetic material carries a *gene duplication*, because the number of copies of gene *C* has been increased from one to two. In most cases, gene duplications happen as rare, sporadic events during the evolution of species. Later in this section, we will consider how multiple copies of genes can evolve to produce a family of genes with specialized functions.

Like deletions, the phenotypic consequences of duplications tend to be correlated with size. Duplications are more likely to have phenotypic effects if they involve a large piece of the chromosome. In general, small duplications are less likely to have harmful effects than are deletions of comparable size. This observation suggests that having one copy of a gene is more harmful than having three copies. In humans, relatively few well-defined syndromes are known to be caused by small chromosomal duplications.

In 1936, Bridges Found that Gene Duplications Produced the Bar and Ultra-bar Eyes Phenotype in *Drosophila*

Early insight into the causes of gene duplications came from studies in *Drosophila* involving a trait that affects the number of facets in the eye. In 1914, Sabra Colby Tice of Columbia University found a fly that had a reduced number of facets. This trait was called bar eyes. A genetic analysis of the trait revealed that it is an X-linked trait that shows incomplete dominance. Females homozygous for the bar allele have a more severe phenotype than heterozygous females that have one bar allele and one wild-type allele.

In 1920, Charles Zeleny at the University of Illinois identified some rare mutants in a stock of flies that was originally homozygous for the *bar* allele. Some rare flies had eyes that were phenotypically normal (*bar revertants*), and other flies had eyes with even fewer facets than the *bar* homozygote. Zeleny called these flies *ultra-bar* (also known as *double-bar*). This trait also shows incomplete dominance, since a female fly that is homozygous for the *ultra-bar* allele has fewer facets than a heterozygous female fly carrying one *ultra-bar* allele and one normal allele.

Based on the results of genetic crosses, Alfred Sturtevant and Thomas Hunt Morgan of Columbia University initially suggested that the ultra-bar phenotype was due to a gene duplication in which two copies of the *bar* allele were on the same X-chromosome. Since it was (erroneously) thought that an ultra-bar female fly had two copies of the *bar* allele, yet displayed a more intense phenotype than a homozygous bar female that also had two copies of the *bar* allele, the term *position effect* was coined. This term refers to changes in phenotype that are caused by the positions of genes on a chromosome. In this case, it was suggested that having two *bar* alleles on the same chromosome produced a different phenotype compared with having two copies of the *bar* allele on separate X-chromosomes.

To gain further insight into the cause of the bar and ultra-bar phenotypes, Calvin Bridges at the California Institute of Technology investigated the bar/ultra-bar phenomenon at the cytological level. In cells of the *Drosophila* salivary gland, the chromosomes are easy to study under the microscope, because they replicate many times to form gigantic *polytene chromosomes*. The structure of polytene chromosomes will be described later in this chapter.

Because polytene chromosomes are so large, their banding pattern is very easy to see in great detail. It is thus possible to detect very small changes in chromosome structure. In the experiment described here, which was conducted in 1936, Bridges investigated changes in chromosome structure associated with the bar and ultra-bar phenotypes.

Hypothesis

The rare formation of the ultra-bar phenotype is due to a duplication of the bar allele. Information concerning the nature of the bar and ultra-bar phenotypes may be revealed by a cytological examination of polytene chromosomes.

Testing the hypothesis

Starting materials: A strain of flies homozygous for the bar allele and a homozygous normal strain.

1. Within the strain of homozygous bar-eyed flies, identify rare these that have normal eyes (bar revertants) or ultra-bar eyes.
2. Dissect the salivary glands from the larva of a normal strain, a homozygous bar strain, a bar-revertant strain, and an ultra-bar strain.
3. Prepare the salivary cells for the microscopic examination of the polytene chromosomes. It involves gently breaking open the cells, staining the chromosomes with dyes, and squashing the preparation on a microscope slide underneath a coverslip.
4. View the banding patterns of the polytene chromosomes under the microscope.

Interpreting the data

The cytological examination of region 16A provided direct insight into the nature of the bar and ultra-bar phenotypes. Bridges found that the bar phenotype itself is caused by a duplication of the genetic material in the 16A region of the chromosome. Furthermore, homozygous bar stocks can occasionally produce phenotypically normal flies (bar revertants) and ultra-bar flies by altering gene number. The polytene chromosomes of bar revertants showed that the 16A region had returned to the wild-type banding pattern. In other words, the duplicated region was re turned to a single copy. By comparison, the opposite situation occurred for the ultra-bar phenotype. In this case, an additional duplication occurred, so that there were three copies of the 16A region. As stated by Bridges, "The Bar-eye reduction [the number of facets] is thus seen to be interpretable as the effect of increasing the action of certain genes by doubling or triplicating their number."

The mechanism for the formation of the bar allele can be explained by unequal crossing over. The formation of *ultra-bar* and *bar-revertant* alleles can likewise be explained by unequal crossing over. As shown here, an unequal crossover between two X-chromosomes carrying the duplicated *bar* allele can result in one chromosome carrying one copy (*bar revertant*) and the other chromosome carrying three copies (*ultra-bar*):

The gene duplication and triplication seen with *bar* and *ultra-bar* alleles are also associated with position effects. A female that is homozygous for the *bar* allele has four copies of this region, and a female that is heterozygous for the *ultra-bar* allele (three copies) and the wild-type allele (one copy) also has four copies. However, a female that is heterozygous for the *ultra-bar* and normal alleles has fewer facets (45) than does a *bar* homozygote (70). This is a position effect—the positioning of three copies next to each other on an X-chromosome increases the severity of the defect. Position effects are caused by the influences of chromosome structure or genetic regulatory regions on the level of gene expression. In this case, each of the three genes in the *ultra-bar* allele is expressed at a higher level than it is when there are just one or two copies on the X-chromosome.

Duplications Provide Additional Material for Gene Evolution, Sometimes Leading to the Development of Gene Families

In contrast to the gene duplication that causes the bar phenotype in *Drosophila*, the majority of small chromosomal duplications have no phenotypic effect. Nevertheless, they are still very important, because they provide raw material for the addition of more genes into a species' chromosomes. Over the course of many generations, this can lead to the formation of a *gene family* consisting of two or more genes that are similar to each other. The members of a gene family are derived from the same ancestral gene. Over time, two copies of an ancestral gene can accumulate different mutations. Therefore, after many generations, the two genes will be similar but not identical. During evolution, this type of event can occur several times, creating a family of many similar genes.

When two or more genes are derived from a single ancestral gene, the genes are said to be *homologous*. Homologous genes make up a gene family. A well-studied example of a gene family illustrates the evolution of the globin gene family found in humans. The globin genes encode polypeptides that are subunits of proteins that function in oxygen binding. For example, hemoglobin is found in red blood cells; its function is to carry oxygen throughout the body. The globins gene family is composed of fourteen homologous genes that were originally derived from a single ancestral globin gene. According to an evolutionary analysis, the ancestral globin gene duplicated about 800 million years ago. Since that time, additional duplication events and chromosomal rearrangements have occurred to produce the current number of fourteen genes on three different human chromosomes.

Gene families have been important in the evolution of traits. Even though all of the globin polypeptides are subunits of proteins that play a role in oxygen binding, the accumulation of different mutations in the various family members has created globins that are more specialized in their function. For example, *myoglobin* is better at binding and storing oxygen, whereas the *hemoglobins* are better at binding and transporting oxygen via the red blood cells. Also, these different globin genes are ex pressed during different stages of human development. The ε- and ζ-globin genes are expressed very early in embryonic life, while the γ-globin gene exhibits its maximal expression during the second and third trimesters of gestation. Following birth, the γ-globin gene is turned off and the β-globin gene is turned on. These differences in the expression of the globin genes reflect the differences in the oxygen transport needs of humans during the embryonic, fetal, and postpartum stages of life.

Inversions Often Occur Without Phenotypic Consequences

We now turn our attention to changes in chromosome structure that involve a re arrangement in the genetic material. A chromosome with an inversion contains a segment that has been flipped to the opposite orientation. Geneticists classify inversions according to the location of the centromere. If the centromere lies within the inverted region of the chromosome, the inverted region is known as a *pericentric inversion*. Alternatively, if the centromere is found outside the inverted region, the inverted region is called a *paracentric inversion*.

When a chromosome contains an inversion, the total amount of genetic material remains the same as in a normal chromosome. For this reason, the great majority of inversions do not have any phenotypic consequences in the individuals who carry them. In rare cases, however, an inversion can alter the phenotype of an individual. Whether or not this occurs is related to the boundaries of the inverted segment. When an inversion occurs, the chromosome is broken in two places and the ends flip around to produce the inversion. If either breakpoint occurs within a vital gene, the function of the gene is expected to be disrupted, possibly producing a phenotypic effect. For ex ample, some people with *hemophilia* (type A) have inherited an X-linked inversion that has inactivated a gene for Factor VIII (a blood clotting protein). In other cases, an inversion (or translocation) may reposition a gene on a chromosome in a way that alters its normal level of expression.

Since inversions seem like an unusual genetic phenomenon, it is perhaps surprising that they are found in human populations in quite significant numbers. About 2% of the human population carry inversions that are detectable with a light microscope. In most cases, these individuals are phenotypically normal and live their lives without knowing they contain this abnormality. In a few cases, however, an individual with an inversion chromosome may produce (one or more) off spring with genetic abnormalities. This event may prompt a physician to request a microscopic examination of the individual's chromosomes. In this way, a phenotypically normal individual may discover that they have a chromosome with an in version. Next, we will examine why an individual carrying an inversion may produce offspring with phenotypic abnormalities.

Inversion Heterozygotes may Produce Abnormal Gametes Due to Crossing Over

An individual who carries one copy of a normal chromosome and one copy of an inverted chromosome is known as an *inversion heterozygote*. Such an individual, though possibly phenotypically normal, may have a high probability of producing abnormal gametes. The frequency of abnormal gametes depends on the size of the inverted segment. An inversion heterozygote with a fairly large inverted segment may produce a sizable number of abnormal gametes (perhaps 1/3 or even higher).

The underlying cause of abnormal gametes is crossing over within the inverted region. During meiosis I of gamete formation, pairs of homologous sister chromatids synapse with each other. For the normal chromosome and inversion chromosome to synapse properly, a loop, termed an *inversion loop*, must form to permit the homologous genes on both chromosomes to align next to each other despite the inverted sequence. If a crossover occurs within the inversion loop, highly abnormal chromosomes will be produced.

The consequences of this type of crossover depend on whether the inversion is pericentric or paracentric. This is a single crossover that involves only two of the four sister chromatids. Following the completion of mitosis, this single crossover yields two abnormal chromosomes. Both of these abnormal chromosomes have a segment that is deleted and a different segment that is duplicated. In this example, one of the abnormal chromosomes is missing genes *A*, *B*, and *C* and has extra copies of genes *H* and *I*. The other abnormal chromosome has the opposite situation; it is missing genes *H* and *I* and has an extra copy of genes *A*, *B*, and *C*. If these abnormal chromosomes are passed on to off spring, they may produce phenotypic abnormalities depending on the amount and nature of the duplicated/deleted genetic material.

The outcome of a crossover involving a paracentric inversion. A single crossover has occurred between two homologous sister chromatids; the other two sister chromatids have not participated in a crossover. This single crossover event produces a very strange outcome. One chromosome is produced that contains two centromeres. This is

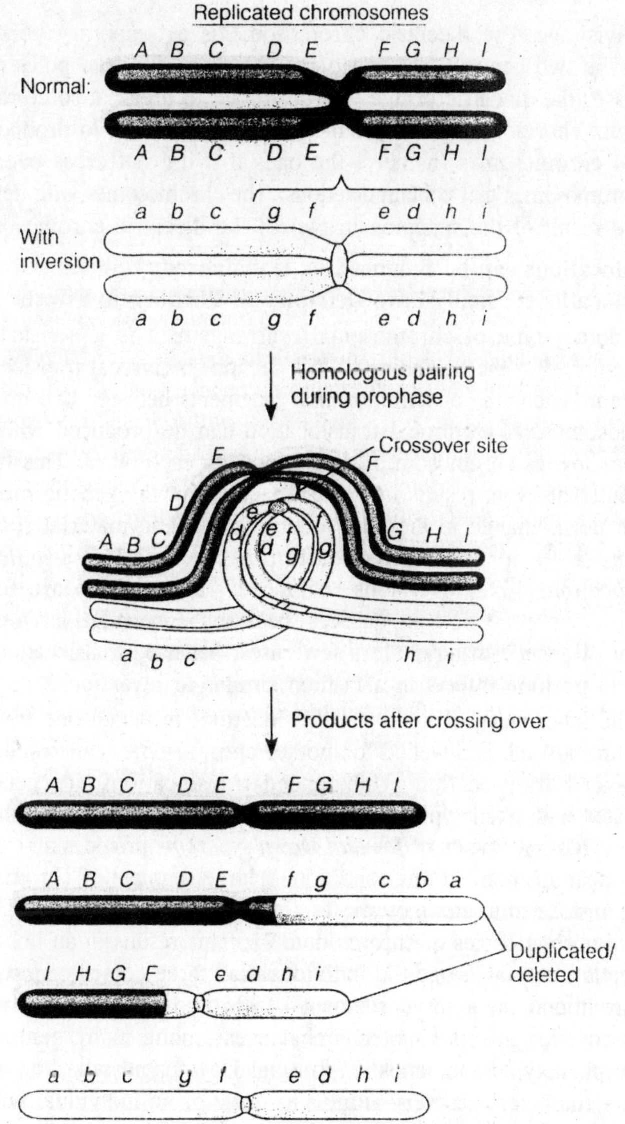

Fig. 5.1. Crossover within a pericentric inversion.

called a *dicentric chromosome*; the region of chromosome connecting the two centromeres is called a *dicentric bridge*. The crossover also produces a piece of chromosome without any centromere. This *acentric* (without a centromere) fragment will be lost and degraded in subsequent cell divisions. The dicentric chromosome is a temporary condition. When the two centromeres try to move toward opposite poles during anaphase, the dicentric bridge will be forced to break at some random location. Therefore, the net result of this crossover is to produce two normal chromosomes (namely, the ones that did not cross over) and two chromosomes that contain deletions. The chromosomes with deletions are the result of the snapping in two of the dicentric chromosome.

Translocations can be Balanced or Unbalanced; Unbalanced Translocations Usually have Detrimental Phenotypic Effects

Another type of chromosomal rearrangement is a translocation. As mentioned at the beginning of this section, reciprocal translocations involve an exchange of chromosomal fragments between two different chromosomes. A reciprocal translocation can be produced when two nonhomologous chromosomes cross over with each other. This type of rare aberrant event results in a rearrangement of the genetic material, though not a change in the total amount of genetic material. For this reason, a reciprocal translocation is also called a *balanced translocation*. Like inversions, balanced translocations are usually without any phenotypic consequences, because the individual has a normal amount of genetic material. In a few cases, balanced translocations can result in position effects in a fashion similar to inversions.

The inheritance of a simple translocation, in which one fragment of a chromosome is attached to another chromosome, can result in an unbalanced translocation. Unbalanced translocations are generally associated with phenotypic abnormalities or even lethality. An inherited human syndrome known as *familial Down syndrome* provides an example of this phenomenon. In this condition, a large segment of chromosome 21 is translocated to chromosome 14. In addition, this individual already has two normal copies of chromosome 21. This results in an imbalance in genetic material, since the individual has three copies of the genes that are found on a large segment of chromosome 21. As will be discussed later in this chapter, imbalances among many genes often lead to phenotypic anomalies. In familial Down syndrome, the person exhibits characteristics very similar to those of an individual with the more common form of Down syndrome, which is due to three entire copies of chromosome 21.

Individuals with Balanced Translocations may Produce Abnormal Gametes due to the Segregation of Chromosomes

Individuals who carry balanced translocations have a greater risk of producing gametes with unbalanced combinations of chromosomes.

Whether or not this occurs depends on the segregation pattern during meiosis I. In this example, the parent carries a balanced, reciprocal translocation and is likely to be phenotypically normal. During gamete formation, the homologous chromosomes attempt to synapse with each other. Because of the translocations, the pairing of homologous regions leads to the formation of an unusual structure that contains four pairs of sister chromatids, termed a *translocation cross*.

To understand the segregation of translocated chromosomes, pay close attention to the centromeres, which are numbered in the figure. During anaphase I of meiosis, each daughter cell will receive one of the two pairs of sister chromatids. For these translocated chromosomes, the expected segregation is due to the segregation of the centromeres; each daughter should receive a centromere located on chromosome 1 and a centromere located on chromosome 2. This can occur in two ways. One possibility is *alternate segregation*, this occurs when the chromosomes on opposite side of the translocation cross segregate into the same cell. One daughter cell receives two normal chromosomes, and the other cell gets two translocated chromosomes. Following meiosis II, four gametes are produced: two are normal and two have reciprocal (balanced) translocations.

Another possible segregation pattern is called *adjacent-1 segregation*. This occurs when adjacent chromosomes (one of each type of centromere) segregate into the same cell. When this occurs, each daughter cell receives one nor mal chromosome and one translocated chromosome. This produces four gametes, all of which are genetically unbalanced because part of one chromosome has been deleted and part of another has been duplicated. If the gametes produced from adjacent-I segregation unite with a normal gamete, the resulting zygote is expected to be abnormal genetically, and possibly phenotypically.

Alternate and adjacent-1 segregation are the likely outcomes when an individual carries a reciprocal translocation. Depending on the sizes of the translocated segments, both types may be equally likely to occur. In many cases, the gametes from adjacent-1 segregation cannot produce viable offspring, thereby lowering the fertility of the parent. On very rare occasions, *adjacent-2 segregation* can occur. In this case, the centromeres do not segregate as they should. One daughter cell has received both copies of the centromere on chromosome 1, the other both copies of the centromere on chromosome 2. This rare segregation pattern also yields four abnormal gametes that contain an unbalanced combination of chromosomes.

Variation in Chromosome Number

As we have seen in the preceding section, chromosome structure can be altered in a variety of ways. The total number of chromosomes, too, can vary. Eukaryotic species typically contain several chromosomes that are inherited as one or more sets. Variations in chromosome number can be categorized in two ways: variation in the number of sets of chromosomes, and variation in the number of particular chromosomes within a set.

Organisms that are *euploid* have a chromosome number that is an exact multiple of a chromosome set. In *Drosophila melanogaster*, a normal individual possesses eight chromosomes. The species is diploid, having two sets of four chromosomes each. We can also recognize that a normal fruit fly is euploid because eight chromosomes divided by four chromosomes per set equals two exact sets. On rare occasions, an abnormal fruit fly can be produced with twelve chromosomes, containing three sets of four chromosomes each. This alteration in euploidy produces a *triploid* fruit fly. Such a fly is also euploid, since it contains exactly three sets of chromosomes. Organisms with three

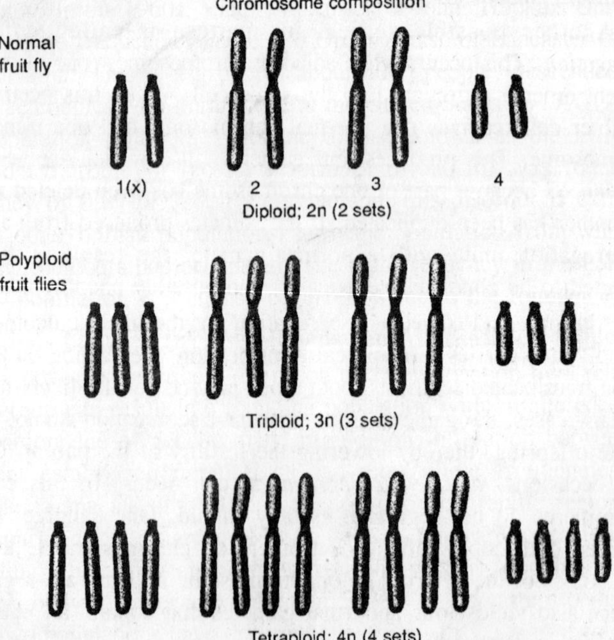

Fig. 5.2. types of variation in chromosome number.

or more sets of chromosomes are also called *polyploid*. Geneticists use the letter *n* to represent a set of chromosomes. A diploid organism is referred to as 2*n*, a triploid organism 3*n*, a *tetraploid* organism as 4*n*, and so forth.

A second way in which chromosome number can vary is by *aneuploidy* (literally, not euploidy). This term refers to a variation that involves an alteration in the number of particular chromosomes, so that the total number of chromosomes it not an exact multiple of a set (not an exact multiple of *n*). For example, an abnormal fruit fly could contain nine chromosomes instead of eight because it had three copies of chromosome 2 instead of the normal two copies. Such a animal is said to have *trisomy*-2 or to be trisomic. Instead of being perfectly diploi (i.e., 2*n*), a trisomic animal is $2n + 1$. By comparison, a fruit fly could be lacking copy of chromosome 1 and contain a total of seven chromosomes ($2n - 1$). The animal would have *monosomy-1*.

In this section, we will examine euploid and aneuploid variation in many eukaryotic species. We will learn that euploid variation among different species occurs occasionally in animals and quite frequently in plants. We will see how natural variation in the number of sets of chromosomes affects phenotypic variation. By comparison, aneuploidy is generally regarded as an abnormal condition We will consider several examples in which aneuploidy has a negative impact of an organism's phenotype.

Aneuploidy Causes an Imbalance in Gene Expression that is often Detrimental to the Phenotype of the Individual

The phenotype of every organism is influenced by thousands of different genes Humans, for example, possess approximately 100,000 different genes. Many genes are only expressed in certain cell types or during particular stages of development To produce a normal individual, intricate coordination has to occur in the expression of thousands of genes. In the case of humans and many other species, the process works correctly when there are two copies of each chromosome per cell. In other words, when a human is diploid, the balance of gene expression among man different genes usually produces a human with a normal phenotype.

Aneuploidy commonly causes an abnormal phenotype. To understand why consider the relationship between gene expression and chromosome number. For many (but not all) genes, the amount of expression is correlated with the number of genes per cell. Compared with a diploid cell, if a gem is carried on a chromosome that is present in three copies instead of two, approximately 150% of the normal amount of gene product will be made. Alter natively, if only one copy of that gene is present due to a monosomy abnormality only 50% of the gene product will be made. Therefore, in trisomic and monosomic individuals, there is an imbalance in the level of gene expression between the majority of chromosomes found in pairs versus those that are not.

At first glance, the difference in gene expression between euploid and aneuploid individuals may not seem terribly drastic. Keep in mind, however, that a eukaryotic chromosome carries hundreds or even thousands of different genes. Therefore, when an organism is trisomic or monosomic, many gene products will occur in excessive or deficient amounts. This imbalance among many genes appears to underlie the abnormal phenotypic effects that aneuploidy frequently causes. In most cases, these effects are detrimental and produce an individual that is less likely to survive than a euploid individual.

The first person to recognize the harmful effects of aneuploidy was Albert Blakeslee at Cold Spring Harbor. In the 1920s, Blakeslee and his colleagues identified many examples of alterations in chromosome number in the Jimson weed (*Datura stramonium*). All 12 trisomies had capsules that were morphologically different from those of the normal diploid strain. These aneuploid plants also had many other morphologically distinguishable traits, including changes in leaf shape, size, and so forth. In many cases, the observed changes in chromosome number produced detrimental traits. For example, Blakeslee noted that the cocklebur plant (trisomy-6) is "weak and lopping with the leaves narrow and twisted."

Aneuploidy in Humans causes Abnormal Phenotypes

One important reason that geneticists are so interested in aneuploidy is its relationship to certain inherited disorders in humans. Even though most people are born with a normal number of chromosomes alterations in chromosome number occur fairly frequently during gamete formation. About 5-10% of all fertilized human eggs result in an embryo with an abnormality in chromosome number. In most cases, these abnormal embryos do not develop properly and result in a spontaneous abortion very early in pregnancy Approximately 50% of all spontaneous abortions are due to alterations in chromosome number.

In some cases, an abnormality in chromosome number produces an offspring that can survive. Several human disorders involve abnormalities in chromosome number. The most common are trisomies of chromosomes 21, 18, or 13, or abnormalities in the number of the sex chromosomes. Most of the known trisomies involve chromosomes that are relatively small (i.e., chromosomes 21, 18, and 13). Trisomies of the other human autosomes and monosomies of the autosomes usually produce a lethal phenotype that causes early spontaneous abortion.

Variation in the number of X-chromosomes, unlike the case for most other large chromosomes, is often nonlethal. The survival of trisomy-X individuals is explained by X-inactivation. In an individual with more than one X-chromosome, all additional X-chromosomes are converted to Barr bodies in the somatic cells of adult tissues. In an individual with trisomy-X, for example, out of the three X-chromosomes are converted to inactive Barr bodies. Unlike the autosomes, the normal level of expression for X-linked genes is from a single X-chromosome. In other words, the correct level of mammalian gene expression results from two copies of each autosomal gene and one copy of each X-linked gene. This explains how the expression of X-linked genes in males (XY) can be maintained at the same levels as in females (XX). It may also explain why monosomy-X and trisomy-X are not lethal conditions. The phenotypic effects involving sex chromosomal abnormalities may be due to the expression of X-linked genes prior to embryonic X-inactivation, or due to the imbalance in the expression of pseudoautosomal genes.

Some human abnormalities in chromosome number are influenced by the age of the parents. Older parents are more likely to produce children with abnormalities in chromosome number. For example, the incidence of Down syndrome rises with the age of the parents, particularly that of the mother. This syndrome was first described by the English physician John Langdon Down in 1866. The association between maternal age and Down syndrome was later discovered by L. S. Penrose in 1933, even before the chromosomal basis for the disorder was identified by the French scientist Jerome Lejeune in 1959. Down syndrome is most commonly caused by nondisjunction at meiosis I in the oocyte. The mechanism of nondisjunction will be described later in this chapter.

Different theories have been proposed to explain the relationship between maternal age and Down syndrome. One popular idea suggests that it may due to the age of the oocytes. Human primary oocytes are produced within the ovary of the female fetus prior to birth and are arrested at prophase I of meiosis I until the time of ovulation. Therefore, as a women ages, her primary oocytes have been in prophase I for a progressively longer period of time. This added length of time may contribute to an increased frequency of nondisjunction. About 5% of the time, Down syndrome is due to an extra paternal chromosome. Prenatal tests can deter mine if a fetus has Down syndrome or some other genetic abnormalities.

Table 5.1. Aneuploid conditions in humans

Condition	*Frequency*	*Syndrome*	*Characteristics*
Autosomal			
Trisomy-21	1/800	Down	Mental retardation, abnormal pattern of palm creases, slanted eyes, flattened face, short stature
Trisomy-18	1/6,000	Edward	Mental and physical retardation, facial abnormalities, extreme muscle tone
Trisomy-13	1/15,000	Patau	Mental and physical retardation, wide variety of defects in organs, large triangular nose
Sex chromosomal			
XXY	1/1,000 (males)	Klinefeltor	Sexual immaturity no sperm, breast swelling
XYY	1/1,000 (males)	Jacobs	Tall
XXX	1/1,500 (females)	Super female	Tall and thin, menstrual irregularity
X0	1/5,000	Turner	Short stature, webbed neck, (females) sexually undeveloped

Variations in Euploidy Occur Naturally in a few Animal Species

We now turn our attention to changes in the number of sets of chromosomes, referred to as variations in euploidy Most species of animals are diploid. In many cases, changes in euploidy are not well tolerated. For example, polyploidy in mammals is generally a lethal condition. In less complex vertebrates and invertebrates, however, a few examples of naturally occurring variations in euploidy occur. Male bees (drones) contain a single set of chromosomes; they are produced from unfertilized eggs. By comparison, female bees are diploid. Geneticists often refer to male bees as being "haploid," although in a strict genetic sense, the term haploid describes gametes that contain half the genetic material of the parent's somatic cells. An organism with a single set of chromosomes within its somatic cells is more accurately called *monoploid*.

A few examples of polyploid animals have been discovered. In some cases, animals that are morphologically very similar to each other can be found as a diploid species as well as a separate polyploid species. This situation occurs among certain amphibians and reptiles. As you can see, they look fairly similar to each other. Their differences in euploidy can only be revealed by an examination of the chromosome number in the somatic cells of the animals.

Variations in Euploidy can Occur in Certain Tissues within an Animal; Polytene Chromosomes are an Extreme example of Repeated Chromosome Doubling

Thus far, we have considered variations in chromosome number that occur at fertilization, so that all the somatic cells of an individual contain this variation. In many animals, certain tissues of the body will display normal variations in the number of sets of chromosomes. In particular, diploid animals sometimes produce tissues that are polyploid. The cells of the human liver, for example, can vary to a great degree in their ploidy. Liver cells can be triploid, tetraploid, and even octaploid (8*n*). This phenomenon is known as *endopolyploidy*. The biological significance of endopolyploidy is not completely understood. One possibility is that the increase in chromosome number in certain cells may enhance their ability to produce specific gene products that are needed in great abundance.

An unusual example of natural variation in the ploidy of somatic cells occurs in *Drosophila* and a few other insects. Within certain tissues, such as the salivary glands, the *chromosomes* undergo repeated rounds of chromosome replication without cellular division. For example, in the salivary gland cells of *Drosophila*, the pairs of chromosomes double approximately nine times ($2^9 = 512$). Repeated rounds of chromosomal replication produce a bundle of chromosomes that lie together in a parallel fashion. This bundle of many chromatids lying side-by-side is a polytene chromosome. They were first observed by E. G. Balbiani in 1881. In the early 1930s, Theophilus Painter and his colleagues at the University of Texas recognized that the size and morphology of polytene chromosomes provided geneticists with unique opportunities to study chromosome structure and gene organization.

Prior to the formation of polytene chromosomes, *Drosophila* cells contain eight chromosomes (two sets of four chromosomes each). In the salivary gland cells, the homologous chromosomes synapse with each other and replicate to form a polytene structure. During this process, the four types of chromosomes aggregate to form a single structure with several polytene arms. The central point where the chromosomes aggregate is known as the *chromocenter*. Each of the four types of chromosome are attached to the chromocenter near their centromeres. The X and Y and chromosome 4 are telocentric, and chromosomes 2 and 3 are metacentric. Therefore, chromosomes 2 and 3 have two arms that radiate from the chromocenter while the X and Y and chromosome 4 have a single arm projecting from the chromocenter.

Because of their considerable size, polytene chromosomes lend themselves to an easy microscopic examination. Ordinarily, we use light microscopy to visualize the highly condensed metaphase chromosomes seen during mitosis or meiosis. Since polytene chromosomes are so large, we can even see them during interphase, when normal chromosomes are not visible. In fact, a polytene chromosome during interphase is actually 100—200 times as large as the average metaphase chromosome. Polytene chromosomes exhibit a characteristic banding pattern. Each dark band is known as a *chromomere*. The structure of the genetic material within a dark band is more compact than in the interband region. More than 95% of the DNA is found within the chromomeres. The banding patterns of polytene chromosomes are much more detailed than those observed in metaphase chromosomes. Cytogeneticists have identified approximately 5000 bands along polytene chromosomes. At one time, each chromomere was thought to correspond to one gene. However, this idea was found to be incorrect. Polytene chromosomes have allowed geneticists to study the organization and functioning of interphase chromosomes in great detail. When a gene is deleted or duplicated in a mutation, researchers can map the change by observing abnormalities in the structure of polytene chromosomes. In addition, since polytene chromosomes can be observed during interphase, the expression of particular genes can be correlated with changes in the compaction of certain bands in the polytene chromosome.

Variations in Euploidy are Common in Plants; this Variation has allowed the Development of Many Agricultural Crops

We now turn our attention to the variations of euploidy that occur in plants. In contrast to animals, plants commonly exhibit polyploidy. Among ferns and flowering plants, about 30—35% of the species are polyploid. Polyploidy is also important in agriculture. Many of the fruits and grains we eat are produced from polyploid plants. For example, the species of wheat that we use to make bread, *Triticum aestivum*, is hexaploid ($6n$). In many instances, polyploid strains of plants display outstanding agricultural characteristics. They are often larger in size and more robust. These traits are clearly advantageous in the production of food. In addition, polyploids tend to exhibit a greater adaptability, which allows them to with stand harsher environmental conditions. Also, polyploid ornamental plants often produce larger flowers than their diploid counterparts.

Polyploids having an odd number of chromosome sets, such as triploids or pentaploids, are usually sterile. The sterility arises because they produce highly aneuploid gametes. To understand why aneuploidy occurs, consider what happens during anaphase Tin a triploid organism. Since there are three copies of each replicated chromosome, they cannot be divided equally between two daughter cells. For each type of chromosome, a daughter cell randomly gets one or two copies. For example, one daughter cell might receive one copy of chromosome 1, two copies of chromosome 2, two copies of chromosome 3, one copy of chromosome 4, and so forth. For a triploid species containing many different chromosomes in a set, it is very unlikely that any gamete will be euploid. If we assume that a daughter cell will receive either one copy or two copies of each kind of chromo some, the probability that a gamete will end up being perfectly haploid or diploid is $(1/2)^{N-1}$, where *N* is the number of chromosomes in a set. As an example, in a triploid species containing 20 chromosomes per set, the probability of producing a haploid or diploid gamete is 1 in 524,288. Thus, a gamete is almost certain to contain one copy of some chromosomes and two copies of the other chromosomes. This high probability of aneuploidy underlies the reason for triploid sterility.

Sterility can be an agriculturally useful trait, since it may result in a seedless fruit. For example, seedless watermelons and bananas are triploid varieties. The domestic banana, which is triploid, was originally derived from a normal diploid species and has been asexually propagated by humans via cuttings. The small black spots in the center of a domestic banana are degenerate seeds. In the case of flowers, the seedless phenotype can also be beneficial. Seed producers such as Burpee are developing triploid varieties of flowering plants. The triploid marigold is sterile and unable to set seed. According to Burpee, "they bloom and bloom, unweakened by seed bearing."

Natural and Experimental Ways to Produce Variations in Chromosome Number

As we have seen, variations in chromosome number are fairly widespread and usually have a significant impact on the phenotypes of plants and animals. For these reasons, researchers have wanted to understand the cellular mechanisms that cause variations in chromosome number. In some cases, a change in chromosome number is the result of nondisjunction. The term *nondisjunction* refers to the event in which the chromosomes do not separate properly during anaphase. As we will see, it maybe caused by an improper separation of homologous pairs in a tetrad, or a failure of the centromere to split during meiosis II or mitosis. *Meiotic nondisjunction* can produce gametes that have too many or too few chromosomes. If such a gamete fuses with a normal gamete during fertilization, the resulting individual will have an abnormal chromosomal composition in all of the cells of the organism. An abnormal nondisjunction event also may occur after fertilization in one of the somatic cells of the body. This second mechanism is known as *mitotic nondisjunction*. When this occurs during embryonic stages of development, it may lead to a patch of tissue in the organism that has an abnormal chromosomal composition.

Finally, a third common way in which the chromosome composition of an organism can vary is by *interspecies matings* to produce an alloploid organism. An *alloploid* organism contains sets of chromosomes from two (or more) different species. This term refers to the occurrence of chromosome sets (ploidy) from different (*allo-*) species.

In this section, we will examine these three mechanisms in greater detail. Also, in the past few decades, researchers have devised several methods to manipulate chromosome number in experimentally and agriculturally important species. As we will learn, the experimental manipulation of chromosome number has had an important impact on genetic research and agriculture.

Meiotic Nondisjunction can Produce Aneuploidy; Complete Nondisjunction can Produce Polyploidy

The phenomenon of meiotic nondisjunction was first described by Calvin Bridges, who compared inheritance patterns in *Drosophila* with the cytological presence of certain chromosomes. Nondisjunction event can occur during meiosis I or meiosis II. If it occurs during anaphase of meiosis I, an entire tetrad will migrate into one of the two daughter cells. All the gametes produced from this event will be abnormal. A second possibility is that nondisjunction can occur during anaphase of meiosis II. If this occurs, the net result will be two normal and two abnormal gametes. In this scenario, the individual produced from an abnormal gamete will be aneuploid (namely, either monosomic or trisomic for one chromosome). In rare cases, all the chromosomes can

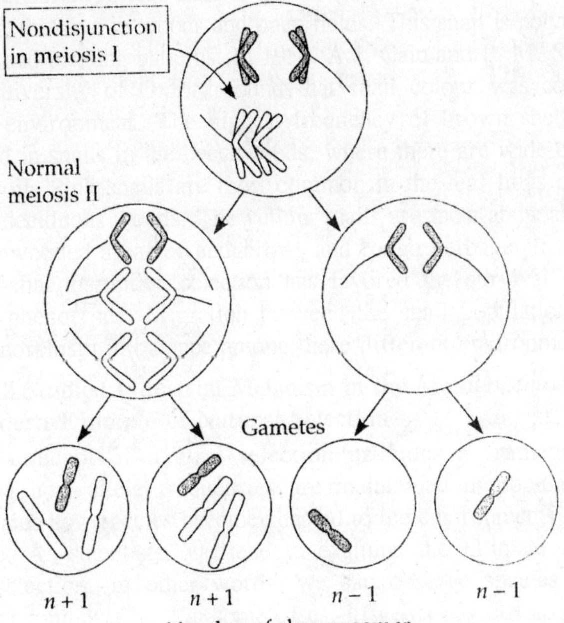

Fig. 5.3. Nondisjunction during meiosis I.

undergo nondisjunction and migrate to one of the daughter cells. The net result of *complete nondisjunction* is a diploid gamete and a gamete without chromosomes. While the gamete without chromosomes is nonviable, the diploid gamete might participate in fertilization with a normal haploid gamete to produce a triploid individual. Therefore, complete nondisjunction can produce individuals that are polyploid.

Mitotic Nondisjunction or Chromosome loss can Produce a Patch of Tissue with an Altered Chromosome Number

Abnormalities in chromosome number occasionally occur after fertilization takes place. In this case, the abnormal event happens during mitosis rather than meiosis. One possibility is that the sister chromatids could separate improperly, so that one daughter cell would have three copies of that chromosome while the other daughter cell would only have one. Alternatively, the sister chromatids could separate during anaphase of mitosis but one of the chromosomes could be improperly attached to the spindle, so that it would not migrate to a pole. If this happens, a chromosome will be degraded if it is left out side of the nucleus when the nuclear membrane re-forms. In this case, one of the daughter cells would have two copies of that chromosome,

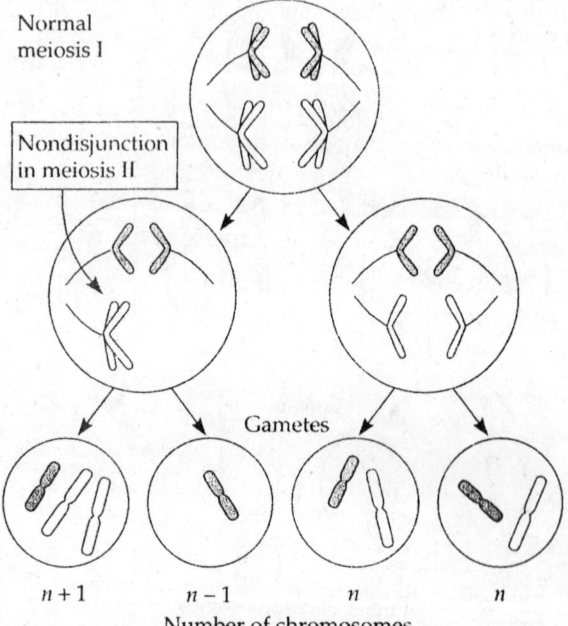

Fig. 5.4. Nondisjunction during meiosis II.

while the other would have only one. When genetic abnormalities occur after fertilization, part of the organism will contain cells that are genetically different from the rest of the organism. This condition is referred to as *mosaicism*. The size and location of the mosaic region will depend on the timing and location of the original abnormal event. If a genetic alteration happens very early in the embryonic development of an organism, the ab normal cell will be the precursor for a large section of the organism, In the most extreme case, an abnormality could take place at the first mitotic division. As an example, consider a fertilized *Drosophila* egg that is XX. One of the X-chromosomes may be lost during the first mitotic division, producing one daughter cell that is XO and one that is XX. XX flies develop into females XO flies develop into males. Therefore, in this example, one-half of the organization will become male and one-half will become female. This peculiar and rare individual is referred to as a *bilateral gynandromorph*.

Changes in Euploidy can Occur by Autopolyploidy, Alloploidy, and Allopolyploidy

Different mechanisms account for changes in the number of chromosome sets among natural populations of plants and animals. As mentioned earlier, complete nondisjunction can produce an individual with one or more extra sets of chromosomes. This individual is known as an *autopolyploid*. The terms *auto-* (meaning self) and polyploid (meaning many sets of chromosomes) refer to the increase in the number of chromosome sets occurring within a single species.

A much more common mechanism for change in chromosome number, called *alloploidy*, is a result of interspecies matings. This event is most likely to occur between species that are close evolutionary relatives. For ex ample, closely related species of grasses may interbreed to produce alloploids. The *allopolyploid* contains a combination of both autopolyploidy and alloploidy. In this case, the allopolyploid contains two complete sets of chromosomes from two different species for a total of four sets. As mentioned earlier in this chapter, domesticated wheat is a hexaploid species. It is actually an allopolyploid that contains two sets of chromosomes from three different, but closely related, wild species.

Experimental Treatments can Promote Polyploidy

Because polyploid plants often exhibit desirable traits, the development of polyploids is of considerable interest among plant breeders. Experimental studies on the ability of environmental agents to promote polyploidy began in the early 1900s. Since that time, various agents have been shown to promote nondisjunction and thereby lead to polyploidy. These include abrupt temperature changes during the initial stages of seedling growth, and the treatment of plants with chemical agents that interfere with the formation of the spindle apparatus.

The drug *colchicine* is commonly used to promote polyploidy. Once inside the cell, colchicine binds to tubulin (a protein found in the spindle apparatus) and thereby interferes with normal chromosome segregation during mitosis or meiosis. In 1937, Alfred Blakeslee and Amos Avery applied colchicine to plant tissue and, at high doses, were able to cause complete mitotic nondisjunction. This strategy can be used to produce new polyploid strains of plants. Colchicine can be applied to seeds, young embryos, or rapidly growing regions of a plant. This application may produce aneuploidy (usually an undesirable outcome), but it often produces polyploid cells, which may grow faster than the surrounding diploid tissue. If the parent plant is diploid, colchicine may cause complete mitotic disjunction, yielding tetraploid ($4n$) cells. As the tetraploid cells continue to divide, they generate a portion of the plant that is often morphologically distinguishable from the remainder. For example, a polyploid stem may have a larger diameter and produce larger leaves and flowers. Since individual plants can be propagated asexually from pieces of plant tissue (i.e., cuttings), the polyploid portion of the plant can be removed, treated with the proper growth hormones, and grown as a separate plant. Alternatively, the polyploid region of a plant may have flowers that produce polyploid seeds; a tetraploid flower will produce diploid pollen and eggs, which will combine to produce tetraploid offspring. In this way, the use of colchicine provides a straightforward method to produce autopolyploid strains of plants.

Alloploids are Often Sterile, but Allopolyploids are More Likely to be Fertile

Geneticists are interested in the production of alloploids and allopolyploids as ways to generate interspecies hybrids with desirable traits. For example, if one species of grass can withstand hot temperatures and a closely related species is adapted to survive through cold winters, a plant breeder may hope to produce an interspecies hybrid that combines both qualities (i.e., good growth in the heat and survival through the winter). Such an alloploid would be desirable in climates with both hot summers and cold winters.

An important determinant of success in producing an interspecies hybrid is the degree of similarity of the different species' chromosomes. In two very closely related species, the number of chromosomes might be identical or very similar. A karyotype of an interspecies hybrid between the roan antelope (*Hippotragus equinus*) and the sable antelope (*H. niger*) is presented. As seen here, these two closely related species have the same number of chromosomes. Moreover, the sizes and banding patterns of the chromosomes show that they correspond to one another. For example, chromosome 1 from both species is fairly large and has very similar banding patterns. This chromosome derived from both species likely carries many of the same genes. Analogous chromosomes from related species are called *homeologous* chromosomes (not to be confused with homologous).

The degree of similarity between the homeologous chromosomes also greatly influences whether an alloploid is fertile. This phenomenon was first recognized by the Russian cytogeneticist G. D. Karpechenko in 1928. He crossed a radish (*Rophanus*) and a cabbage (*Brassica*), both of which are diploid and contain 18 chromosomes. Each of these organisms produces haploid gametes containing 9 chromosomes each. Therefore, the alloploid produced from this interspecies mating contains 18 chromosomes. Since the radish and cabbage are not closely related, the nine *Raphanus* chromosomes are distinctly different from the nine Brassica chromosomes. During meiosis I, the radish and cabbage chromosomes cannot synapse with each other, resulting in a high degree of aneuploidy. Therefore, the radish/cabbage hybrid is sterile.

Karpechenko discovered a way to overcome the sterility of the radish/cabbage hybrid. The idea is to produce an allopolyploid (rather than an alloploid) with a diploid number of chromosomes from each of the two species. In the example here, a radish/cabbage allopolyploid would contain 36 chromosomes instead of 18. During metaphase I, the homologous chromosomes from each of the two species can synapse properly When anaphase I occurs, the pairs of synapsed chromosomes can disjoin equally to produce gametes with 18 chromosomes each (a haploid set from the radish and a haploid set from the cabbage). These gametes with 18 chromosomes can combine with each other to produce an allopolyploid containing 36 chromosomes. In this way, the allopolyploid is a fertile organism. Unfortunately, it is not agriculturally useful, because its leaves are like the radish and its roots are like the cabbage!

Modern plant breeders employ several different strategies to produce allopolyploids. One is to start with two different tetraploid species. Since these make diploid gametes, the hybrid from this cross would contain a diploid set from each species. Alternatively, if an alloploid containing a monoploid set from each species already exists, a second approach is to create an allopolyploid by using agents such as colchicine.

Cell Fusion Techniques can be Used to Make Hybrid Plants

Thus far, we have discussed several mechanisms that produce variations in chromosome number. Some of these processes can occur naturally and have figured prominently in speciation and evolution. In addition, agricultural geneticists can administer treatments such as colchicine to promote nondisjunction and thereby obtain useful strains of organisms. More recently, researchers are devising cellular approaches to produce hybrids with altered chromosomal composition. As de scribed here, these cellular approaches have important applications in research and agriculture.

In the technique known as *cell fusion*, individual cells are mixed together and made to fuse. This method can be used in gene mapping. In agriculture, cell fusion can create new strains of plants. Cell fusion may allow the crossing of two plants that normally cannot interbreed. There are several reasons that interbreeding cannot occur. For example, one of the parents may be sterile. Also, two distantly related species may not be able to cross-pollinate. Similarly, within a single species there may exist" incompatibility alleles" that prevent certain strains from cross-pollinating or self-fertilizing. In many cases, cell fusion techniques have successfully circumvented such interbreeding problems.

The parent cells were derived from tall fescue grass (*Festuca arundinacea*) and Italian rye-grass (*Lolium multiflorum*). Prior to fusion, the plant cells from these two species were treated with agents that gently digest the cell wall without rupturing the plasma membrane. A plant cell without a cell wall is called a *protoplast*. The protoplasts are mixed together and treated with agents that promote fusion. Immediately after this takes place, a cell containing two separate nuclei is formed. This cell, known as a *heterokaryon*, will spontaneously go through a nuclear fusion process to create a *hybrid cell* with a single nucleus. After nuclear fusion, the hybrid cells can be grown on laboratory media and eventually regenerate an entire plant. The allopolyploid has phenotypic characteristics that are intermediate between tall fescue grass and Italian ryegrass. This type of approach is sometimes used by agricultural geneticists to produce intraspecies and interspecies hybrids.

Monoploids Produced in Agricultural and Genetic Research can be Used to Create Homozygous and Hybrid Strains

Plant breeders sometimes wish to have diploid strains of crop plants that are homozygous for all of their genes. One true-breeding strain can then be crossed to a different true-breeding strain to produce an F_1 hybrid that is heterozygous for many genes. Such hybrids are often more vigorous than the corresponding homozygous strains. This phenomenon, known as *hybrid vigor* or *heterosis*.

Seed companies often use this strategy to produce hybrid seed for many crops, such as corn and alfalfa. To achieve this goal, the companies must have homozygous parental strains that can be crossed to each other to produce the hybrid seed. One way to obtain these homozygous strains involves inbreeding over many generations. This may be accomplished after several rounds of self-fertilization. As you might imagine, this can be a rather time-consuming endeavor.

As an alternative, the production of monoploids can be used as part of an experimental strategy to develop homozygous diploid strains of plants. In 1966, Sipra Guha and Satish Maheswari at the University of Delhi produced monoploid plants directly from pollen (containing the haploid male gamete). Monoploids produced experimentally have been used to improve many agricultural crops such as wheat, rice, corn, barley, and potato. Experimental technique called *anther culture*, which can be used to produce a diploid strain homozygous for all of its genes. It involves alternation between monoploid and diploid generations. The parental plant is diploid but not homozygous for all of its genes. The anthers from this diploid plant are collected, and the haploid pollen grains within them are induced to begin development by a cold shock treatment. After several weeks, monoploid plantlets will emerge, and these can be grown on agar media in a laboratory. Eventually, the plantlets can be transferred to small pots. After the plantlets grow to a reasonable size, a section of the monoploid plantlet can be treated with colchicine to convert it to diploid tissue. A cutting from this diploid section can then be used to generate a separate plant. This diploid plant is expected to be homozygous for all of its genes, be cause it was produced by the chromosomal doubling of a monoploid strain.

In certain animal species, monoploids can be produced by experimental treatments that induce the eggs to begin development without fertilization by sperm. This process is known as *parthenogenesis*. In many cases, however, the haploid zygote will only develop for a short period of time before it dies. Nevertheless, a short phase of development can be useful to research scientists., the zebrafish (*Brachydanio rerio*), a common aquarium fish, has gained recent popularity among researchers interested in vertebrate development. The haploid egg can be induced to begin development by exposure to sperm rendered biologically in active by UV-irradiation.

We classify variations in chromosome structure as *deletions* (or *deficiencies*), *duplications*, *inversions*, and *translocations*. Deletions and duplications involve changes in the total amount of genetic material; inversions and translocations are genetic rearrangements. Deletions tend to be detrimental to the phenotype of the individual, although this depends on the size and location of the deletion. In some cases, a deletion heterozygote will exhibit *pseudodominance* of (normally recessive) genes located in the corresponding region of the non-deleted chromosome. In comparison, duplications tend to be less harmful. Gene duplications provide the raw material for the evolution of *gene families* in which a group of genes encode proteins that carry out similar yet specialized functions.

Inversions and translocations are chromosomal rearrangements; they often do not affect the phenotype of the individual who carries them.

Reciprocal translocations are sometimes called *balanced translocations*, because the individual has a normal amount of genetic material. In a few cases, inversions and translocations can affect phenotype because a breakpoint is within a vital gene or the rearrangement results in a *position effect* that alters gene expression or regulation. Inversions and translocations are often associated with fertility problems and a higher probability of producing abnormal offspring. Even though an individual carrying an inversion or balanced translocation may be phenotypically normal, crossing over and chromosomal segregation may lead to the production of gametes that do not have a balanced amount of genetic material. Such gametes are likely to produce phenotypic abnormality or even lethality.

Chromosome number is another critical factor in determining the phenotype of an organism. In the condition known as *aneuploidy*, an individual may have an extra chromosome (trisomy) or may be missing a chromosome (monosomy). This usually is detrimental to the phenotype. For example, various human genetic diseases, such as Down syndrome (trisomy-21), are due to irregularities in chromo some number. Likewise, in plants, aneuploidy also significantly affects the phenotype of an individual.

Variations in the number of sets of chromosomes are also common, particularly in the plant kingdom. This has dramatically influenced agriculture. Many of our modern crops are *polyploid* plants. These frequently exhibit characteristics superior to those of their diploid counterparts. Polyploids with an odd number of chromosome sets are usually sterile, and these are useful in producing seedless varieties of plants. In animals, some cells of the body may be *endopolyploid*—that is, they have more sets of chromosomes than other somatic cells. An extreme example are the *polytene chromosomes* found in *Drosophila*.

Changes in the chromosome number can arise via several different mechanisms. In natural populations, the most important of these are *nondisjunction* and *interspecies matings*. Nondisjunction produces gametes with alterations in chromosome number, which can lead to aneuploidy or even polyploidy. Also, nondisjunction or chromosome loss can occur in cells after fertilization to produce an individual that is a genetic *mosaic*. Another way to produce new combinations of chromosomes is by interspecies matings. These can produce *alloploids*, which have one set of chromosomes from each species, or *allopolyploids*, which have two or more sets from each species. Allopolyploids are more likely to be fertile when the chromosomes are found in homologous pairs.

The primary experimental strategy for studying variation in chromosome structure and number is the cytological examination of chromosomes. A karyotype is a photographic representation of the chromosomes from actively dividing eukaryotic cells. When preparing a karyotype, the chromosomes are usually stained with a dye such as Giemsa, which gives individual chromosomes their own characteristic banding pattern. By observing the banding patterns and number of chromosomes under a microscope, a cytogeneticist can determine if an individual carries a change in chromosome structure or number.

Experimental methods are also available to cause changes in chromosome number. For example, drugs such as colchicine can alter chromosome number by promoting nondisjunction. In addition, laboratory methods can produce hybrids by cell fusion techniques. This can be useful in making fertile interspecies hybrids. Also, monoploid plants and animals can be produced by the experimental activation of haploid gametes.

6

CHROMOSOME ABERRATIONS

It has already been stated that chromosomes occasionally break and that the rate of breakage can be increased by the use of various mutagens such as ionizing radiation(s), ultraviolet light, and certain chemicals. When a chromosome breaks, the two broken ends usually undergo *restitution* or healing, and everything is as before. But this does not always happens, and the result can be the formation of (1) deficiencies, (2) duplications, (3) inversions, or (4) translocations. The first three of these ordinarily involve a single chromosome, while the fourth involves two or more chromosomes.

There is very strong evidence the DNA in chromosomes that extends as a double helical thread throughout the entire length of the structure. Furthermore, we know that the genes are linearly arranged in the DNA and that at least in some instances groups of genes under the control of common regulatory units are grouped together. It is of great interest then to ask about the genetic consequences of breaks in specific regions of chromosomes.

One interesting observation made by George Lefevre on Drosophila chromosomes is that perhaps as many as 70 percent of chromosome break points that have successfully reunited with new neighbouring genes have no detectable genetic effect at all. In other words, only about 30 percent of healed breaks points have mutant effects associated with them. This could mean that only a portion of the DNA of chromosomes is informational, and breaks within the regions without information are inconsequential. It could also means that these regions may be more susceptible to breakage and rejoining than others. We do not yet know, but C. S. Lee has suggested that those break points that manage to reunite are likely to be in regions of the DNA that are represented repeatedly in the genome. The reasoning behind this suggestion has to do with the healing process itself rather than with a differential fragility of chromosome regions. The union of broken ends may involve some degree of base pairing at the joint in order for the DNA strands to be properly positioned and stabilized for ligase to be effective in forming covalent bonds between the two chains. It is of course possible that unions can be formed without a perfect match between the bases of the two strands. Experiments, with a λ phage by Wu and Taylor have shown that the ends of the λ chromosome in its linear form have complementary sequences of nucleotides at the protruding cohesive ends. The λ chromosome goes from the linear to the circular form by the formation of hydrogen bonds between the complementary bases as shown in the figure. The same sort of protruding single strand ends with complementary or mostly complementary sequences may cause the joining of broken ends of eukaryotic chromosomes. Broken chromosome ends have been described for many years by cytogeneticists as being sticky. This sickness may be nothing more than the complementarily that exists between repeated DNA sequences.

It is known that in some tissues of many organisms, such as the endosperm of maize kernels, where the broken end soft chromosomes remain unhealed at the time of chromosome replication, that there is often a fusion of the broken ends of the sister chromatids. The results in a dicentric chromosome that will produce a chromosome bridge at the anaphase of the next mitotic division. In plant cells, such a bridge will often break as the cell wall is laid down between newly formed daughter cells. This produces another broken end in each of the nuclei, and the cycle may begin again. This breakage-fusion-bridge cycle may be carried on for several cell generations, and it can be used to great advantage for manipulating gene dosage in such cells. As the bridge breaks, it may not do so at the original fusion point and the result will be a duplication of some genes in one cell and their deletion in the other. McClintock used such a scheme to study chromosome organization and gene regulation in the endosperm of maize.

For most cells, however, the rule is that aberrations that persist are those that result from rejoined ends. This is because a broken chromosome, unrejoined, will generally not persist through the zygote stage. The segment without the centromere, called the acentric fragment, will be lost. The segment that has the centromere may act as a dominant lethal, depending on how much and what is lost in the acentric fragment.

DELETIONS

Deletions are also referred to as *deficiencies*. Some geneticists, following the historical origins of the terms, use deletion to describe an intercalary (involving two breaks in the chromosome) loss of chromatin and deficiency to describe a terminal loss. But here we use the terms deficiency and deletion as synonyms, because in either case an acentric fragment is lost. A spindle fiber cannot attach to an acentri fragment during mitosis or meiosis and guide it to one of the poles.

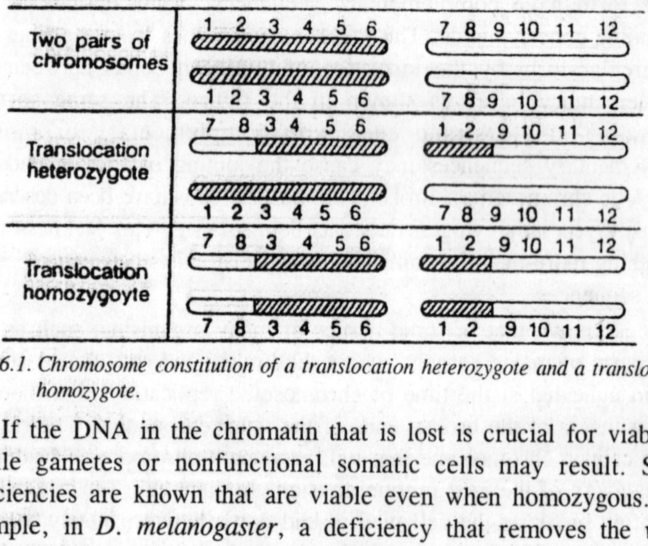

Two pairs of chromosomes	1 2 3 4 5 6	7 8 9 10 11 12
	1 2 3 4 5 6	7 8 9 10 11 12
Translocation heterozygote	7 8 3 4 5 6	1 2 9 10 11 12
	1 2 3 4 5 6	7 8 9 10 11 12
Translocation homozygoyte	7 8 3 4 5 6	1 2 9 10 11 12
	7 8 3 4 5 6	1 2 9 10 11 12

Fig. 6.1. Chromosome constitution of a translocation heterozygote and a translocation homozygote.

If the DNA in the chromatin that is lost is crucial for viability, sterile gametes or nonfunctional somatic cells may result. Some deficiencies are known that are viable even when homozygous. For example, in *D. melanogaster*, a deficiency that removes the *white* locus produces, when homozygous, fully developed flies that appear normal except for their white eye colour. Most deficiencies, however, are lethal when homozygous, and these may involve no more loss of chromatin than in the example of white locus. A loss of a large segment of a chromosome may result in a dominant lethal effect even if a normal homologue is present with it in the heterozygote. If the dominant effect of a deletion is not lethal, it may result in an abnormal phenotype.

Genetic Significance of Deficiencies

The chromosomal deficiencies have following genetic effects :

(i) Lethal effect

Organisms with homozygous deficiency usually do not survive to an adult stage because a complete set of genes is lacking.

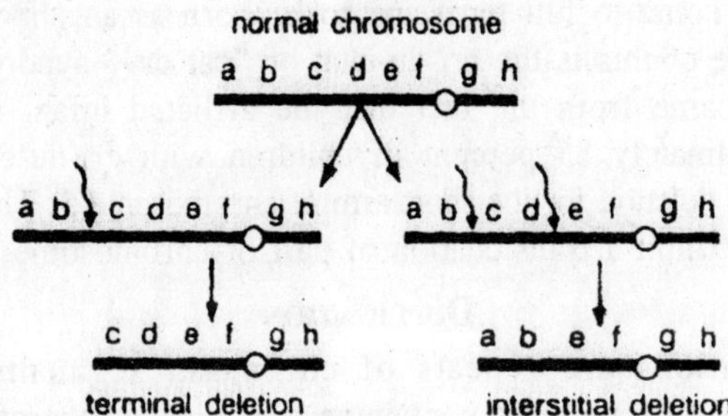

Fig. 6.2. Production of terminal and interstitial deletion. Chromosomes can be broken when struck by ionizing radiation (wavy arrows).

Examples

A mutant sex-linked condition called "notch" is lethal in *Drosophila* when hemizygous in males or when homozygous in females. This "notch phenotype is a sex-linked deletion (deficiency) which acts like a dominant mutation ; a deletion at another sex-lined locus behaves as recessive mutation, producing yellow body colour when homozygous.

(ii) Pseudodominance

When an organism heterozygous for a pair of alleles, A and a, loses a small portion of the chromosome bearing the dominant allele (A), the recessive allele (a) on the other chromosome being hemizygous will become expressed phylogenetically. The phenotypic expression of recessive allele (a) due to deficiency of dominant allele (A) called ***pseudodominance.***

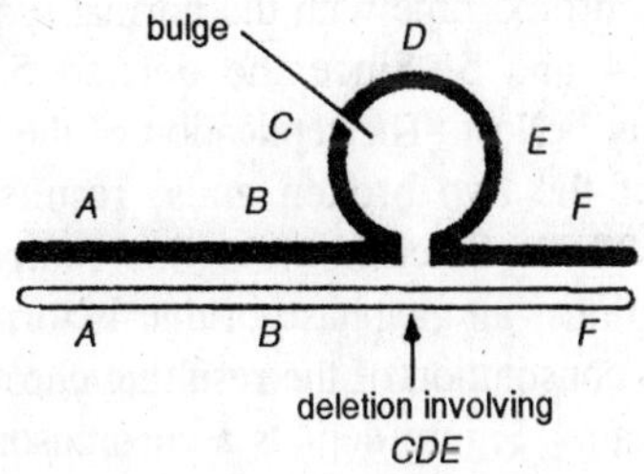

Fig. 6.3. Formation of a deletion loop during synapsis in a deletion heterozygote.

With a deficiency exchange (crossing over) becomes impossible. Disturbance of pairing causes a reduction of exchange in a segmen around the deficiency.

Occurrence of Deletions in Human Beings

Deletions are expected to occur in all organisms including humans. Some do not cause early spontaneous abortions, but unfortunately allow the fetus to come to full term and to be born as an abnormal child. One example of this is the cri-du-chat or "cat cry" syndrome, which derives its name from the fact that the afflicted infant cries like a cat. Approximately 85 percent of children with cri-du-chat show a heterozygous deletion for the short arm of chromosome 5. The remaining 15 percent exhibit a translocation of part of chromosome 5.

DUPLICATION

Duplications are repeats of chromatin segments within a chromosome. Chromosome containing a duplication therefore possess more than the normal amount of DNA. There are several possible mechanisms for generating duplicate DNA segments in chromosomes.

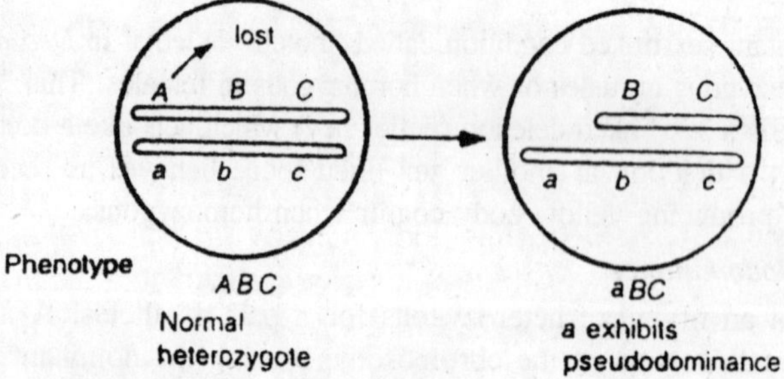

Fig. 6.4. Pseudodominance. A deficiency in the segment of chromosome bearing the dominant gene A allows the recessive allele a to become phenotypically expressed.

Duplications and deletions may also be generated by the breakage-fusion-bridge cycle. A chromosome with the normal sequence of 1.23456 has a break between 4 and 5. Since the deleted 5,6 is an acentric fragment it will probably be lost. The replication of the 1.2 3 4 fragment, followed by joining of the two broken ends, results in the dicentric chromosome 1. 234-432. 1. Since homologous centromeres generally move in opposite directions, an anaphase bridge is formed and breakage will occur. The genetic constitution of the resulting chromosomes depends on where the break occurs. Figure depicts a chromosome with deletions of regions 4,5 and 6. Also it shows a chromosome with both deleted (5,6) and duplicated (3,4) regions.

The breakage-fusion-bridge cycle that has been found in maize results in a wide variety of duplications and deletions. Note that duplications and deletions. Note that duplications created in this manner often are displaced, that is, duplicated segments are not positioned side by side on the chromosome. Duplications play an important role in the regulation of gene activity because the multiple copies of a

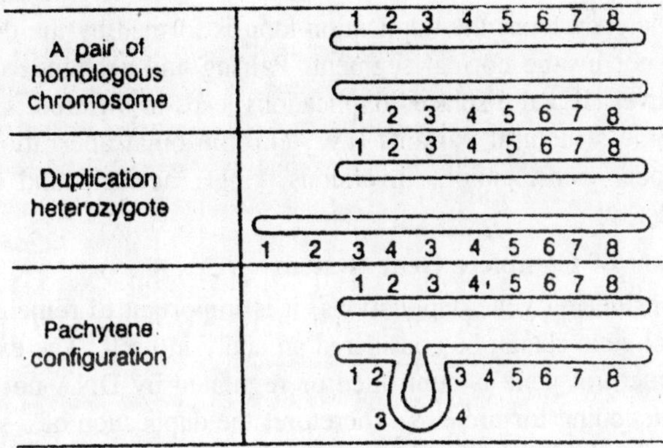

Fig. 6.5. Loop formation during chromosome pairing in a duplication heterozygote.

gene allow large quantities of RNA to be generated quickly. Examples of this type of reiteration are the cistrons coding for the ribosomal RNA's and those coding for histones. Table 6.1. gives some examples of proteins and RNA's thought to be coded by repetitive DNA sequences.

Table 6.1. Examples of repetitive DNA sequences

Group	*No repetitive DNA sequences*	*Degree of DNA or protein homology, %*
Satellite DNA	103–106	80–100
18S-28S Ribosomal RNA	100–600	97–100
5S Ribosomal RNA	100–200	97–100
tRNA	6-400	
Histones	10–1200	87–99
Hemoglobins	=10	<75–100

Satellite DNA's contain large numbers of repeating sequences which apparently are never transcribed. The nature of the mechanism(s) that generate large numbers of DNA segments in tandem linkage is unknown but it may well be by unequal crossing over. Also unknown are the factors that protect the repeated sequences from mutational changes.

Duplication occurs when a segment of the chromosome is represented two or more times in a chromosome of a homologous pair. This extra-chromosomal segment may be a free fragment with a centromere or chromosomal segment of the normal complement. During meiotic pairing the chromosome bearing the duplicated segment forms a loop to maximize the juxtraposition of homologous regions. In contrast to the deficiency loop, the duplication loop is formed by the duplicated segment, not by the normal segment. Pairing and exchange (crossing over) is inverted and displaced duplications leads to different secondary chromosome structural variants (i.e., chromosomal aberrations) such as reciprocal translocation, inversions, rings, acentric and dicentric chromatids.

Duplication of Complete Gene System

In considering gene duplications, it is important to remember that a structural gene cannot be considered an entity in itself. The expression of any structural gene is controlled or regulated by DNA not directly involved in coding for proteins. Therefore, the duplication of a structural gene without the corresponding duplication of its regulator may lead to a large array of new phenotypes including lethality.

Gene Duplication as a source of New Genetic Material

In addition to permitting a cell or organism to vary its genetic capabilities through gene dosage effects, duplications may also provide a cell with a reservoir of extra DNA. The duplicated genes could be considered as spare parts in the sense that one gene is sufficient to enable normal cell function. With one gene functioning in the normal manner, the duplicate would be free to undergo mutation, since its gene product could no longer be essential to maintain normal function.

By taking this reasoning a step further, it may be postulated that the duplicate gene may mutate and eventually evolve into a different gene coding for a protein with a new function. If so, gene duplications offer a mechanism for the evolution of new genes. There are many examples in nature that appear to fit the hypothesis that gene duplication is followed by gene changes through point mutation. Base nucleotide additions or deletions and frameshift mutations may result in genes coding for new phenotypes. The following two examples illustrate this concept.

Some of the most extensively studied human proteins that are thought to have arisen through gene duplications are the multiple hemoglobins of vertebrates. Hemoglobins are tetramers composed of two different polypeptide chains. Adult humans possess hemoglobins A and A_2. These two hemoglobins, represented by the formulas $\alpha_2\beta_2$ and $\alpha_2\delta_2$, respectively, share a common α-chain, but have distinct β and δ-chains. Thus, the hemoglobins found in the red blood cells of adults are coded by three distinct genes. Hemoglobin studies in certain families with α-chain mutants indicate that there is probably more than one α-chain gene per haploid genome. In addition, during development, embryonic hemoglobins and fetal hemoglobins occur. The Gγ, Aγ, δ and β genes have been shown to be closely linked. As we have shown the amino acid sequences of the beta and delta chains are very similar, suggesting that the genes coding for these chains arose by means of gene duplication of a common ancestral gene. Furthermore, the gene that codes for the α polypeptide is structurally related to the other hemoglobin genes, but is not linked to them. The genes coding for the δ, γ, and β chains probably arose through gene duplications from the original a gene and subsequently moved to a different linkage group by means of translocation.

Since the myoglobin chain is also structurally related to the hemoglobin polypeptides, it is thought that its genes and all the hemoglobin genes are derived from a common ancestral gene by a series of duplications over a long period of time.

Haptoglobin, one of the iron-binding proteins in vertebrate serum, is another example of a protein that probably arose by gene duplication. Haptoglobin, like hemoglobin, is a tetramer. Apparently, two genes α and β are involved. The α locus has two alleles, α_1 and α_2. Three phenotypes occur in the population Hp1, Hp2, and Hp1-2. The sera of humans exhibit either the Hp1 type (α, β) or the Hp2 type (α_2,β_n), or the Hp–2 type (α, α_2, β_n). The Hp2 haptoglobin is unique to humans, suggesting that a recent gene duplication gave rise to the Hp2. Studies of amino acid sequences clearly show that a partial gene duplication (intracistronic duplication) us have arisen resulting in a large polypeptide chain, different from the original α gene product in size and ability to combine with several beta chains to yield the polymeric haptoglobin Hp1–2.

These two examples—hemoglobin and haptoglobin–clearly indicate that gene duplication has probably played an important role in evolution.

Inversions

Inversions result when two breaks occur within a chromosome and the broken fragment is rejoined to the original chromosome in a reversed order. In this example, a chromosome with a normal sequence of 12345678910 undergoes breakage where the arms are in the contact while folded over on itself, say between 3 and 4 and 8 and 9. An inversion takes place when the 3 end fuses with the 8 end and the 4 end fuses with the 9 end to give the order 12387654910. Inversions are of two types, pericentric, with breaks flanking the centromere, and paracentric, with both breaks on the same side of the centromere. Remember that the same polarity must be maintained along a strand of DNA. Proper integration of an inverted segment could never occur in a single strand of DNA. The proper integration of a double-stranded segment of DNA occurs if the inverted segment undergoes a 180° rotation.

TYPES OF INVERSIONS

The inversions are of following types :

(i) Pericentric Inversion

When the inverted segment of chromosome includes or contains centromere, then such inversions are called heterobranchial or pericentric inversions. If crossing over occurs within the loop of a pericentric inversion, the resulted chromatids include half of the chromatids with duplications and deficiencies forming nonfunction. The other half of the chromatids form functional gametes : 1/4 gametes have normal chromosome order, 1/4 gametes have the inverted arrangement.

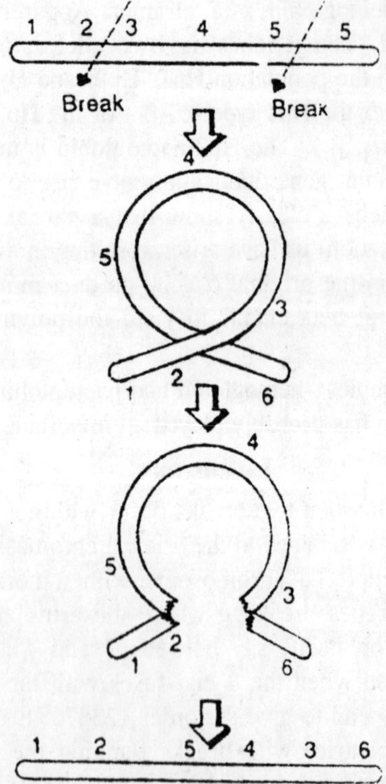

Fig. 6.6. Origin of an inversion in a chromosome.

(ii) Paracentric Inversions

When the inverted segment includes no centromere and the centromere remains located outside the segment, then such type of inversion is called homobranchial or paracentric inversion. Crossing over within the inverted segment of a paracentric inversion, produces a dicentric and an acentric chromosome. The dicentric chromosome contains two centromeres and forms a bridge from one pole to the other during first meiotic anaphase. When anaphase chromosomes separate towards poles, this bridge breaks somewhere along its length and the resulting fragments contain duplications and/or deficiencies. The acentric chromosome because lacks in centromere and fails to move to either pole and so, is not included in the meiotic products. Such, breakage-fusion bridge cycles of crossing over of paracentric inversions are most common in maize. The meiotic products includes half non-function, 1/4 functional normal and 1/4 functional inverted chromosomes.

Cytological and Genetical Methods for Detection and Identification of Inversions

Since the banding patterns of polytene chromosomes are visible, is an obvious way of detecting inversions. Figure shows a polytene chromosome in Drosophila with an inversion paired with a normal chromosome. The development of techniques that differentially stain metaphase chromosomes has also made the cytological characterization of inversions possible in species other than Drosophila.

The detection of inversions also involves the study of pairing relationships between a chromosome containing inverted DNA and its noninverted homologue during the prophase stage of meiosis. Pairing would homologous chromosomes on a point-by-point basis. However, homologous pairing between an inverted arrangement and its normal arrangement is difficult because they cannot line up properly. Therefore, in order for pairing to occur, an inversion loop must be formed.

Inversions have been dubbed crossover suppressors. The results of recombination within paracentric and pericentric inversions will clearly illustrate the reason for this name, which is, in fact, not entirely appropriate. Keeping in mind that crossing over occurs at the four-strand stage, it is probable that only two of the nonsister chromatids will be involved in the crossover while the other two remain unaffected. Let us consider two chromosomes that pair, one chromosome containing the normal gene sequence, ABCD and the other chromosome containing an inverted segment, ABCD. After replication and pairing occur, and while they are still in the four-strand stage, a single crossover occurs between nonhomologous chromatids within the inversion loop (e.g,. between genes C and B). Such crossing over results in the formation of a dicentric chromosome ABCD and an acentric fragment, DCBD, both of which contain duplications and deficiencies. Dicentric chromosomes undergo anaphase bridge formation which, in some types of cells, may be followed by breakage. The result will be chromosomes with deletions and duplications. The acentric treatment probably will remain at the metaphase plate and subsequently be lost. All of the recombination products that result from crossovers that occur within a paracentric inversion will form deficient gametes. Crossing over within the inversion loop region of a pericentric inversion also produces duplication and deficiency products.

Inversions are crossover suppressors, not because crossovers fail to occur within them in heterozygotes, but because the recombinant chromosome resulting from a crossover does not permit the offspring receiving it to survive if the crossover occurs within the inversion.

As crossover suppressors, inversion. Provide useful tools for geneticists. Inversions maintain particular seta of alleles of different loci within linkage groups by preventing recombinant offspring from surviving. Therefore, as for example in the isolation of sex-linked genes in Drosophila with the Basic stock. Inversions have also proved useful in the detection of linkage groups. A reduced number of recombinant types in a particular cross indicate that the genes are located within or near an inverted segment. And, conversely, a normal number of recombinant types indicate that the genes are located outside the inversion.

The number and type of inversions and the effects that they have in humans have not been extensively investigated until recently. Investigation has been limited by the difficulties in detecting inverted segments in human chromosomes. However, improved chromosome staining procedures now allow geneticists to detect paracentric and pericentric inversions in humans, but much work remains to be done in the field. One target area involves investigating the role of inversions in certain diseased states, such as heritable cancerous conditions.

TRANSLOCATIONS

Translocation is a broad term including all types of unilateral or bilateral transfer of chromosome segments from one chromosome to another. An important class of translocation having evolutionary significance is known as reciprocal translocations or segmental interchanges, which involve mutual exchange of chromosome segments between non-homologous chromosomes.

Cytology of a Translocation Heterozygote

If a translocation is present in one of the two sets of chromosomes, this will be a translocation heterozygote. In such a plant, the normal pairing into bivalents will not be possible among the chromosomes involved in translocation. Due to pairing between homologous segments of chromosomes, a cross-shaped (+) figure involving four chromosomes will be observed at pachytene. These four chromosomes at metaphase I can have one of the following three orientations:

1. *Alternate*. In alternate orientation, the alternate chromosomes will be oriented towards the same pole. In other words, the adjacent chromosome will orient towards opposite poles. This will be possible by the formation of a figure of eight.
2. *Adjacent I*. In adjacent I orientation, adjacent chromosomes having non-homologous centromeres will orient toward the same pole. In other words the chromosomes having homologous centromeres will orient towards opposite poles. A ring of four chromosomes will be observed.
3. *Adjacent II*. In adjacent II orientation, the adjacent chromosomes having homologous centromeres will orient towards the same pole. A ring of four chromosomes will be observed.

The alternate disjunction will give functional gametes. Adjacent I and adjacent II disjunctions will form gametes, which would carry duplications or deficiencies and as a result would be non-functional or sterile. Therefore, in a plant having translocation in heterozygous condition, there will be considerable pollen sterility.

A ring of four chromosomes as described above is found under conditions when a single interchange is found. If two interchanges are involving three non-homologous chromosomes, a ring of six chromosomes is found, and the size of the ring can increase with additional interchanges. More than one rings can also be found if two or more interchanges are independently found, each involving two different non-homologous chromosomes.

The first case of translocation was found in *Oenothera*, which was originally described as mutation by de Varies while working for his Mutation Theory. Oenothera, *Tradescantia* and *Rhoeo* are such cases, where translocations in the heterozygous condition are frequently found in nature. In many other crop plants they have been artificially induced.

Breeding Behaviour of a Translocation Heterozygote

The presence of translocation heterozygosity can be detected by the presence of semi-sterility and low seed set. This can then be confirmed at meiosis by quadrivalent formation. As shown above only two types of functional gametes are forced which result from alternate disjunction.

The functional gametes will give rise to three kinds of progeny namely (i) normal (ii) translocation heterozygote and (iii) translocation homozygote. These three types would be obtained in 1 : 2 : 1 ratio.

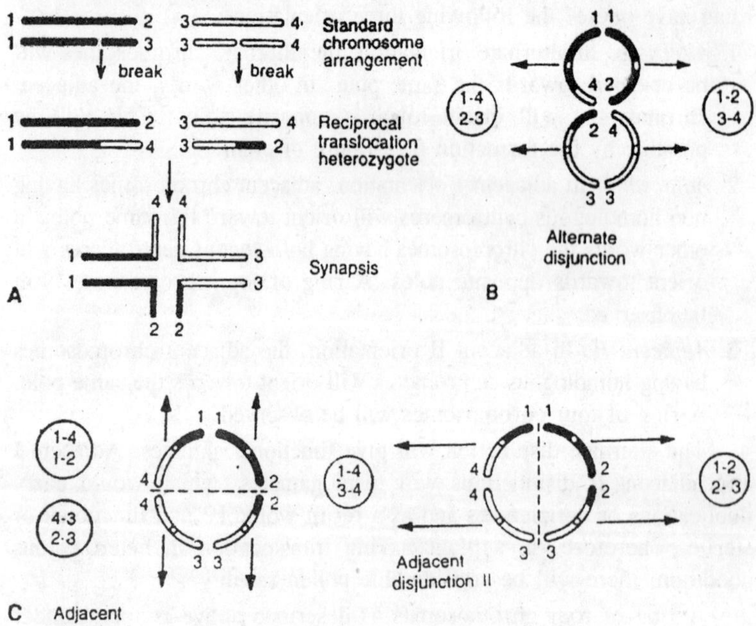

Fig. 6.7. A—Cross shaped figure formation during heterozygous reciprocal translocation; B—Alternate disjunction; C—Adjacent I and II disjunctions.

Interchange Heterozygosity in Oenothera

Subgenus *Euoenothera* of the genus *Oenothera* has been studied during 1920-1930 and cytogenetic structure leading to evolution in this group was examined. This group has 2 n = 14 and all 7 chromosomes of haploid complement have median chromosomes. There are three classes. (i) First is represented with species showing bivalents or small rings at meiosis (e.g. *O. hookeri*, *O. grandiflora*, *O. argillicola*). (ii) Second class is represented by species forming rings of various sizes at meiosis indicating the presence of interchanges. These rings are not permanent but are maintained due to their superiority in adaptive value (e.g., *O. irrigua*). (ii) The third class is represented by those having permanent translocation heterozygosity involving all chromosomes, so that a ring of 14 chromosomes is regularly formed (e.g., *O. biennis*, *O. strigosa*, *O. parviflora*).

The three classes described above also differ in phenotypes like flower size etc. and can be identified. The members of third category behave like pure lines and are actually permanent heterozygotes.

Balanced lethals and gametic complex-permanent hybridity in Oenothera. Permanent hybridity is maintained due to operation of a balanced lethal system, which may function duc to gametic lethality or zygotic lethality. Since complete rings are formed and alternate disjunction is a rule, only two types of gametes are formed showing complete linkage between chromosomes. The gametic and zygotic lethality leads to survival of only heterozygotes.

Chromosome Aberrations in Humans

Chromosome deletions are usually lethal even as heterozygotes, resulting in zygotic loss, stillbirths, or infant deaths. Sometimes infants with small chromosome deficiencies, however, survive long enough to permit observation of some of the abnormal phenotypes they express. Lejeune and his colleagues, for example, discovered a chromosome deficiency in humans that has been associated with the *cri-du-chat* (cat cry) *syndrome*. The name of this syndrome came from a plaintive catlike mewing cry from small weak infants with the disorder. Other characteristics microcephaly (small head), broad face and saddle nose, widely spaced eyes with epicanthic folds, unique facial features and physical and mental retardation IQs of cri-du-chat children studied are in the range of 20–40. The chromosome deficiency is in the short arm of chromosome 5 and is designated 5p.

Cri-du-chat patients die in infancy or early childhood and do not transfer the chromosome deletion to offspring. This chromosome deficiency, however, has been shown by Lejeune and others to become involved sometimes in a reciprocal translocation and thus to be transmitted. When the short arm of chromosome 5 became translocated to chromosome 15, the heterozygous translocation was carried in a normal healthy parent. Some gametes, however, carried only the deficient member of the translocation pair. Children inheriting the 5p chromosome expressed the cri-du-chat syndrome.

Another human disorder that is associated with a chromosome abnormality is chronic myelocytic (myelogenous) leukemia. A deletion of chromosome 22 was first described by Nowell and Hungerford and was called the Philadelphia (Ph) chromosome after the city in which the discovery was made. It was observed consistently in bone marrow preparations of patients with chronic myelocytic leukemia. Later, a translocation discovered by J. Rowley in a leukemia patient provided the correct chromosome rearrangement. A part of the long arm of chromosome 22 was translocated to another chromosome, usually chromosome 9 (46,XX.9q + 22q–), leaving a deficiency in the long arm of No.22.

More chromosomal anomalies are now being associated with malignancy. New staining methods permit more precise comparisons of (1) chromosome arms and (2) differentially stained sister chromatids. With balanced chromosomes, Burkitt lymphoma was related to a translocation of chromosome 14 [t(8q–; 14q+) (q24;q32). A deletion in chromosome 13q (band 14.1) is associated with the human embryonic tumor retinoblastoma. Williams tumor, an embryonic kidney tumor, is associated with a deletion in band 11p13. Sister chromatid exchanges, exchanges occurring between sister chromatids, occur with high frequency in somatic cells of patients with Bloom syndrome, indicating chromosome instability. Most people with this syndrome die with some form of cancer before age 30. Irregular chromosome numbers are also observed in malignant cells, particularly in later stages of cancer.

Like deficiencies, chromosome duplications are usually lethal even as heterozygotes, but sometimes they are sufficiently viable to permit observations of the abnormalities they produce. Like deficiencies, they may be associated with translocations and thus transmitted by normal healthy parents. Remember that a translocation is an exchange of parts between nonhomologous chromosomes or a transfer from one chromosome to a nonhomolog. Broken parts may be further divided and some may be lost or gained in a transfer. Thus, deficiencies and duplications may accompany a translocation carried in a parent. Such a parent may be a translocation heterozygote, with the long arm of one chromosome 21 attached at the centromere to the long arm of chromosome 14 along with normal chromosomes 14 and 21. Some gametes will contain a normal 21 and translocation 14q21q. When such a gamete is fertilized with a gamete from a normal individual, a translocated 21, along with two normal chromosomes 21, results in a viable infant with trisomy-21 and the Down syndrome.

With fluorescene microscope techniques, I.A. Uchida and C.C. Lin discovered a partial trisomy-12 that could not be identified by conventional methods. A body with some clinical features resembling the Down syndrome had been studied with conventional chromosome techniques. Because of continued lack of motor development at seven months age, further chromosomal investigations were carried out on lymphocyte cultures and slides stained with quinacrine dihydrochloride. With this technique, an additional band was identified in the short arm of chromosome 8. The mother's chromosomes were normal; but in the father, the short arm of No. 21 had become translocated to the short arm of No. 8 [46,XY, (8p+,12p–)]. The same chromosomal rearrangement was found in his daughter, an older sister of the patient. Since both father and daughter were clinically normal, the translocation was presumed to be reciprocal. The patients's No. 8 pair consisted of one normal and one translocation chromosome, similar to that of his father and sister, but both of his No. 12 chromosomes were normal. Thus, the boy had a duplication of part of the short arm of No. 12 and a deficiency of the tip of No. 8. Many translocations, addition to those cited above, have been detected in studies of human chromosomes, but most were apparently reciprocal and produced no phenotypic anomalies. At the Yale-New Haven Hospital, for example, cytological studies were conducted on 4500 infants born consecutively during one year. Lymphocytes from umbilical cord blood of each infant were grown in vitro and prepared for microscopic observation of chromosomes. Six translocations were detected. None of these was associated with a distinctive phenotypic anomaly. It is not the translocation process per se that produces abnormalities, but the imbalance of genetic material reflected in chromosome deficiencies and duplications that are produced by segregation of translocated chromosomes.

P. W. Allerdice et al. described a syndrome called the *chromosome 3 duplication-deletion syndrome*. Phenotypically, this syndrome includes a group of morbid symptoms; stillbirths, neonatal deaths, and spontaneous abortions. Most pregnancies are lost, but two children survived and became probands for the investigation. He cannot sit up, turn over, or eat solid food. Facial malformation of the living children includes a distorted head shape; thick, low eyebrows; low hair-line; long eyelashes; persistent lanugo; distended veins on scalp; hypertelorism; oblique palpebral fissures; a very short nose with a broad, depressed bridge and anteverted nares; protruding maxilla; thin upper lip; micrognathia; low-set ears; and short, webbed neck. Port-wine stains, congenital glaucoma, cloudy corneas, cleft palate, and harelip also occur frequently. Each infant had difficulty sucking and swallowing. Internal physical abnormalities were noted in infants who died neonatally.

Giemsa- and quinacrine-banded karyotypes from a parent of an affected child revealed an inversion [inv(3)(p25q21). Fetal cells cultured for prenatal diagnosis from a subsequent pregnancy of this couple carried a recombinant chromosome 3, with the long arm described as rec (3) del p, dup q, inv(3)(p25q21). With a banded chromosome 3, the inversion was analyzed. The inverted segment included the centromere (pericentric). Breaks had occurred in the short arm (p) at band 25 and in the long arm (q) at band 21. The central part of the chromosome including the centromere had rotated 180° and become reinserted into chromosome 3. Inv 3(p25q21) had been carried in the kindred for at least four generations and 35 kindered members were presumed to carry it. Karyotype of Case 1 is 46,XYrec(3), dup q, inv(3) (p25q21). He was the second child born to a 23-year-old carrier mother with karyotype 46,XX, inv(3)(p25q21) and a 26-year-old father, karyotype 46,XY.

Crossing over within the inversion loop is presumed to produce the imbalance of genetic material associated with duplications and deficiencies and to cause the symptoms of the syndrome. Odd numbers of crossovers within the inverted are of a chromosome 3 would be expected to result in genetic imbalance. Earlier in this chapter, crossing over in *Drosophila* inverted chromosomes was shown to create irregular combinations resulting in lethals. This, in turn, resulted in "suppressing" crossing over. In humans, unbalanced zygotes from inverted chromosome crossovers sometimes survive. Some infants carrying unbalanced chromosomes are kept alive, but with varying degrees of birth defect handicaps. The degree of imbalance of genetic material may be determining factor for life or death and degree of abnormality.

7

POLYPLOIDY

Somatic cells of higher plants and animals usually have chromosomes in pairs (2n); that is, two of each kind of chromosome are present in each cell. Mature germ cells, having undergone reduction division, normally have one number of each pair (n). Many individual plants and animals, however, have local areas of somatic tissue characterized by a multiple of the basic chromosome number. A doubling process in cell division is the usual explanation for these deviations.

With the exception of sex differences, somatic doubling, and minor variations that occur in natural and experimental populations, all members of a species of plants or animals have the same basic chromosome number. Chromosome number can be overemphasized as a criterion for species identification, but it represents a valid characteristic to be used along with others for distinguishing species. The range or reported chromosome numbers in animals extends from 2 pair in a rhabdocoel Gyratrix hermaphroditus and some mites, midges, and scale insects, to more than 100 in some butterflies and Crustacea. The Crustacean, *Paralithodes camtschatica*, for example, has 208 chromosomes or 104 pairs. The reported range in plants is from 2 pairs in the small composite plant Haplopappus gracilis to several hundred in some ferns. A species of fern-like plants of the genus Ophioglossum is reported to have 768 chromosomes.

Where all individuals within a species, with the exceptions noted above, have the same chromosome number, different species within a genus often have different numbers. Cytological investigations of chromosomes help to unravel problems of species formation. The evolutionary path of a certain species can be followed in some cases by comparing numerical and structural relations of its chromosomes with those of other species within the genus.

The essential genetic material is a major factor in determining evolutionary patterns, but chromosome number itself represents merely the number of packages into which the genetic material is divided. Chromosome number is probably more constant, however, than any other single morphological characteristic that is available for species identification.

Table 7.1. Chromosome number of some common animals and plants

Species	*Number of chromosomes pair*
Plants	
Garden pea, *Pisum sativum*	7
Sorghum, *Sorghum vulgare*	10
Maize, *Zea mays*	10
Johnson Grass, *Sorghum halepense*	20
Alfalfa, *Medicago sativa*	16
Barley, *Hordeum vulgare*	7
Oats, *Avena sativa*	21
Tomato, *Lycopersicon esculentum*	12
Tobacco, *Nicotiana tabacum*	24
Trillium, *Trillium erectum*	5
Animals	
Gypsy moth, *Lymantria dispar*	31
Mouse, *Mus musculus*	20
Rabbit, *Oryctolagus cuniculus*	22
Cow, *Boss tarus*	30
Horse, *Equus caballus*	32
Donkey, (ass) *Equus asinus*	31
Dog, *Canis familaris*	39
Monkey, *Macaca rhesus*	21
Chimpanzee, *Pan troglodytes*	24

Changes in the number of chromosomes may be reflected in phenotypic variations, which constitute a useful tool for identifying the influence of individual chromosomes. If, for example, phenotypically distinguishable individuals with different chromosome numbers can be identified in natural populations or produced experimentally, it is sometimes possible to determine the effect of adding or removing certain chromosomes. Some plants with increased chromosome numbers have phenotypic changes in morphological or physiological characteristics which are of practical importance to man. Grape and tomato plants with chromosome numbers above 2n are larger and produce more desirable fruit than do corresponding varieties with the usual 2n number.

Classifications of chromosome changes are arbitrary and superficial because these changes are necessarily interpreted in terms of obvious additions or eliminations of parts of chromosomes, whole chromo-somes, or whole chromosome sets. The presently accepted classification system, therefore, is merely a working tool. Two main classes are euploidy and aneuploidy ("ploid," Greek for unit; "eu," true or even; and "aneu," uneven). Euploids have chromosome complements consisting of whole sets or genomes. The chromosome number of euploid organisms is basically represented by the monoploid (n). Euploids with chromosome numbers above the monoploid level may be diploid (2n), triploid (3n), tetraploid (4n), or have some other "polyploid" number.

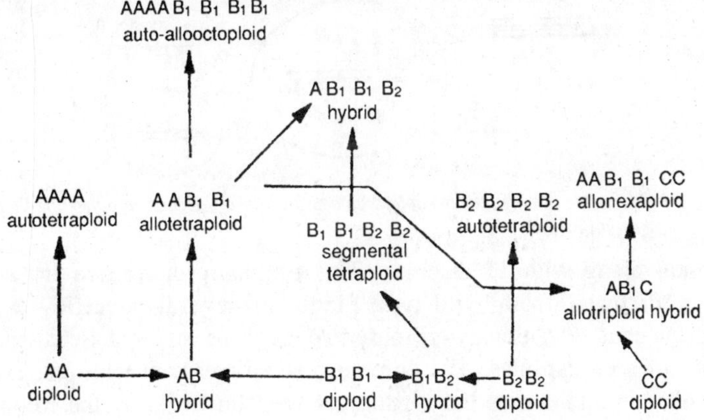

Fig. 7.1. Mode of formation of different kinds of ploypoloid.

ANEUPLOIDY

The first critical study of aneuploid plants was made by Blakeslee and Bellingusing the common Jimson weed Datura stramonium, which normally has 12 pairs of chromosomes in the somatic cells. In 1924, these investigators announced the discovery of a "mutant type" having 25 rather than 24 chromosomes. At the meiotic metaphase, one of the 12 pairs was found to have an extra member; that is, one trisome was

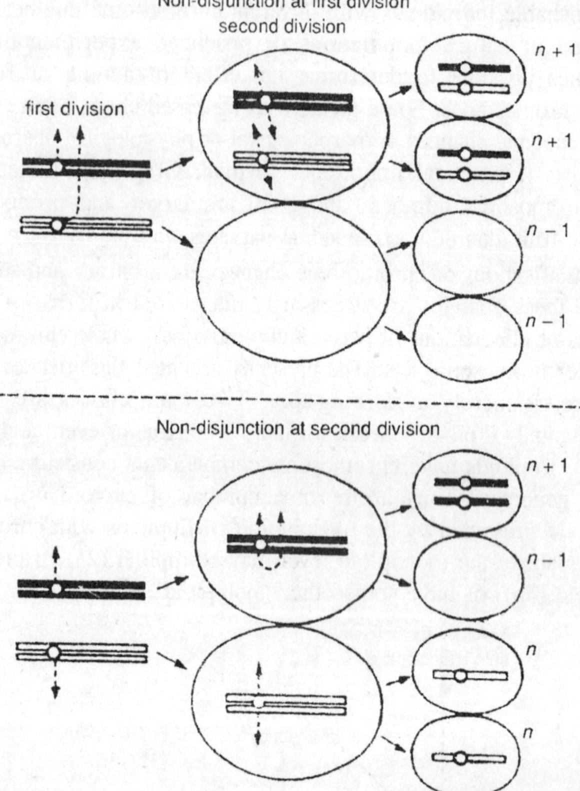

Fig. 7.2. The mode of origin of aneuploid gamete by non-disjunction at either the first or second meiotic division.

present along with 11 disomes. This originally discovered trisomic plant differed from wild-type plants in several specific ways. Conspicuous deviations were observed in shape and spine characteristics of seed capsules. The chromosome complement with one extra member in addition to the regular set was illustrated by the formula 2n + 1. Theoretically, because the some have, however, been divided successfully into 7 groups identified with letters A to G. Numbers 1 to 22 are associated with the autosomes in descending order by length. All autosomes can be placed satisfactorily within a group but the numbering within the groups is more or less tentative. The X chromosome is difficult to distinguish from members of the C group and the Y chromosome shows considerable variability in different preparations.

Nullisomics (2n – 2)

Although nullisomy is a lethal condition in regular diploids, an organism like wheat (which "pretends" to be diploid but is fundamentally hexaploid) can tolerate nullisomy. In fact, all of the possible 21 wheat nullisomics have been produced. Their appearances are different from normal wheat; furthermore, most of them show less vigorous growth.

Monosomics (2n – 1)

Monosomic chromosome complements are generally deleterious for two main reasons. First, the balance of chromosomes that is necessary for a finely tuned cellular homeostasis, carefully put together during evolution, is grossly disturbed. For example, if a genome consisting of two of each of chromosomes, a, b, and c becomes monosomic for c (that is 2a + 2b + 1c), the ratio of these chromosomes is changed from 1c:1 (a+b) to 1c : 2(a + b). Second, any deleterious recessive on the single remaining chromosome becomes hemizygous and may be directly expressed phenotypically. (Note that these are the same effects as those produced by deletions.)

Monosomics, trisomics (2n +1), and other chromosome aneuploids are probably produced by nondisjunction during mitosis or meiosis. In meiosis it can happen at either the first or the second division. If an n – 1 gamete is fertilized by an n gamete, a monosomic (2n – 1) zygote is produced; an n + 1 and an n gamete give a trisomic 2n +1; and an n + 1 and n + 1 give a tetrasomic if the same chromosome is involved or a double trisomic if different chromosomes are involved, and so on.

In Neurospora (a haploid), the n – 1 meiotic products abort, and do not darken like the normal ascospore; so MI and MII nondisjunctions are detected as asci with 4:4 and 6:2 ratios of normal to aborted spores. For loci on the aneuploid chromosomes, what ascus genotypes are produced?

In humans, the sex-chromosome monosomic (44 autosomes + 1X) produces a phenotype known as Turner's syndrome. Affected people have a characteristic, easily recognizable, phenotype: they are sterile females, are short in stature, and often have a web of skin extending between the neck and shoulders. Their intelligence is near-normal, although some specific cognitive functions are defective. Their frequency is about 1 in 5000 female births. Monosomics for all autosomes die in utero.

If viable, nullisomics and monosomics are useful in locating newly found recessive genes on specific chromosomes in plants. In one such method, different monosomic lines lacking a different chromosome in each line are obtained. Homozygotes for the new gene are crossed with each monosomic line, and the progeny of each cross are inspected for expression of the recessive phenotype. The cross in which the phenotype appears identifies its chromosomal location. In nullisomics and monosomics, of course, n - 1 gametes are produced. In general, these gametes tend to be more viable in a female parent than in a male. It is the union of these n – 1 gametes within n gametes, bearing the new mutation, that provides the crucial progeny types for the linkage test.

A similar approach can be used in humans. For example, two people whose vision is normal may produce a daughter who has Turner's syndrome and is also red-green colour-blind. This shows that the allele for red-green colour blindness is recessive, that it was on the X chromosome of the mother, and the nondisjunction must have occurred in the father.

Trisomics (2n + 1)

In trisomics, trivalents are regularly seen. For genes that are tightly linked to the centromere of a trisomic chromosome set, the random segregations can be represented a trisomic Aaa. All types occur equally frequently, and a gametic ratio of 1A:2Aa:2a:1aa is produced. Trisomics are sometimes recongnised by these ratios, which are also useful in locating genes on chromosomes. We have already observed a complete set of trisomic lines in Datura. Once again, note that chromosome imbalance produces highly chromosome-specific deviations from the normal appearance.

In humans there are several examples of viable trisomics. The combination XXX (1/1000 male births) results in Klinefelter's syndrome, producing males that have lanky builds, are mentally retarded, and sterile. Another combination, XYY, also occurs in about in 1000 male births. A lot of excitement was aroused when an attempt was made to link the XYY condition with a predisposition toward violence. This is still hotly debated, although it is now clear that an XYY condition in no way guarantees such behaviour. Nevertheless, several enterprising lawyers have attempted to use the XYY genotype as grounds for acquittal or compassion in crimes of violence. The XYY males are usually fertile. We have already looked at the generation of Down's syndrome through adjacent segregation in translocation heterozygotes.

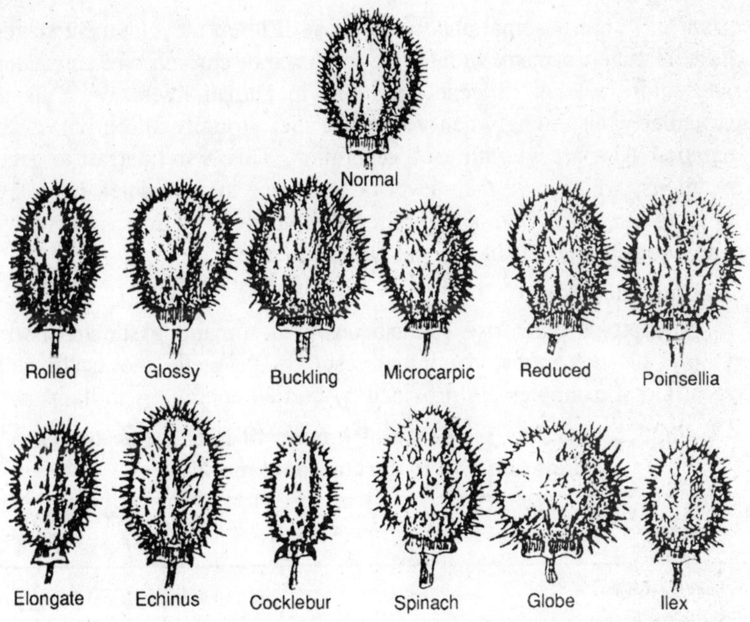

Fig. 7.3. Fruit capsules of the 12 primary trisomics of Datura stramonium, each with its particular phenotype.

Down's syndrome also occurs much more commonly as a result of nondisjunction during meiosis, and it is then called trisomy 21. In this form of Down's syndrome, there is generally no family history of the phenotype; however, the frequency of this form is dramatically higher among children born to older mothers. The overall incidence of this abnormality is about 0.15 percent of births.

Down's syndrome is a severely incapacitating condition. Affected individuals are mentally retarded, and about one-third die by the age of 10 years. Recent advances in mapping the human genome will allow the identification of precisely those genes on the long arm of chromosome 21 that must be trisomic to produce this syndrome; these advances offer some hope of a more precise understanding of its chemical nature and possible therapy. In humans, the only other two autosomal trisomics known to survive past birth are individuals with trisomy 13 and trisomy 18. Affected children are even more severely handicapped, both mentally and physically, and rarely survive to 1 year of age.

Chromosome mutation in general plays a prominent role in determining genetic ill health in humans. The surprisingly high levels of various chromosomal abnormalities at different development stages of the human organism. In fact, the incidence of chromosome mutations ranks close to that of gene mutations in human livebirths. This is particularly surprising when we realize that virtually all chromosome mutation arise a new with each generation. This is in contrast to gene mutations, which owe their level of incidence to a complex interplay of mutation rates and environmental selection, acting over many generations of the history of the human species.

Somatic Aneuploids

Aneuploids can arise spontaneously in somatic tissue or tissue culture. In such cases, the initial result is a genetic mosaic of cell types. Good examples are provided by certain conditions in humans.

Table 7.2. Relative incidence of human ill health due to gene mutation and to chromosome mutation

Type of mutation	*Percentage of live birth*
Gene mutation	
Autosomal dominant	0.90
Autosomal recessive	0.25
X-linked	0.05
Total gene mutation	1.20
Chromosome mutation	
Autosomal trisomies (mainly Down's syndrome)	0.14
Other unbalanced autosomal aberrations	0.06
Balanced autosomal aberrations	0.19
Sex chromosomes	
XYY, XXY, and other YY	0.17
XO, XXX, and other MM	0.05
Total chromosome mutation	0.61

Sexual mosaics provide the first example. These are people whose bodies are a mixture of male and female tissue. One type of sexual mosaic is XO/XYY. This mosaic can be explained by postulating an XY zygote in which an early mitotic division involved a nondisjunction of the Y chromosomes, so that both went to one pole:

The phenotypic sex of such individuals depends upon where in the body the male and female sectors end up. In this case, if nondisjunction occurred at a later mitotic division, there would be a three-way XY/XO/XYY.

Aneuploidy in Man

More recent studies on individuals have related chromosome numbers below and above 46 with intersex conditions and other irregularities manifesting physical, reproductive, and mental abnormalities. Participants in the Chicago Conference agreed on a system of nomenclature for identifying numerical and structural chromosome alterations. In a description of a karyotype, the first item to be recorded is the total number of chromosomes, including the sex chromosomes, followed by the sex chromosome constitution and any autosomal aberrations. A complement of 44 autosomes and one X, for example, is symbolized 45, X. This monosomic chromosome complement has been associated with an abnormal female condition known for many years as Turner's syndrome and which occurs in one of about 5000 people in the general population.

The Turner Syndrome (45,X)

This monosomic has a chromosome complement of 44 autosomes and one X chromosome. The chromosome anomaly is associated with an abnormal female phenotype described in 1938 by H. H. Turner and associates and known as the Turner syndrome. It occurs in about 1 per 2500 live female births. More than 90 percent abort spontaneously. A rough estimate for 45,X adults in the general population is in 1 in 5000. These adults have virtually no ovaries, have limited secondary sexual characteristics, and are sterile. Microscopic sections of the ovaries show fibrous streaks of tissue representing remnants of ovaries. Affected females have short stature, low-set ears, webbed neck, and shieldlike chest. Mental deficiency is not usually associated with this syndrome. Epithelial cells of 45,X patients are X chromatin negative, as expected when only X chromosome is present.

X monosomic probably originate from exceptional eggs or sperm with no sex chromosome or from the loss of a sex chromosome in mitosis during early cleavage stages, after an XX or XY zygote has been formed. This latter probability is supported by the high frequency of mosaics that result from postzygotic events in patients with the Turner syndrome. Mosaics with X/XX sex chromosomes show symptoms of the Turner syndrome but are usually taller than X and have fewer anomalies that nonmosaic 45,X females. They show more feminization, more normal menstruation, and may be fertile. Many cases of the somatic Turner phenotype without the typical 45,X chromosome constitution are now known. Most of these have one normal X chromosome and a fragment of a second X chromosome. Both arms of the second X chromosome are apparently necessary for normal ovarian differentiation. Individuals with only the long arm of the second X are short in stature and show other somatic symptoms of the Turner syndrome, whereas those with only the short arm of the second X have normal stature and do not show as many signs of the Turner syndrome. This indicates that the Turner phenotype is mostly controlled by genes on the short arm of the X chromosome.

Patients with partial deletion of one X chromosome are X chromatin positive and therefore may be misdiagnosed if a buccal smear is the only test used. The deficient X chromosome always forms the X chromatin body. A Y chromosome also occurs in some individuals with the Turner phenotype. These patients are usually mosaic for 45,X/46,XY, with a normal Y. People with one X and a Y fragment, not including the Y short arm, have only streak ovaries but are normal in phenotype. This suggests that male-determining genes are in the short arm of the Y chromosome, and those that prevent Turner phenotype are in the Y long arm as well as the X short arm. Major features of the Turner phenotype occur in some males as well as in females. The male Turner syndrome is characterized by defective development of the testes, sterility, and limited male secondary sexual characteristics, along with somatic features of the Turner phenotype. These people have normal male karyotypes.

The Klinefelter Syndrome (47,XXY)

An extra X chromosome in addition to the usual male (XY) chromosome complement (47,XXY) has been associated with the abnormal male syndrome described (in 1942) by H.F. Klinefelter and known as the *Klinefelter syndrome*. It is estimated to occur in 1 per 500 live male births. Individuals with this syndrome are phenotypically males but with some tendency toward femaleness, particularly in secondary sex characteristics. Such features as enlarged breasts, underdeveloped body hair, small testes, and small prostate glands are a part of the syndrome. Presumably, the XXY constitution originates either by fertilization of an exceptional orginates either by fertilization of an exceptional XX egg by a Y sperm or of an X egg by an exceptional XY sperm. Studies of Klinefelter syndrome and Turner syndrome indicate that the Y chromosome in human beings, unlike that in *Drosophila*, determines male sex.

The most common karyotype (about three-quarters of the cases) for the Klinefelter syndrome is 47,XXY, but the symptoms of the syndrome will usually occur whenever more than one X chromosome is present along with a Y chromosome. More complex karyotypes associated with the Klinefelter syndrome include: XXYY, XXXY, XXXYY, XXXXY, XXXXYY, and XXXXXY. All patients with the Klinefelter syndrome have one or more X chromatin bodies in the their cells. Mental retardation is usually found when there are more than two X chromosomes. The XY/XXY mosaicism in patients with Klinefelter syndrome is associated with less severe physical and reproductive anomalies.

Aneuploidy of X Chromosomes and Mental Deficiency

Other irregular combination of X chromosomes have also been recognized among females with X chromosome aberrations. About 1 percent of all mentally defective women in institutions have been shown to have one or more extra X chromosomes. This chromosome abnormality occurs in about 1 in 700 live births in the general population. Individuals with the "triple X syndrome" are comparable in some ways with Drosophila metafemales (XXX). In Drosophila, however, such individuals are usually lethal, and those that survive are strikingly abnormal and sterile. Human XXX individuals are sometimes visibly undistinguished from normal XX females, but there is considerable range in phenotypic expression. They may be mentally abnormal.

The best-known symptoms in this syndrome are abnormalities associated with functional processes such as menstruation. One patient cited by P. A. Jacobs was a 37-year-old female who reported that the first suggestion of an abnormality was highly irregular menstruation. When the abdominal wall was opened, the ovaries appeared as if they were postmensopausal. Microscopically, they showed deficient ovarian follicle formation. Of 63 cells observed, 51 had 47 chromosomes; the extra chromosome was an X. Nondisjunction in the production of the egg from which this woman developed was postulated as the mechanism for the occurrence of extra chromosome. Buccal smears showed two sex chromatin bodies in the epithelial cells as expected when three X chromosomes are present. Individuals with tetrasomic X chromosomes (48,XXXX) are all mentally defective. The degree of mental deficiency increases with the number of X chromosomes present.

47,XYY and Behaviour

P. A. Jacobs and her associates reported in 1965 that seven XYY males were detected in a population of 197 male, mentally subnormal inmates of a penal institution in Scotland. The XYY men were unusually tall, with an average height of 73.1 inches, compared with 64 inches for XY men in the same prison. Numerous other studies, mostly in institutionalized populations, have since confirmed that a high proportion of XYY individuals are tall, subnormal in intelligence (with IQs individuals are tall, subnormal in intelligence (with IQs ranging from 80 to 95), and antisocial. The aggressive behaviour that brought them into conflict with the law was usually against property rather than people.

XYY trisomy occurs about once in 1000 live male births in the general European population. Only a few of these can be accounted for in the criminal population. Furthermore, most XYY men have been described as perfectly normal in behaviour. Hook has shown that only 3.6 percent of all XYY men are institutionalized for any reason.

Environmental factors are presumed to be involved in the development of aggressiveness. Since some XYY men are subnormal in intelligence and excessively tall in stature, particular environmental situations in childhood or adulthoood may lead to withdrawal from society or aggressive behaviour. Unfavourable social conditions such as frustration in personal accomplishment and taunting from associates may encourage physical aggression as a means of adaptation. Males with this sex trisomy have not been found to transmit the extra Y chromosome to their sons. This extra chromosome seems to be weeded out in gametogenesis. A wide range of physical and mental abnormalities has been detected in institutionalized XYY men, but most of these are irregular in occurrence and do not form a syndrome. Tallness of stature and mental dullness, however, are fairly constant characteristics among institutionalized XYY men.

Chromosomal Mosaics

Individuals who have at least two cell lines, with different karyotypes derived from one zygote, originate from nondisjunction in a cleavage division after fertilization. One daughter cell would thus receive one too many and the other would be one deficient. Each cell would give rise to cell line with its irregular chromosome number. Proportions of cells representing the different cell lines would vary in different tissues, making the extent of the mosaicism and the effect on the organism difficult to predict.

Many sex chromosome mosaics have been detected in human beings. The main phenotypic characteristic is extreme variability. Some sex chromosome mosaics have been reported —X/XX, X/XY, XX/XY, XXY/XX, XX/XXX, XXX/X XXX/XXXXY—and several other combinations reflecting two or three cell lines. Mid to severe phenotypic symptoms have been associated with these sex chromosome mosaics.

Abnormal Euploidy

Monoploids

In this section we shall consider monoploidy as an unusual condition. Monoploid individuals can arise spontaneously in natural populations as rare aberrations, but in several forms (such as bees, wasps, and ants) the males are normally monoploid, having derived from unfertilized eggs.

In the germ cells of a monoploid, meiosis cannot occur normally because the chromosomes having no pairing partners. Thus monoploids are characteristically sterile. (However, meiosis can be bypassed in some monoploid animals, such as male honeybees, which produce gametes essentially by mitotic division). If meiosis occurs and the single chromosomes segregate randomly, then the probability of their all going to one pole is $(1/2)^{x-1}$, where x is the number of chromosomes. This will determine the frequency of viable (whole-set) gametes, obviously a vanishingly small number if x is large.

Monoploids have a major role in modern approaches to plant breeding. Diploidy is an inherent nuisance in the induction and selection of new favourable plant mutations and of new selection of new favourable plant mutations and of new combinations of genes already present. Monoploids provide a way around some of these problems. In some plants, monoploids may be artificially derived from the products of meiosis in the plant's anthers. A cell destined to become a pollen grain may be induced by cold treatment to grow instead into an embryoid, a small dividing mass of cells. The embryoid may be grown on agar to form a monoploid planter, which can then be potted in soil to mature.

Monoploids may be exploited in several ways. In one method, they are first examined for favourable traits or gene combinations. Thcsc may arisc from heterozygosity already present in the parent or induced in the parent by mutagens. The monoploid can then be subjected to chromosome doubling to achieve a completely homozygous diploid with a normal meiosis, capable of providing seed. How is this achieved? Quite simply, by the application of a compound called colchicine to meristematic tissue. Colchicine, an alkaloid drug extracted from the autumn crocus, inhibits the mitotic spindle, so that cells with two chromosomes sets are produced. These may proliferate to form a sector of diploid tissue that can be identified cytologically.

Another way the monoploid may be used is to treat its cells, basically like a population of haploid organisms, in a mutagenesis-and-selection procedure. The cells are isolated, their walls are removed by enzyme treatment, and they are treated with mutagen. They are then plated on selective medium— perhaps a toxic compound normally produced by one of the plant's parasites, or an insecticide— to select resistant cells. Resistant plantlets grow eventually into haploid plants, which can then be doubled (using colchicine) into a pure-breeding resistant type. These are potentially powerful techniques that can circumvent the normally slow process of what is basically meiotic plant breeding. The techniques have been successfully applied in several important crop plants, such as soybeans and tobacco. This is, of course, another aspect of somatic-cell genetics in higher organisms.

The another technique of or producing monoploids does not work in all organism or in all genotypes of an organism. Another useful technique has been developed in barley, an important crop plant. When diploid barley, Hordeum vulgare, is pollinated using a diploid wild relative called Hordeum bulbosum, fertilization occurs, but during the ensuing somatic cell divisions, the chromosomes of H. Bulbosum are preferentially eliminated from the zygote, resulting in a haploid embryo. (The haploidization process appears to be caused by a genetic incompatibility between the chromosomes of the different species.) The resulting haploids can be doubled with colchicine. This approach has led to the rapid production and widespread planting of several new barley varieties. It is being used successfully in other species too.

Polyploids

Once into the realm of polyploids, we must distinguish between autopolyploids and allopolyploids. *Autopolyploids* are composed of multiple sets from within one species, whereas *allopolyploids* are composed of sets from different species. Allopolyploids form only between closely related species; however, the different chromosome sets are homeologous (only partially homologous), not fully homologous as they are in autopolyploids.

Triploids

Triploids are usually autopolyploids. They are constructed from the cross of a 4x (tetraploid) and a 2x (diploid). the 2x and the x gametes unite to form a 3x triploid.

Triploids also are characteristically sterile. The problem again involves pairing at meiosis. Although pairing can take place in several ways, it usually occurs between only two chromosomes at a time. The

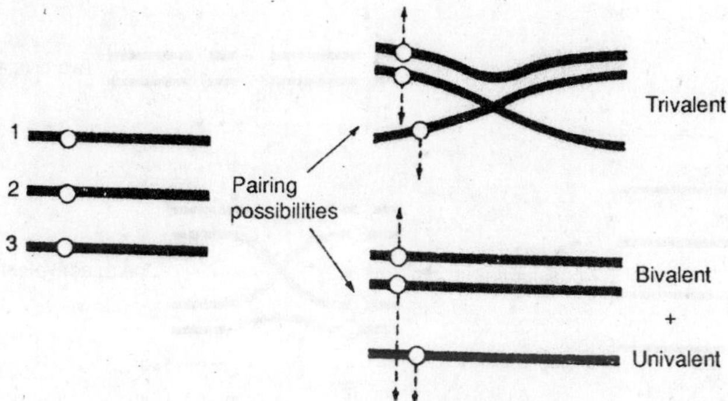

Fig. 7.4. Meiotic pairing possibilities in a triploid.

net result is always the same, an unbalanced segregation of one of the following types:

$$\frac{1+2}{3} \text{ or } \frac{1+3}{2} \text{ or } \frac{2+3}{1}$$

This happens for every chromosome threesome, and the probability of obtaining either a 2 x or x gamete is $(\frac{1}{2})^{x-1}$, where x is the number of chromosomes in a set. The other will be unbalanced gametes, having two of one chromosome type, one of another, two of another, and so on, and most will be nonfunctional. Even if the gametes are functional, the resulting zygotes will be unbalanced. A practical application of the sterility associated with triploidy lies in the production of seedless varieties of watermelons and bananas.

Autotetraploids

Autotetraploids occur either naturally, by the spontaneous accidental doubling of a 2x genome to 4x, or artificially, through the use of colchicine. Autotetraploids are evident in many commercially important crop plants because, as with other polyploids, the larger number of chromosome sets is often associated with increased size of the plant. This is manifested in increased cell size, fruit size, stomata size, and so on. Because 4 is an even number, autotetraploids can have a regular meiosis, although this is by no means always the case. The crucial factor is how the four chromosomes of one type pair and segregate. The two bivalent and the quadrivalent pairing modes tend to be most regular in segregation, but even here there is no guarantee of a 2 → 2 segregation. If a regular 2 → 2 segregation is achieved at each chromosome type, as is the case in some species, than a formal genetic analysis can be developed for autotetraploids.

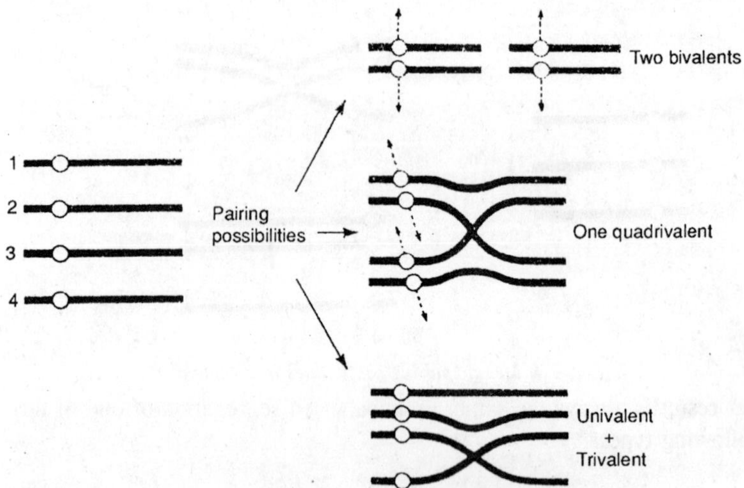

Fig. 7.5. Meiotic pairing possibilities in tetraploids.

Let's hypothesize an experiment in which colchicine is used to double the chromosomes of an *Aa* plant into an *AAaa* autotetraploid, which we will assume shows 2 → 2 segregation. We now have a further worry because autotetraploids give different genetic results in their progeny, depending on whether or not the locus concerned is tightly linked to the centromere. First we consider a centromeric gene. The three possible pairing and segregation patterns are presented in Figures, these occur by chance and with equal frequency. As the figure shows, the 2x gametes will be *Aa, AA, or aa,* and these will be produced in a ratio of 8:2:2, or 4:1:1. If such a plant is offspring is obviously 1/6 × 1/6 = 1/36. In other words, a 35:1 phenotypic ratio will be observed if *A* is fully dominant over three *a* alleles.

If, in the same kind of plant, a genetic locus *B/b* is very far removed from the centromere, crossing-over must be considered. This forces us to think in terms of chromatids instead of chromosomes, and we have for *B* chromatids and for *b* chromatids. Because the number of crossovers in a such a long region will be large, the genes will become effectively unlinked from their original centromeres, and the packaging of genes two at a time into gametes is very much like grabbing two balls at random from a bag of eight balls, four of one kind and four of another. The probability of picking two *b* genes is then

4/8 (the first one) × 3/7 (the second one) = 12/56
= 3/14

So, in a selfing, the probability of a *bbbb* phenotype will be equal $3/14 \times 3/14 = 9/196 = 1/22$. Hence there will be a 21:1 phenotype ratio of *B* – – –:*bbbb*. For genetic loci of intermediate position, intermediate ratios will, of course, result.

Allopolyploids

The "classical" allopolyploid was synthesized by G. Karpechenko in 1928. He wanted to make a fertile hybrid between the cabbage (*Brassica*) and the radish (*Raphanus*) that would have the leaves of the former and roots of the latter. Each of these species has 18 chromosomes, and they are related closely enough to allow intercrossing. A variable hybrid progeny individual was produced from seed. However, this hybrid was functionally sterile because the nine chromosomes from the cabbage parent were different enough from the radish chromosomes that homology was insufficient for normal synapsis and disjunction.

However, one day a few seeds were in fact produced by this (almost) sterile hybrid. On planting, these seeds produced fertile individuals with 36 chromosomes. These individuals were allopolyploids. They had apparently been derived from spontaneous accidental chromosome doubling in the sterile hybrid, presumably in tissue that eventually became germinal and underwent meiosis. Thus, in $2n_1 + 2n_2$ tissue, there is a pairing partner for each chromosome, and balanced gametes of the type $n_1 + n_2$ are produced. These fuse to given $2n_1 + 2n_2$ allopolyploid progeny, which are in turn fertile also. This kind of allopolyploid is sometimes called an amphidiploid. (Unfortunately for Karpechenko, his amphidiploid had the roots of a cabbage and the leaves of a radish.)

If the allopolyploid is crossed to either parent species, sterile offspring result. In the case of the cross to radish, these offspring would be $2n_1 + n_2$, constituted from an $n_1 + n_2$ gamete from the allopolyploid, and an n_1 gamete from the radish. Obviously, the n_2 chromosomes will have no pairing partners, so sterility will result. Consequently, Karpechenko had effectively created a new species, with no possibility of gene exchange with its parents. He called his new species *Raphanobrassica*.

Nowadays, allopolyploids are routinely synthesized as a major tool in plant breeding. The goal obviously is to combine some of the worthwhile features of both parental species into one type. This kind of endeavor is very uncertain, as Karpechenko found out. In fact, only one amphidiploid has ever been intentionally produced that is of potentially widespread use. This is *Triticale*, an amphidiploid between wheat (*Triticum, 2n = 6x = 42*) and rye (*Secale,* 2n = 2x = 14). *Tricale* combines the high yields of wheat with the ruggedness of rye. A massive international *Triticale* testing program is now under way, and many breeders have great hopes for the future of this artificial amphidiploid.

In mature, allopolyploidy seems to have been a major force in speciation of plants. There are many different examples. One particularly satisfying one is shown by the genus *Brassica*. Here three different parent species have been hybridized in all possible pair combinations to form new amphidiploid species. This has all taken place in nature, but *Brassica* amphidiploids also have been artificially synthesized.

A particularly interesting natural allopolyploid is bread wheat, *Triticum aestivum* (2n = 6x = 42). By a study of various wild relatives, it has been possible to reconstruct a probable evolutionary history of breat wheat. In a wheat meiosis, there are always 21 pairs of chromosomes. Furthermore, it has been possible to establish that any given chromosome has only one specific pairing partner (homologous pairing)—not five other potential ones (homologous pairing). The suppression of such homeologous pairing (which would lead to much reduced stability of the species) is maintained by a gene *Ph* ensures a diploid-like genetics for this basically hexaploid species. Without *Ph*, bread wheat could probably never have arisen. It is interesting to speculate on whether Western civilization could have arisen or progressed without this species—in other words, without *Ph*.

Somatic Allopolyploids from Cell Hybridization

Another innovative approach to plant breeding is to try to make allopolyploid-like hybrids by asexual methods. Theoretically, such a technique would permit combination of widely differing parental species. The technique does indeed work, but so far the only allopolyploids that have been produced are those that can also be made by the sexual methods we have considered already. The procedure is as follows. Cell suspensions of the two parental species are prepared and stripped of their cell walls by special enzyme treatments. The stripped cells are called *protoplasts*. The two suspensions (protoplast suspensions) are combined with polyethylene glycol, which enhances protoplast fusion. The parental cells and the fused cells will proliferate to form colonies (in much the same way as microbes) on agar medium. If these colonies, or calluses, are examined, a fair percentage of them are found to be allopolyploid-like hybrids with chromosome number equal to the sum of the parental types. Thus, not only do the protoplast cell membranes fuse to form a kind of heterokaryon, but the nuclei fuse too.

A good example of an allopolyploid-like hybrid is commercial tobacco, *Nicotiana tabacum*, which has 48 chromosomes. This species of tobacco was originally found in nature as a spontaneously occurring amphidiploid. The two probable parents are N. Sylves and N. Tomentosiformis, each of which has 24 chromosomes. A sexual cross between N. Tabacum and either of the other two gives a 36-chromosome hybrid n which there are 12 chromosome pairs plus 12 unpaired chromosomes. A cross between N. Sylvestris and N. Tomentosiformis gives a 24-chromosome hybrid n which there is no pairing at all. Hence, it appears that part of the N. Tabacum genome is from N. Sylvestris and part from N. Tomentosiformis. This amphidiploid can be re-created either sexually, by processing involving colchicine as described previously, or somatically by cell fusion. When cells or the prospective parental species are fused, a 48-chromosome hybrid cell line is produced from which may be grown plants whose behaviour is identical to that of N. Tabacum. (Note that in the latter method, colchicine is not required, since the fusion product is already amphidiploid.) The recovery of somatic hybrids may be enhanced if a selective system is available. For example, two different monoploid lines of *N. Tabacum* had light-sensitive yellowish and light-resistant, as a result of complementation between the parental genotypes. The calluses can be grown into plantlets, which then either are grafted onto a mature plant to develop or are themselves potted.

Application of Polyploidy

Among the cultivated varieties of wheat, three different chromosome numbers are represented: 14, 28, and 42 (x = 7). For example, the primitive small-grained einkorn type of Europe and Asia, Triticum monococcum, has 14 chromosomes in its vegetative cells. Its yield is low and it is of comparatively little value. An emmer wheat (durum), T. dicoccum, grown chiefly in southern Europe but also in the United States, has 28 chromosomes. It has thick heads with large hard kernels and issued mainly for macaroni, spaghetti, and stock feed. The bread wheats, T. aestivum, with 42 chromosomes, were postulated by J. Percival in England to have come from a cross between emmer wheat and goat grass (Aegilops), both of which are native to the Babylonian region where bread wheat originated.

When techniques for artificial chromosome doubling became established, investigations of the origin of bread wheat confirmed

Percival's theory. Experimental evidence obtained by E. S. McFadden and E. R. Sears and separately by H. Kihara traced the pathway for the origin of one type of bread wheat, T. Spelta.

Aegilops squarrosa (n = 7) was found to carry a group of major characteristics that distinguish the hexaploid (n=21) T. Spelta from the tetraploids (n = 1) T. dicoccum and T. dicoccoides. Hybrids between these tetraploid species of wheat and A. squarrosa proved to have all of the major taxonomic characters of T. spelta but the hybrids were completely or nearly sterile. When the F_1 hybrids of T. dicoccoides × A. squarrosa were treated with colchicine, highly fertile allopolyploids with 42 chromosomes were obtained. These synthetic hexaploids closely resembled the cultivated. T. spelta, and they produced highly fertile hybrids with that species and with T. vulgare, known to be in the ancestry of the bread wheats. This demonstrated that the genome of the hexaploid wheats corresponded to one chromosome set of A. squarrosa. It was postulated that T. spelta is the ancestral hexaploid wheat of Europe, having arisen, possibly in fairly recent times, in southeastern Europe or southwestern Asia following chromosome doubling of natural hybrids of T. dicoccoides (or its cultivated close relative, T. dicoccum) × A. squarrosa. T. spelta is believed to have been carried over the northerly route into central and western Europe. Experiments of McFadden, Sears, and Kihara reconstructed the pathway through which a moderately useful wheat and a goat grass hybridized in nature and produced forerunners of a most valuable crop, bread wheat.

New World Cotton

Crosses can be made between distinct species of cotton, members of the benes Gossypium. The hybrids show a wide range of vigor and fertility, making the material favourable for studies of origins. Three cytological groups have been found to correspond with the major world distributional areas. Old World cotton had 13 pairs of large chromosomes. American cotton, which originated in Central or South America, has 13 pairs of small chromosomes. New world cotton (the cultivated long-staple type) has 26 pairs, 13 large and 13 small. Evidently, hybridization and chromosome duplication occurred somewhere in the ancestry of the New World Cotton.

J. O. Beasley used the colchicine technique and succeeding in doubling the chromosomes of a hybrid between the Old World and American cotton. The resulting hybrids, with four set of chromosomes (amphidiploids), crossed readily among themselves and produced fertile plants resembling New World cotton. The process by which the valuable polyploid cotton may have originated in nature was thus duplicated in the laboratory.

Primrose Hybridization

The primrose, Primula kewensis, is an allotetraploid with 36 (2n) chromosomes. It was derived from a cross between two diploids, P. floribunda (x= 9) and P. verticillata (x= 9). Plants from these two species crossed readily, producing hybrids with 18 chromosomes in their vegetative cells. 9 from P. floribunda and 9 from P. verticillata, but the hybrids were sterile. Eventually, however, a branch on a hybrid plant developed from a cell in which the chromosome number was doubled (36), so that each chromosome had a homologous partner. This branch was propagated and gave rise to a fertile primrose plant with cells containing 36 chromosomes of the two diploid parents, the sterile diploid hybrid, and the fertile allotetraploid are shown.

Tobacco Resistance

Induced polyploidy has been exploited to a great extent. Practical applications may become more common as additional data are accumulated. By artificially induced polyploidy, disease resistance and other desirable qualities have been incorporated into some commercial crop plants. Tobacco, Nicotiana tabacum, for example, is susceptible to the tobacco mosaic virus (TMV), whereas N. glutinosa appeared at first observation to be resistant. Further investigation, however, showed that in N. glutinosa the virus killed the cells that were invaded and the virus particles became isolated in the dead cell. The apparent resistance thus was attributable to hypersensitivity. When the two tobacco species were crossed, the hybrid was found to be "resistant" to the virus, but totally sterile. When the chromosomes were doubled, it was possible to secure a fertile polyploid "resistant" to the virus.

Polyploid Fruits, Flowers, and Wheat

Some varieties of plants that serve human needs more effectively than others have now been identified as polyploids. Many polyploids were selected and cultivated because of their large size, vigor, and ornamental values, before their chromosome numbers were known. Giant "sports" from twings of McIntosh apple trees that were found to be tetraploid (4n) were propagated into whole trees, which produce extra-large fruit. The texture of the giant apples is as fine as that of diploids, but the yield is inferior. Mass selection of seedlings may overcome this difficulty. Bartlett pears, several varieties of grapes, and cranberries have also produced sports with giant fruits. Some of these show promise of practical usefulness. With colchicine treatment, a number of polyploids have been developed artificially. This technique has provided a way to explore the mechanism involved in polyploid formation and to make use of the good qualities of polyploids. Tetraploid (4n) maize is more vigorous than the ordinary diploid and produces some 20 percent more vitamin A. Its fertility is somewhat reduced, but this drawback responds to selection. Polyploid watermelons have been developed from colchicine treatment by Kihara and others. The tetraploid with 44 chromosomes is large and has practical value. Triploid watermelons with 33 chromosomes are especially desirable because they are sterile and have no seeds. Among the flower garden varieties, 4n marigolds and snapdragons are widely cultivated.

Polyploid plants respond to artificial selection and hybridization, as do diploid species. The recent history of plant breeding has been characterized by a marked improvement in many polyploid plant crops. The yield of wheat, for example, has increased appreciably. This has been accomplished by developing disease-resistant strains and breeding for increased hardiness and greater efficiency so that crops may survive under the various environmental conditions found in wheat-growing areas. A constant threat to the wheat crop is rust—a fungus that attacks the stems and leaves of the growing plants and destroys the ripening grain. Spores are borne by wind and, when conditions are right, they spread like fire through wheat fields. The disease can be combated by developing rust-resistant strains and by eradicating barberry bushes, which are hosts to the spores during the spring months. But new varieties of rust that destroy previously resistant grain keep evolving, thus perpetuating the job of plant breeders. The larger kernels at the left are from a new strain of rust-resistant spring wheat. At the right are shown kernels of wheat, similar in other respects but not resistant, that are dwarfed from infection with stem rust. The number of kernels of grain per plant as well as the size of the kernels is decreased by rust infection. Investigators in agricultural experiment stations are constantly alert for new rusts. When a new one is found, the standard wheat varieties are tested against it. If they are not resistant, breeding programs are initiated immediately to develop new strains resistant to that particular rust.

Chromosome Anomalies in Spontaneous Abortions in Humans

A wide variety of chromosome numbers and chromosome structural aberrations is found in spontaneoulsy aborted human fetuses. The types of aberrations found vary according to differences in the age of the fetus at the time of abortion. For example, 40 percent of spontaneously aborted fetuses under 90 days of gestational age (i.e., length of time since the last menstrual period of the mother) exhibit chromosome anomalies. For 91-to 120-day-old fetuses, 25 percent show chromosomal anomalies, and only 5 percent of fetuses over 120 days old exhibit chromosome anomalies. Thus, chromosomal anomalies cause most fetuses to die and be aborted in early developmental stages.

Before the percentage of spontaneous abortions resulting from chromosomal anomalies can be calculated, one must first define what constitutes an aborted fetus. Apparently a large number of conceptions occur in humans, and the rechronic myelogenous leukemia possess two cell lines; one cell line has a normal chromosome complement, whereas the other appears to be missing the long arm of chromosome 2. This condition was originally thought to be a monosomy; however, using band staining techniques to identify the long arm, it was learned that the long arm was translocated to one of the larger chromosomes, usually chromosome 9 (referred to as the Philadelphia chromosome). The role of this chromosomal aberration in the induction, development, and progression of cancer is unknown.

Inherited autosomal recessive disorders, such as Bloom's syndrome, Fanconi's anemia, ataxia-telangiectasia, and xeroderma pigmentosum, have been associated with chromosomal instability and or deficiency in mutation repair mechanisms that result in chromosomal aberrations. These individuals have a high incidence of cancer.

8

Cytological Genetics

The field of cytological genetics has grown up at the interface of molecular genetics and cytology, or cell biology. The structures of cells as they relate to cell division and gene expression are studied by *cytogeneticists*. With regard to the first main gene function, self-duplication or reproduction, the structure and function of chromosomes and the spindle are particularly relevant. With regard to the second main gene function, phenotypic expression, the structure and function of interphase chromatin and of the nucleolus are important.

Chromosome Banding

Several techniques for staining mitotic or meiotic chromosomes, so that characteristic transverse bands can be seen, were developed in the early 1970s. The methods and results of some of the major techniques are given in Three of the techniques shown (*G banding*, *C banding*, and *R banding*) are usually accomplished by staining pretreated chromosomes with the same dye mixture, Giemsa stain. This dye results in coloured (dark) bands and light interbands. The fourth technique, *Q banding*, is performed using quinacrine, a fluorescent dye. The stained bands are then fluorescent when observed using UV microscopy.

The mechanisms underlying these techniques have been investigated, and, although there is not complete agreement, many cytogeneticists now think that both Giemsa and quinacrine stain with DNA rather than the chromosomal proteins. G banding seems to result from stacking of dye molecules along the sides of certain regions of the DNA. The interbands may stain less because more of their DNA is covered by proteins and/or because more of the interbands DNA is extracted before stain is applied. The G-bands have been identified as regions that are

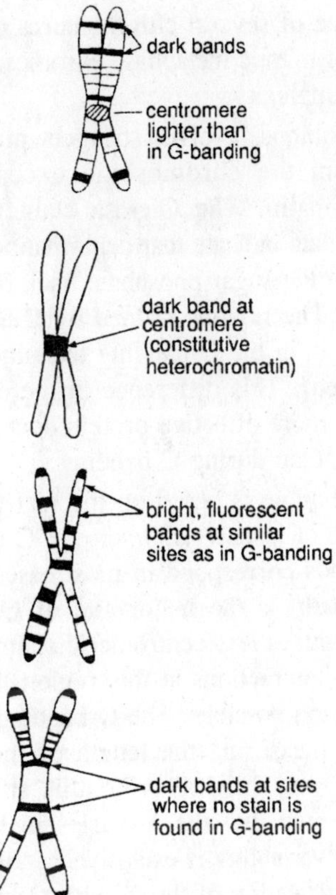

Fig. 8.1. Effects of major mitotic chromosome banding techniques on a hypothetical chromosome.

heterochromatic; these same regions replicate late in the S period of the cell cycle and thus can be differentially labelled using 3 added at the end of S.

Another interesting correlation is between G bands of mitotic chromatin and the dark-staining clumps of meiotic chromatin called *chromomeres*. During the early prophase of meiosis 1 (pachytene), the chromosomes are condensed but not maximally contracted. The arrangement of chromomeres in such chromosomes is the same as the arrangement of C bands in mitotic chromosomes of the same organism. In meiotic prophase, the extended human chromosomes can be seen to have even more chromomeres than G bands (2000 to 300 chromomeres rather than 350 G bands listed in the Paris nomenclature), but as the chromosomes condense further, the 350 characteristic bands are produced by coalescence of several chromomeres to form each band. Certain manipulations of mitotic metaphase chromosomes have revealed a similar underlying complexity.

The C-banding technique involves stringent pretreatments, which extract the DNA from the chromosome except for regions of constitutive heterochromatin. The Giemsa stain then binds to this remaining DNA. Some data indicate that depurination (release of purine bases, A and G, leaving the sugar-phosphate back bone intact) may be essential to C banding. The regions stained by C banding are usually stained less intensely than in the G-banding technique (which uses the same stain, as mentioned). This difference in the effects of the same stain may result from a more effective protein cover over centromeric DNA during G banding than during C banding.

The Q banding technique is based on the fact that AT-rich DNA enhances the fluorescence of quinacrine, whereas GC-rich DNA quenches the fluorescence. Q bands correspond in most cases to G bands, thus suggesting heterochromatin is the major site of Q banding. Like G banding, Q banding produces less centromeric staining than expected. Probably, protein-DNA interactions at this region limit dye access or limit the type of interaction possible. The Q banding technique is very useful in detecting and measuring the length of the Y chromosome, since the Y fluoresces very brightly with this stain. A number of observations have suggested that the Y varies in length in different men, and that this length variability is exclusively in the heterochromatin of the Y. The easy identification of the Y with Q banding has greatly facilitated such studies.

R banding or reverse banding is thought to result in Giemsa staining of GC rich DNA, especially noticeable in phase-contrast microscopy. Other stains (for example, acridine orange) work equally well in producing R bands after the same pretreatment. One possible mechanism for R banding involves denaturation and extraction of the AT-rich DNA, leaving behind the GC-rich DNA, which is more resistant to thermal denaturation. R bands are, in general, those regions that do no stain well in G banding or Q banding. They are thus euchromatin. These bands have been shown to replicate early in S periods.

The characteristics of DNA from the R bands (euchromatin, expressed genes), the G bands (*facultative heterochromatin*, inactive genes), and the C bands (*constitutive heterochromatin*, highly repetitive non Genic DNA). The model shows how the chromosome might be organized into G bands and inter bands (which would stain during R

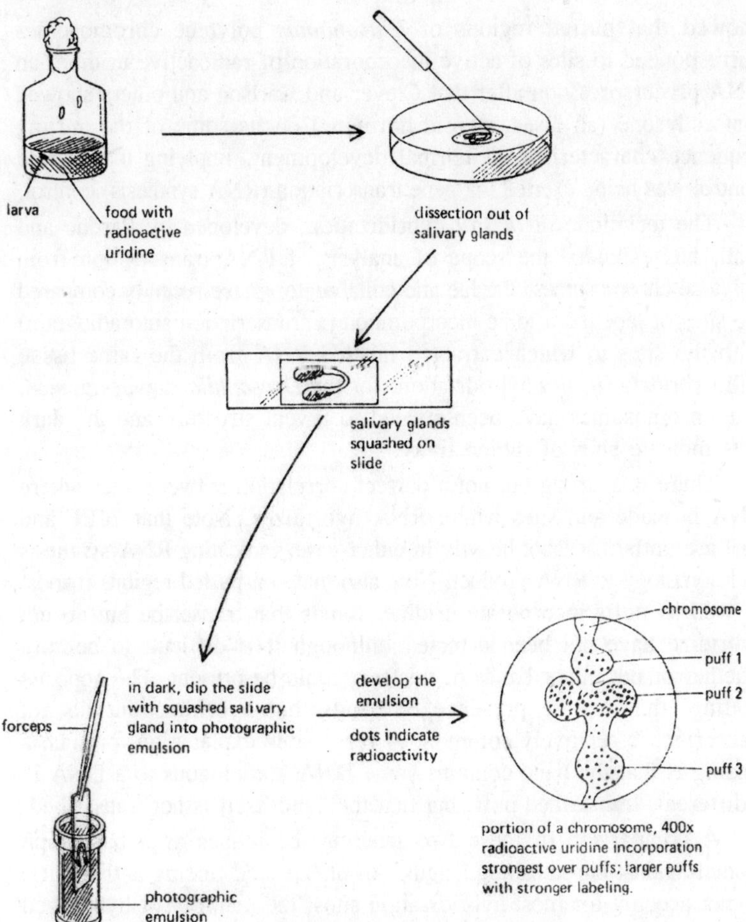

Fig. 8.2. The process of demonstrating enhanced transcription of RNA in puffs via autoradiography after radioactive uridine incorporation.

banding). Notice how both helical and clumped levels of organization are used to build up the final condensed chromosome, according to this model.

STRUCTURE AND FUNCTIONS OF SPECIAL TYPES OF CHROMATIN

The role of nonhistone proteins in opening our the structure of these chromosomes for gene expression was also covered. At that point, *puffs* along the polytene chromosomes were merely identified as sites of RNA transcription. This finding was worked out by cytogeneticists in a series of experiments that will now be described. In 1959, Pelling showed that puffed regions of *Chironomus* polytene chromosomes corresponded to sites of active incorporation of radioactive uridine, an RNA precursor. Soon after that Clever and Karlson and others showed that ecdysone (an insect steroid hormone) elicits some of the puffing sequence characteristic of normal development, implying that steroid control was being exerted via gene transcription (RNA synthesis) control.

The technique of *in situ* hybridization, developed by Pardue and Gall, has extended the scope of analysis of RNA transcription from polytene chromosomes. Pardue and collaborators have recently compared the sites of labelled uridine incorporation (a transcription autoradiogram) with the sites to which extracted labeled RNA from the same tissue will hybridize (*in situ* hybridization) for the *Drosophila* salivary glands. The chromosomes have been stained to reveal structure and the dark dots indicate sites of radioactivity.

There is a strong but not a perfect correlation between sites where RNA is made and sites where RNA hybridizes. Note that 74EF and 75B are puffs that label heavily in either case, indicating RNA synthesis and homology to RNA product. Note also that non-puffed regions (bands) as well as puffs incorporate uridine. Bands that transcribe but do not hybridize have not been detected, although it is difficult to be sure whether or not minor bands of this sort could be present. The opposite finding, that is, the presence of bands that hybridize but do not transcribe, is relatively common. A reasonable explanation for such a finding is that the band contains some DNA homologous to a DNA in a different, transcribed puff, but that the band itself is not transcribed.

A comparison of these two labeling techniques as a bar graph plotted against chromosome length. An unexpected finding is that puffs do not account for most hybridization sites; the majority of hybridized sites are in unpuffed bands. It is significant that high levels of *in situ* hybridization in puffs are found upon hybridization with RNA probes labeled during transcription of those same puffs. For bands that puff, puffing thus does seem to go hand-in-hand with transcription.

A different sort of refinement of our understanding of puff structure comes from the observation that in vivo hybridization does not always label the entire puff uniformly, as ^{3}H-uridine incorporation does. This sort of finding in the 75B region, where the ^{3}H-uridine covers the puff but the labeled RNA hybridizes at only one end of the puff. This finding may indicate that much of the RNA product of the puff is lost during RNA processing an idea consistent with regulation via RNA processing.

The use of fluorescent antibodies to localize proteins in the polytene chromosomes. This technique has also revealed that RNA polymerase is heavily concentrated in puffs, although it is also present in some bands. Another special type of chromosome is the lamp brush chromosome. This chromosomal structure is seen during the late pachytene and diplotene of meiotic I prophase in oocytes. RNA synthesis occurs along the loops of these chromosomes, while the DNA along the axis is inactive. RNA synthesis is characteristically observed during meiotic prophase I, even when the lamp brush configuration cannot be detected. A typical pattern of events is that autoradiography after ^{3}H-uridine incorporation in zygotene and pachytene stages shows the whole chromosome complement to be involved in RNA synthesis, whereas later, in the diplotene stage of meiotic I prophase, rRNA is almost the only type of RNA made, By metaphase I, RNA synthesis is undetectable. It is thus reasonable to assume that lampbrush structure is the reflection of a normal activity of meiotic prophase in many organisms, namely RNA transcription.

A paired set of lampbrush chromosomes, held together by two chiasmata, and a view of the RNA synthesis in progress along one of the loops. In *Triturus cristatus* the *lampbrush loops* persist about 200 days. Gall and Callan showed that in one giant loop, transcription begins where the end of the loop reaches the axis of the chromosome, and progresses around the loop. It takes about 10 to 14 days of continuous labeling with ^{3}H-uridine for the label to proceed all the way around the loop. When the labeled RNA is found at the thick end of the loop, it is evidently about to be released from the chromosome as ribonucleoprotein. The thick (RNA rich) and thin (RNA poor) ends of such a loop, as well as evidence for the progression of ^{3}H-uridine labelling around the loop. It is probable that such RNA is an important set of mRNAs needed during early development of the fertilized egg.

A third special type of chromosome, not previously described, is the B chromosome. *B chromosomes* are extra chromosomes that are not vital to the organism. They may be present or absent in different members of a population without conferring any obvious advantages or disadvantages. Such chromosomes exist in hundreds of species of plants and animals, yet since they have little or no effect on the phenotype. They have not been extensively studied. In mealy bugs, radioactive RNA was synthesized from organisms having only the normal (A set) of chromosomes. Then, cells with both A and B chromosomes were incubated with the labeled RNA for *in situ* hybridization. The B chromosomes were lightly labeled or unlabeled, whereas the A chromosomes were heavily labeled. This experiment suggests that B chromosomes have little DNA homology with A chromosomes.

In corn, the B chromosome is denser than the regular A chromosomes at mitotic metaphase. In addition, it replicates late in the mitotic S period, probably because it is highly heterochromatic. This B chromosome controls a system of non-Mendelian inheritance, involving non disjunction at the second pollen mitosis followed by preferential fertilization of the egg by sperm containing B chromosomes. (*Non-disjunction* is the failure of a pair of daughter chromatids to separate during cell division.) Evidently, the corn B chromosome also enhances recombination in certain regions of the A chromosomes. It has been suggested that the corn B chromosome may have structural features in common with abnormal 10, an unusual corn chromosome consisting of a very heterochromatic region attached to chromosome 10.

Detection of Changes in Chromosome Number and Structure

Change in Number Chromosomes

Changes in chromosome number may create aneuploid organisms. *Aneuploids* have changed in chromosome number to a number of chromosomes not divisible by a, the haploid number. In some organisms there is a supernumerary B set of chromosomes, as described in the previous section. In other organisms (and in the A set of chromosomes, if both A and B sets are present) there is a necessity for a balanced set of chromosomes. The loss of a chromosome, representing loss of several hundred to several thousand genes, represents a loss of gene copies which may be needed at times of maximal gene expression. Such loss is expected to have a deleterious effect. It is more surprising to discover that gain of a chromosome has deleterious effects, although they are usually less severe than in chromosome loss. The explanation must lie in the necessity to have an optimum balance between the gene products encoded by the different types of chromosomes.

One mechanism producing aneuploidy is non-disjunction. If non disjunction occurs during meiosis, a gamete with two copies of that chromosome is formed as well as a gamete with no copies. Fertilization by such gametes leads (barring developmental arrest) to entire aneuploid organisms. Non-disjunction can also happen later in development, during a mitotic cell division. This situation produces a mosaic organism, in which most cells are normal but those descended from the cell with abnormal mitosis are aneuploid. If sex chromosomes are involved in the non disjunction, the genotypes produced are X0, XXY, and so on. Deficiency of one autosome is called *monosomy*, whereas the presence of an extra autosome is called *trisomy*, by contrast with the normal diploid state with two copies of each autosome. Non-disjunction occurs in many organisms, including humans.

In 1959, Lejeune, Gautier, and Turpin showed that Down's syndrome (mongolism) due to trisomy of one of the two shortest human chromosomes (21 or 22). Down's syndrome had been recognized medically since the mid-1800s. Symptoms of this trisomy include mental deficiency, short stature, a round facc with cpicanthal eyelid folds, and a simian fold pattern of lines in the palm of the hand. Since identical twins in which Down's syndrome occurred were nearly always either both mongoloid or neither mongoloid, it was assumed to be hereditary. The puzzling feature of the inheritance was that only very rarely was more than one person affected in a particular family. Establishing that the condition was due to an extra autosome convincingly explained these findings. Using G banding, one can easily distinguish the two shortest pairs of autosomes. When chromosome banding became widely used, the extra chromosome of Down's patients was called number 21, in accordance with the literature traditions.

Few individuals with other trisomies survive until birth in humans. Trisomies of 13 and 18 have been described, and both produce a variety of effects including mental retardation. Nonviable individuals with trisomies for other chromosomes do occur, however, as has been shown by examining the chromosomes of spontaneously aborted features.

The sex chromosomes can also undergo non-disjunction, and many persons with abnormal doses of sex chromosomes are nearly normal in most phenotypic qualities except for sexual differentiation. The mildness of such effects is probably due to the inactivity of all but one X in adult organisms and the presence of very few genes on the Y chromosome. Two of the most common syndromes involving sex chromosome aneuploidy are Klinefelter's (XXY or other rarer forms such as XXXY) and Turner's (X0) syndromes. Patients with *Klinefelter's syndrome* are phenotypically male but with underdeveloped genitalia, occasional breast development and less body hair than usual. The incidence is approximately one (XXY) male per 1000 births. They may or may not be mentally deficient. The XXY configuration poses problems in reproduction, and most such people are sterile. Patients with Turner's syndrome are phenotypically female but with underdeveloped ovaries. The incidence is about one (X0) per 20,000 births, but 99% of X0 conceptions. Persons with *Turner's syndrome* are usually short and have underdeveloped secondary characteristics, but are often normal mentally. They may have reduced mathematical and spatial ability.

A third common abnormality is the XYY occurring approximately once per 1000 births. This syndrome is rather controversial, there having been some studies suggesting increased aggressiveness from XYY men and others refuting this notion. The XYY males are fertile and seem normal in most phenotypic characteristics. They have some tendency to be extra tall and possibly somewhat reduced in intelligence. Other syndromes, much rarer, have been described in humans with the following sex chromosome sets: XXX, XYYY, XXXX, XXXXX, XXXXY, XXYY, and XXXYY.

Detection of Chromosomal Rearrangements by Different Chromosome Shapes, Sizes, or Banding Patterns

In addition to changes in the number of chromosomes, a number of macrolesions have been described that can be detected (a) by different banding patterns from normal, suggesting that chromosomal rearrangements have occurred, (b) by meiotic synapsis of particular chromosomes showing that homologous regions have been rearranged, or (c) by polytene chromosome synapsis showing different arrangements of the same bands.

A great deal can be told about normalcy of a chromosome complement by simply examining a *karyotype* (all the metaphase chromosomes from a cell). A human genetic disease called the cri-du-chat syndrome is characterized by severe mental deficiency, a very round face, microcephaly and a very plaintive cry in infancy. In 1963, Lejeune and coworkers showed that a partial deletion of the short arm of chromosome 5 was responsible for this syndrome. This deletion essentially converts the submetacentric chromosome to an acrocentric, so it is easily visible in the karyotype.

The *Bar* eye mutation in *Drosophila*, which was shown, by examining the salivary gland chromosomes, to result from a duplication of a short set of bands and interbands. This mutation is only one of many duplications and deficiencies which have been detected by examining *Drosophila* chromosome bands.

In addition, the mitotic banding techniques for human chromosomes have permitted detection of rearrangements. Q banding of chromosomes of a person with Down's syndrome but with only the normal number

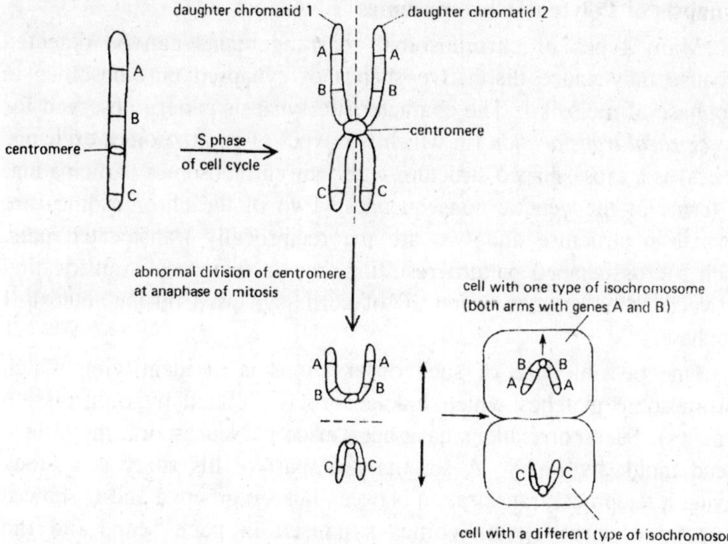

Fig. 8.3. Formation of isochromosomes by division of the centromere in the wrong plant.

of chromosomes, 46. Careful examination of this figure will show that a chromosome 21 has been translocated to the short arm of one of the chromosomes 14. In addition, 15-21 translocations are often implicated in cases of Down's syndrome. Another human abnormality distinguishable by banding is a variant of Turner's syndrome (Xiso X), caused by chromosomal aberration in a person with 46 chromosomes. One of the X chromosomes in such a patient is an *isochromosome* (both arms carry the same genes) for the long arm of the X chromosome. Consequently, no genes commonly found on the short arm of X are present on this chromosome; thus, the patient has only one copy of the X short arm genes. Isochromosomes are thought to form by division of the centromere in the wrong plane.

In view of the currently accepted unineme model of the chromosome, breakage and reunion of the DNAs involved would presumably be necessary. The Philadelphia chromosome associated with chronic myelocytic leukemia is another translocation that can be detected by banding. The usual type of aberrant chromosome observed is a translocation from chromosome 22 to chromosome 9. The persons affected are essentially always mosaic for the very short 22, the Ph′ chromosome, suggesting that the translocation occurred in a somatic cell rather late in development.

Detection of Rearrangements by Meiotic Synapsis or Interphase Synapsis of Polytene Chromosomes

Many types of chromosomal rearrangements can be detected because they cause distinctive shapes of synapsed chromosomes in prophase of meiosis I. The characteristic synapsis pattern observed for a *reciprocal translocation* (in which two types of chromosomes exchange pieces) is a cross-shaped structure with four chromosomes participating, in terms of the genetic consequences. Two of the chromosomes are normal in structure and two are the reciprocally translocated pair. Such a cross-shaped pattern resulting from a reciprocal translocation between chromosomes 8 and 10 of corn, synapsed during meiosis I prophase.

One possible use of such observations is in identifying which chromosome matches which linkage group (defined by transmission genetics). Such correlations have been made in Neurospora, the orange bread mold, by Barry. A summary of part of his study is a stock having a reciprocal translocation between linkage groups I and II showed a group of four chromosomes synapsed in pachytene, and the cytologically identifiable chromosomes involved were 1 and 6, ranked in order of size. These and other similar data using other translocation heterozygotes established linkage group I to be chromosome 1, linkage

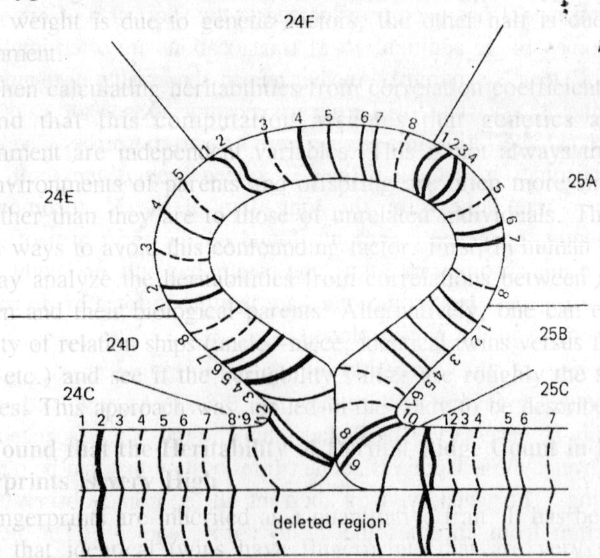

Fig. 8.4. Diagram of a loop formed in somatically paired polytene chromosome in a region of a deletion in chromosome II of Drosophila.

group II to be chromosome 6, and linkage group VII to be chromosome 7. In addition to details of banding and meiotic synapsis, a third way of cytologically defining chromosomal rearrangements is via the somatic synapsis found for the polytene chromosomes. In a heterozygote for a chromosomal rearrangement, such pairing is quite characteristic in shape, depending on the type of aberration. For example, shows a looped synapsis in an inversion heterozygote. What happens in a deletion heterozygote. (Note that the loop starts and ends at the same place along the deleted chromosome, but has been stretched for facility in diagramming. Synapsis is thus a powerful tool for detecting rearrangements, whether it is observed in meiotic prophase or in polytene somatic cells.)

Ultrastructure of Mitotic and Meiotic Chromatin

Mitotic Chromatin

Electron microscopy of dividing cells has given us a mental image of a single chromosome at the start of metaphase. In chromosomes sectioned and observed by transmission electron microscopy (TEM) at metaphase, not much structure is usually apparent in the DNA-protein complex, but the points of attachment of the spindle fibers can be seen. The scanning electron microscopy (SEM) process has generated a rather different view of mitotic chromosomes, also represented in the drawing of Wolfe. This view shows a knobby structure (with *microconvules* or knobs consisting of supercoiled loops) typified.

An interesting question with regard to mitotic chromatin is how it is condensed from the extended nucleosome state to the 300Å

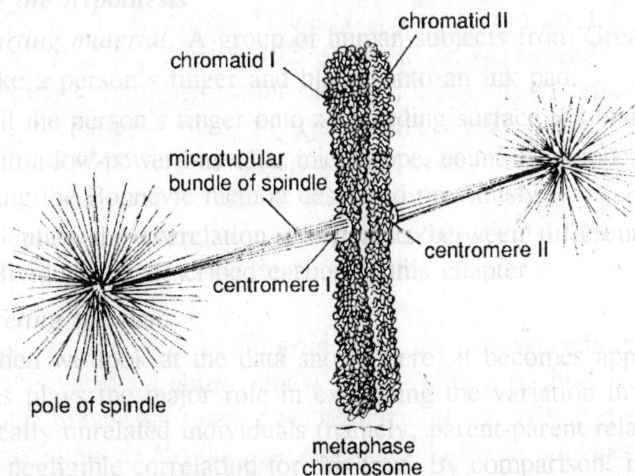

Fig. 8.5. Mitosis, showing a chromosome at mitotic metaphase.

nucleohistone fiber and finally to the knobby chromosome. Laemmli and coworkers have approached this question by removing the histones from metaphase chromatin on the hypothesis that higher order structure is stabilized by means of nonhistone proteins, and one might find out how DNA is associated with structural nonhistone proteins if those two components were observed in the absence of histones. A chromosome of this type, isolated from a HeLa cell. The dark material in the center is a nonhistone protein *scaffold*, and the DNA is spilling out on all sides. The DNA has no visible free ends, suggesting that DNA loops might have both ends rooted at the scaffold. The view in such loops in a region of less dense DNA, convincingly demonstrating that the scaffold has associated with it both ends of these loops, at least in the region shown.

In the absence of histone extraction, the looped DNA would have been in the nucleosome or nu-body structure. This structure is packed in some way to form the 300-Å diameter unit fiber of chromatin. Until recently, the suggestion had been, almost universally, that the nucleosomes formed a regular spiral of 300 Å diameter. Presently, there is some evidence that this unit fiber may be an association of clumps of nucleosomes, and that these clumps may vary in size. The 200- to 300 Å (20- to 30-nm) fibers from a chicken erythrocyte nucleus, showing apparent variation along the fiber axis in the degree of nucleosome packing.

Meiotic Chromatin

In order to discuss meiotic chromatin, it is necessary to describe the stages of meiotic divisions I and II especially the substages of prophase I. As meiosis begins, each chromosome shortens by coiling. Its centromere, the region for spindle attachment, may become visible as a constriction. This earliest stage of meiotic prophase is called *leptotene*. During this phase, a specific mixture of RNA and proteins called *synaptinemal complex* (SC) forms between the two sister chromatids of each chromosome. Just before the next phase, this SC moves from between the chromatids to cover one surface of each chromatid.

You might visualize each chromosome at this point as an opened out hot-dog bun, spread over the whole inside surface with peanut butter. Each half of the bun represents a *sister chromatid*, the whole bun represents the whole chromosome, and the peanut butter represents the SC in this analogy. At the *zygotene* phase of prophase, the two homologs of each chromosome stick together by means of their SC

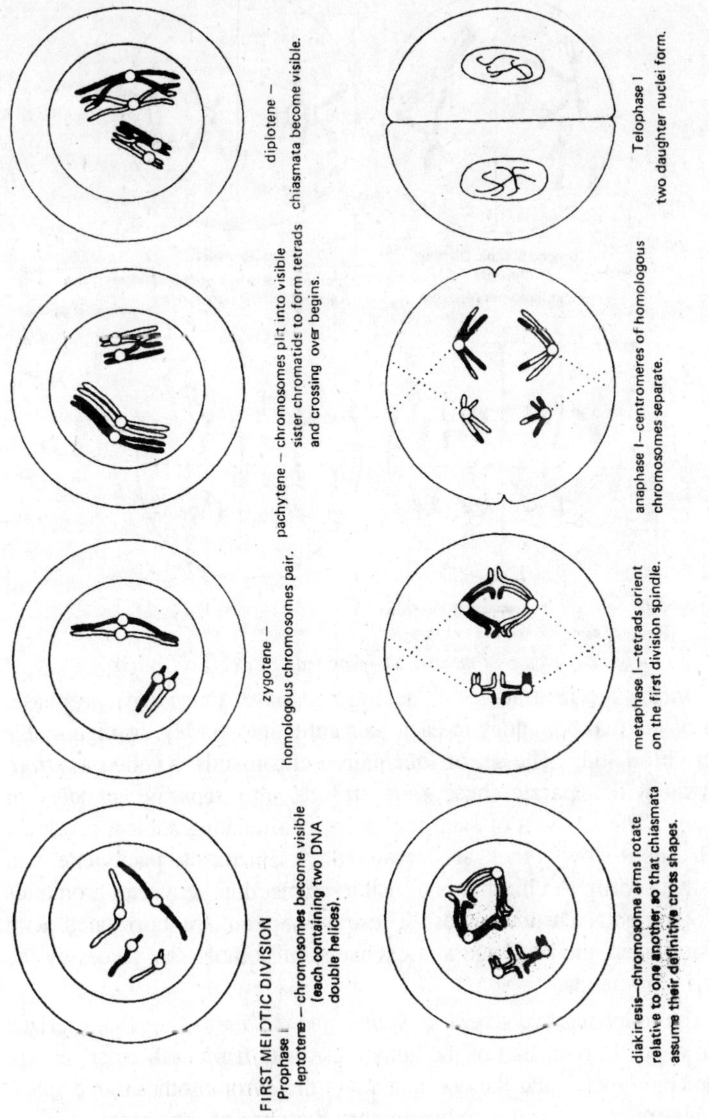

Fig. 8.6. Stages of meiosis, showing major cytological events.

surfaces. In this analogy, we would take another exactly similar peanut-butter-spread, opened-up hot-dog bun and stick the two buns together by their peanut butter surfaces. This situation of lengthwise pairing of homologous maternal and paternal chromosomes is called *synapsis*. Notice that both chromatids of each homolog are paired with their matching chromosome all along their lengths.

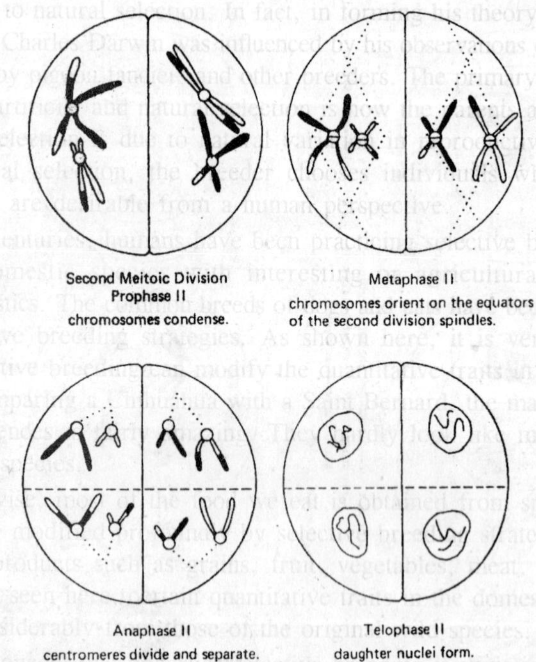

Fig. 8.7. Stages of meiosis, showing major cytological events.

During the next stage of meiotic prophase, the *pachytene* phase, each of the two homologs in each pair splits into visibly distinguishable sister chromatids. The set of four paired chromatids is called a *tetrad*. Meiosis will separate these four strands into separate gametes or progeny. The analysis of such tetrads is a fascinating subject revealing much about how genes are arranged. It is during pachytene that chiasmata form. A chiasma is a visible connection between chromatids of homologous chromosomes. These *chiasmata* are correlated with crossing over, the breakage and exchange of homologous pieces by the paired chromatids.

The following two stages, *diplotene* and *diakinesis*, are characterized by an apparent repulsion of the homologs away from each other, except at the centromeres and the chiasmata. As the chromosomes move apart, the chiasmata *terminalize* (move toward the ends of the chromosomes). Then prophase of meiosis I is complete. Some cells (for example, cells defined as human egg cells) remain frozen at this stage for years. At metaphase I, the chromosomes align themselves along the equator of the spindle so that the maternal homolog's centromere faces one pole of the spindle and the paternal homolog's centromere faces the other pole. Notice that there is no sorting mechanism to make all of the maternal chromosomes face one pole and the paternal chromosomes face the other.

Now, in *anaphase I*, the homologs separate. In telophase, the two cells separate. Commonly, each of these cells receives some chromosomes from each parent. We have shown each cell resulting from meiosis I receiving a paternal homolog of one chromosome and a maternal homolog of the other type of chromosome, but which homolog goes where is merely a matter of chance. In an organism with only two types of chromosome, often one might see complete separation of maternal from paternal chromosomes at this point. In humans, with 23 pairs of chromosomes, it is almost inconceivable ($p = (0.5)^{23} = 1/8,388,608$) that a cell produced by meiosis would get a complete set of chromosomes from only one of the two parental sets.

The next two phases, *telophase I* and *prophase II*, may be abbreviated or absent. They are mainly concerned with decondensation and recondensation of the chromosomes. In *metaphase II*, the chromosomes orient on the metaphase plate so that the centromeres of the sister chromatids of each of the chromosomes face opposite poles. The centromeres then separate in *anaphase II* and each chromatid ends up in a separate cell from its sister chromatid. In *telophase II*, the chromosomes decondense and meiosis is complete. Four cells result, but all four may not be viable entities. In egg development only one of the four is typically viable, and this one receives almost all of the cytoplasm.

Meiotic prophase I (recall from earlier in this section that the stages are leptotene, zygotene, pachytene, diplotene, and diakinesis) has been examined extensively by cytologists in order to find out the nature of synapsis and the nature of crossing over. As already described, the synaptinemal complex (SC) appears to be a necessary part of synapsis. Evidence for this conjecture comes from studies of *Drosophila*, where the males do not exhibit meiotic crossing over but the females do have crossing over. The females in this case form SC and have chiasmata along their chromosomes during meiosis I prophase, but the males have no SC and are achiasmatic. A mutant, $c(3)G^{17}$ has no crossing over in the females and also has no SC. We can infer that chromosomes without SC probably cannot undergo crossing over.

Although the SC is important in chiasma formation, it is not sufficient to insure chiasmata. In the silk moth (*Bombyx mori*), both males and females have SC, but females are achiasmatic. The structure of the SC in females becomes modified during pachytene, and these modifications may prevent recombination. Another example of a developed SC without the presence of chiasmata is the Black Beauty hybrid lily, in which Hotla and Stern have shown that an important endonuclease, necessary to break the homologous DNA strands for recombination, is missing.

A synapsed pair of chromosomes, held together by the SC, is usually anchored to the nuclear envelope on both ends. In some organisms, these attached ends are widely scattered over the nuclear envelope, whereas in others they form a bouquet at a single, small area on the nuclear envelope. An electron micrograph of the end of a pachytene synapsed pair attached to the nuclear envelope, and a study tracing exactly where all the chromosome ends were anchored in a *Locusta* pachytene spermatocyte (note that all ended on one side of the nuclear envelope). Although this type of structure is found in most organisms examined, there are exceptions. In *Drosophila* for example, the chromosomes are not attached to the nuclear envelope.

When pairing and synapsis begin, there are few coils holding the pair together along its length; however, the process of synapsis during zygotene is accompanied by a coiling, possibly produced by rotation of one or both of the anchored ends. By mid-pachytene, these coils are observed throughout the length of the pair, unless aberrations (for example, translocations) interfere with their formation. In a rat spermatocyte, by late pachytene the total number of such counter-clockwise 1800 twists, counting all the paired chromosomes, can approach 100 turns. In very late pachytene, the number of these coils usually decreases. This figure also shows that the SC is forming between the homologs at a region removed from the centromere.

Zickler has studied electron microscopic images of synapsed chromosomes in order to determine what the appearance of a chiasma is at a higher level of resolution. In the fungus, *Sordaria macrospora*, there are seven chromosomes. Each long chromosome pair has three to four chiasmata, each middle-sized pair has two to three chiasmata, and the short pair has one chiasma. The average number of chiasmata per nucleus, in 20 nuclei scored using light microscopy, was 18. The structure on the electron microscopic level, which corresponded in quantity, was the *synaptinemal complex node* (SC node). There were 17 of these in one pachytene nucleus and 21 in another. There were from one to four nodes present in any synapsed pair of chromosomes. Such nodes have been described in a number of other organisms as well, particularly by Moens. An electron micrograph of synapsed rat chromosomes in which nodes appear as definite, discrete, very darkly staining regions, often extending across the space occupied by the SC itself. In other organisms, the size, shape, and location of such structures may differ. In all cases studied, though, a modification of the SC structure appears a likely candidate for recombination sites.

The special lampbrush chromosomes of meiotic prophase in many oocytes have been characterized by cytologists in terms of the centromere placements, the placement of certain large loops, the placement of knobs of chromatin, the placement of chromatin-associated proteinaceous granules, and the location of nucleoli. An example of such a map, as worked out for *Triturus marmoratus*. It is generally possible to recognize such chromosomes by use of relative lengths and these working maps of conspicuous features.

The main feature of meiosis subsequent to prophase I that is of interest is the centromere placement insuring proper alignment at metaphase I and metaphase II a drawing of the alignment at metaphase I, and electron micrographs showing the contrast in centromere alignments in metaphases I and II.

DNA Replication and Transcription at the Cytological Level

Data implying that bacterial DNA replicates semiconservatively, as obtained by autoradiography of entire *E. coli* chromosomes. In addition, the eyelike openings where DNA replication occurs in eukaryotic chromosomes have been. The semiconservative nature of whole chromosome replication remains to be described. In 1957, Taylor and coworkers labeled chromosomes with 3 during one cell generation, then followed the distribution of label in subsequent generations. Some results from their study, demonstrating that the label was passed on exactly as predicted, assuming unineme chromosomes with semiconservative replication of DNA. Notice that some new and old segments have been interchanged. The extent of sister-strand exchange was not known at that time, but it is rather extensive.

Transcription of RNA has been studied extensively in polytene and in lampbrush chromosomes. In addition to transcription from special chromosomes, transcription from RNA- and mRNA-coding regions of DNA has been observed via electron microscopy.

Foe, for example, studied the electron microscopic appearance of chromatin in milkweed bug embryos after various lengths of

First generation, all chromatids labeled

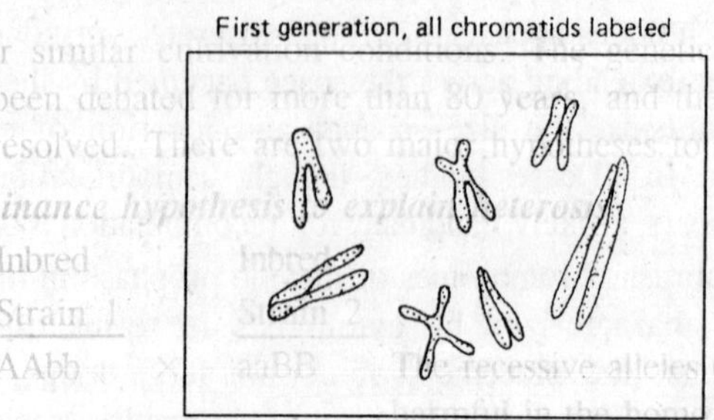

Second generation

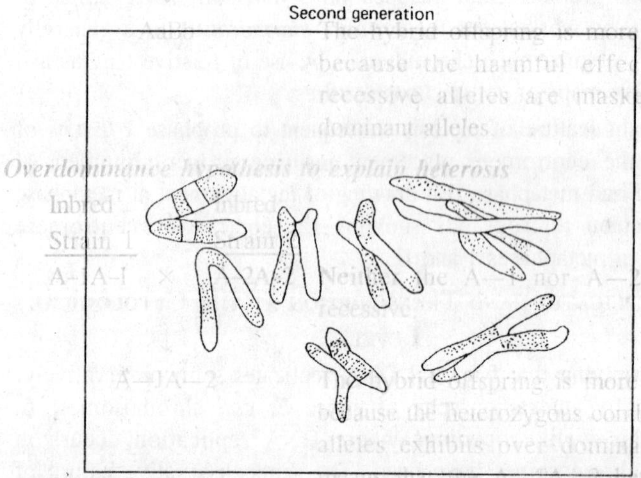

Fig. 8.8. Chromosome semiconservative DNA replication in Bellevalia romona.

development. At 68 hr of development, the neurula stage, rRNA is being synthesized very actively. A typical tandemly repeated array of rRNA genes at this stage contained 51 fibers (transcription product) with transcribed regions of the DNA averaging 2.4 μm long, separated by untranscribed spacers averaging 0.6 μm long. The entire group of fibers being transcribed from one 2.4 μm long region of DNA was termed an rTU, *ribosomal transcription unit*. By 116 hr of development, rRNA transcription has been partially deactivated. There are only about 13 fibers per rTU, and, in fact active rTUs occur less frequently. Since this lower fiber number results in a less densely packed "brush" on the rTU, it is possible to see whether or not the chromatin is beaded into nucleosomes under the transcription products. The answer seems to be that it is not beaded.

The rDNA has an average width of 73 Å, where DNA is 20 Å in diameter and a nucleosome is 100 Å. Since micrococcal nuclease-

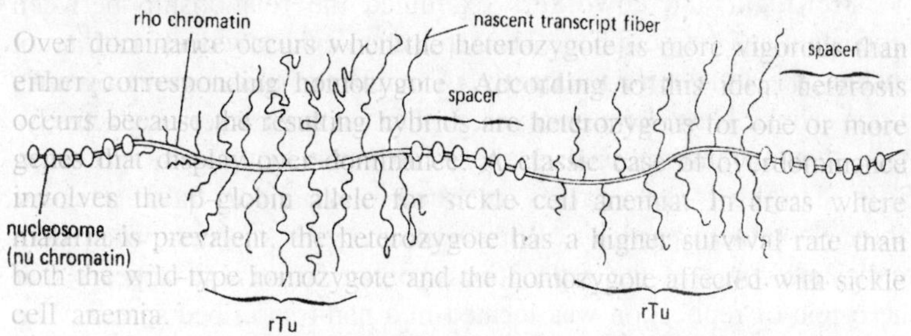

Fig. 8.9. Diagram of ribosomal transcription units (rTU), showing nascant transcript fibers and the spacers between rTU.

sensitive sites are spaced similarly to those in nucleosomes, it is probable that the non-beaded chromatin represents a different packing of the general nucleosome components. Foe has proposed that this state be termed *rho* chromatin. Her observations of the progress of development suggested that *rho* chromatin is necessary but not sufficient for rDNA transcription, since *rho* chromatin's smooth, nonbeaded 70-A fibers are present in some cases before any nascent fibers can be detected. Both the number of active rTUs and the level of transcription of single units appear to be modulated during development.

McKnight and associates, on the other hand, studied the *Drosophila melanogaster* embryo nonribosomal gene transcription as well as the ribosomal gene transcription. In preblastoderm embryos, a few short regions of nonribosomal DNA (0.5 to 3 μm long) are transcribed. In general, RNA polymerases on these genes are very numerous, leading to dense brushes of nascent RNA or ribonucleoprotein fibrils. The rDNA is not active in preblastoderm embryos of the fruit fly.

The rDNA is transcribed as the embryo reaches the cellular blastoderm stage (when a syncitial structure is partitioned into cells). These rTUs are not activated simultaneously, and may be regulated independently by different signal molecules. Non-rDNA cistrons, similar to those present before cellular blastoderm formation, continue to be transcribed. In addition, there is transcription of regions with very low RNA polymerase frequency and without clear cut gradients in lengths of nascent fibrils. McKnight and coworkers found both sparsely and densely RNA polymerase populated non-ribosomal regions had beaded structure like the nucleosomes. Antibodies to the histones reacted to similar extents with the beads of transcribed and nontranscribed chromatin, implying that both types of structure contain histones. But techniques would not permit saturation level antibody reactions, so the meaning of this observation is not certain.

McKnight and coworkers examined the relationship between replication and transcription. The replication evidently does not disrupt nucleosome structure significantly. A number of electron micrographs of the cellular blastoderm embryos, replicating and transcribing actively, have shown transcriptional gradients on newly replicated chromatin. Both daughter chromatids in such a gradient have nascent ribonucleoprotein particles. With regard to rRNA genes, when both replication forks of the replication eye can be seen, it appears that the origin of replication was located in a non-transcribed spacer. The replication complex in some cases dislodges RNA polymerase and nascent rRNA or ribonucleoprotein, but in other cases it may follow the RNA polymerase. Patterns that can be interpreted in either of these fashions have been seen.

Sister Chromatid Exchange

The two chromatids that are produced during chromosome replication are called sister chromatids. Until recently, it was not possible to distinguish the sister chromatids from each other, so the level of exchanges between sister chromatids in somatic cells could not be assessed cytologically. A debate raged for years over the extent of sister strand exchange in synapsed pairs of chromosomes during meiosis. Conclusive proof that ring chromosomes underwent *sister chromatid exchanges* (SCE) in both meiosis and mitosis was presented by McClintock and by Schwartz before the 1960s. Evidently, ring chromosomes were felt to be sufficiently atypical that results regarding them might not apply to all chromosomes.

In the late 1950s and early 1960s, Taylor studied chromosome replication, and found that it was possible to detect SCEs via ^{3}H-thymidine labeling techniques. When he prevented the separation of cells by adding coichicine and examined the tetraploid cells produced, he could detect the level of SCEs in two different cell cycles. These frequencies turned out to be the same. A problem in interpreting these experiments was that, in 1972, Gibson and Prescott discovered that tritium incorporation evidently increased the frequency of SCEs.

More recently, a number of staining techniques have been developed that allow the investigator to distinguish between chromosomes with BU substituted for T in one strand of DNA and chromosomes with BU substituted for T in both strands. Zakharov and Egolina showed in 1972 that Giemsa stained the two chromatids differently with BudR (deoxyribonucleoside with BU substituted for T) was substituted for TdR for two rounds of DNA replication. Giemsa stained the sister chromatid with BU in both strands darkly but the sister chromatid with BU in only one strand lightly. Latt subsequently showed that the fluorescent dye, Hoechst 33258, stained the chromosomes very brightly when T was present, less brightly when T was in one strand and BU in the other, and barely perceptibly if both strands contained BU. Other techniques similar to one or both of these approaches have been worked out subsequently.

The studies of SCEs have given rise to a number of interesting observations. The SCEs have a very strong tendency to occur in junctions between euchromatin and heterochromatin, at least in muntjac and in kangaroo rat. These organisms have certain distinctive distributions of euchromatin and heterochromatin that make it possible to distinguish locations of these chromatin categories in chromosomes stained to reveal SCEs. Another observation has been that there are more SCEs than normal in a genetic disease, Bloom's syndrome, in which cancer incidence is high and there are many chromosomal aberrations. This result does not necessarily mean that SCEs and chromosomal aberrations have a common basis, however, since two other genetic diseases with the same symptoms (Fanconi's anemia and ataxia telangiectasia) lack SCE increases.

Even when no differences in basal levels of SCEs are found, it is sometimes possible to see differences between cells where the DNA has been damaged. Mitomycin C, a bifunctional alkylating agent that crosslinks DNA, causes fewer SCEs in young tissue culture cells than in aged ones, for example, although the basal SCE levels for the two types of cells are similar. Cells of patients with Fanconi's anemia are less sensitive than normal to Mitomycin C as an SCE inducer, but are more likely than normal cells to have chromatid breaks where an SCE might have been expected. With regard to chemically induced increases in SCEs, it has been proposed that mutagenic carcinogens can be detected by in creases in SCEs.

Concluding Remark

Cytological genetics is concerned with cell structures related to DNA replication, nuclear division, and gene expression. Mitotic chromosomes or meiotic chromosomes can be stained to reveal bands. Depending on the stain and the pretreatment, different regions of each chromosomes are stained. Characteristic band patterns can be used to identify particular chromosomes. Currently, C-bands are thought to be constitutive heterochromatin, G-bands are thought to be intercalary heterochromatin, and R-bands are thought to be euchromatin.

The polytene chromosomes generally, but not always, have visible puffs at the sites of RNA synthesis. The transcribed regions contain DNA that hybridizes with the RNA produced; other bands which are not transcribed also can hybridize with the product RNA. Transcription involves entire puffs, but much of the RNA product may be quickly degraded so that hybridization occurs only at one portion of the puff. The location of the mRNA-specific RNA polymerase can be shown via fluorescent antibody binding; it is found in puffs but also in bands that are not puffed.

Lampbrush chromosomes of meiotic I prophase in oocytes are in the process of transcription of their loops. Both mRNAs and rRNAs are synthesized. RNA transcription proceeds around each loop, starting at one end of the loop and terminating at the other end.

B chromosomes are supernumerary chromosomes, having little or no homology with necessary A chromosomes. In corn, the B chromosome is highly heterochromatic. It tends to enhance recombination in certain regions of the A chromosomes.

Aneuploids have either lost or gained one or more chromosomes from the normal set. Non-disjunction is a major cause of aneuploidy. Down's syndrome and other autosomal aneuploidies have far-reaching effects on the phenotypes of the affected persons. Sex chromosomes aneuploidies are also found in viable persons.

Chromosomal rearrangements can be detected cytologically. The human cri du-chat syndrome results from a short deletion, for example. Translocations and inversions can be detected via examination of synapsed homologs.

Mitotic and meiotic chromosomes have a fine structure consisting of microconvules, which are condensed loops. The loops are attached at each end to a scaffold of non-histone proteins; these loops are made of aggregated nucleosomes of the 250- to 300-Å fiber.

Meiosis consists of a long prophase I, subdivided into leptotene, zygotene, pachytene, diplotene, and diakinesis phases. Synapsis via synaptinemal complex occurs in zygotene; crossing over can be seen at pachytene. This prophase is followed by metaphase I; anaphase I and telophase I separate the homologs into different nuclei. Prophase II is often absent; metaphase II ensues, then anaphase II, and telophase II separate the sister chromatids.

DNA replication and transcription can be studied at the cytological level, via light microscopical autoradiography or via electron microscopy. Results are consistent with unineme chromosome structure.

Transcription requires rearrangement of nucleosomal beads into a smoother structure, but transcription need not begin as soon as this rearrangement occurs. Evidently chromosomes are divided into transcription units (blocks of adjacent DNA that are turned on or off simultaneously). Electron microscopical studies have shown that DNA replication and transcription can occur simultaneously.

Sister chromatid exchange (SCE) is more common than it was once believed to be, and can be detected via differential labeling or staining of the sister chromatids. Abnormally high levels of SCE are associated with some genetic diseases, notably Bloom's syndrome.

9

QUANTITATIVE GENETICS

Qantitative genetics is the study of traits that can be described in a quantitative way. In humans, quantitative traits include height, the shape of our noses, the rate at which we metabolize food, and our I.Q, to name a few examples. The features of these traits can be characterized and analyzed with numbers.

Quantitative genetics is an important branch of genetics for several reasons. In agriculture, most of the key characteristics considered by plant and animal breeders are quantitative traits. These include traits such as weight, resistance to disease, and the ability to withstand harsh environmental conditions. As we will see later in this chapter, quantitative genetic techniques have improved our ability to develop strains of agriculturally important species with desirable quantitative traits.

Another important reason to study quantitative genetics is its relationship to evolution. Many of the traits that allow a species to adapt to its environment are quantitative. Examples include the long neck of a giraffe, the swift speed of the cheetah, and the sturdy branches of trees in windy climates. In this chapter, we will examine how genes and the environment contribute to the phenotypic expression of quantitative traits.

QUANTITATIVE TRAITS

When we compare characteristics among members of the same species, the differences are often quantitative rather than qualitative. Humans, for example, all have the same basic anatomical features (two eyes, two ears, etc.), but they differ in quantitative ways. People vary with regard to height, weight, the shape of facial features, pigmentation, and many other characteristics. As shown in Table 9.1, quantitative traits can be categorized as anatomical, physiological, and behavioral. In addition, a few human diseases exhibit characteristics and inheritance patterns analogous to those of quantitative traits.

Table 9.1. Types of quantitative traits

Trait	*Examples*
Anatomical traits	Height, weight, number of bristles in *Drosophila*, ear length in corn, the degree of pigmentation in flowers and skin
Physiological traits	Metabolic traits, speed of running and flight, tolerance of harsh temperatures, milk production in mammals
Behavioural traits	I.Q., mating calls, courtship rituals, ability to learn a maze, the ability to move toward light
Complex diseases	Diabetes, hypertension, arthritis, obesity

In many cases, quantitative traits are easily measured and described numerically. For example, height and weight can be measured in centimeters and kilograms. Speed can be measured in kilometers per hour, and metabolic rate can be assessed as the grams of glucose burned per minute. Behavioural traits can also be quantified. For example, a mating call can be evaluated with regard to its duration, sound level, and pattern. The ability to learn a maze can be described as the time and/or repetitions it takes to master the skill. Finally, complex diseases such as diabetes can also be studied and described via numerical parameters. For example, the severity of the disease can be assessed by the age of onset or by the amount of insulin needed to prevent adverse symptoms.

From a scientific viewpoint, the measurement of quantitative traits is essential when comparing individuals or evaluating groups of individuals. It is not very in formative to say that two people are very tall. Instead, we are better informed if we know that one person is 6'4" and the other is 6'7". In this branch of genetics, the measurement of a quantitative trait is how we describe the phenotype.

In the early 1900s, Francis Galton in England and his student Karl Pearson showed clearly that many traits in humans and domesticated animals are quantitative in nature. To understand the underlying genetic basis of these traits, they founded what became known as the *biometric field* of genetics. During this period, Galton and Pearson developed various statistical tools for studying the variation of quantitative traits within groups of individuals; many of these tools are still in use today. In this section, we will examine how quantitative traits are measured, and how statistical tools are used to analyze their variation within groups.

Quantitative Traits Exhibit a Continuum of Phenotypic Variation that may Follow a Normal Distribution

In Part II of this text, we considered many traits that fell into discrete categories. For example, fruit flies might have white eyes or red eyes, and pea plants might have wrinkled or smooth seeds. The alleles that govern these traits affect the phenotype in a qualitative way. In analyzing crosses involving these types of traits, each offspring can be put into a particular phenotypic category. Such attributes are called *discontinuous traits*.

In contrast, *quantitative traits* show a continuum of phenotypic variation within a group of individuals. For such traits, it is often impossible to place organisms into a discrete phenotypic class. Though there is a minimum and maximum height, the range of heights between these minimum and maximum values is fairly continuous.

Since quantitative traits do not naturally fall into a small number of discrete categories, an alternative way to describe them is a *frequency distribution*. To construct a frequency distribution, the trait is divided arbitrarily into a number of convenient discrete phenotypic categories. For example, the range of heights is partitioned into 1-inch intervals. Then a graph is made that shows the numbers of individuals found in each of several phenotypic categories.

The measurement of the height is plotted along the x-axis, and the number of individuals who exhibit that phenotype is plotted on the y-axis. The values along the x-axis are divided into the discrete 1-inch intervals that define the phenotypic categories, even though height is essentially continuous within a group of individuals. For example, 22 students were between 64.5 and 65.5 inches in height, which is plotted as the point (65 inches, 22 students) on the graph. This type of analysis can be conducted on any group of individuals that vary with regard to a quantitative trait.

The dotted line in the frequency distribution depicts a *normal distribution curve*, a distribution for an infinite sample in which the trait of interest varies in a symmetric way around an average value. The distribution of measurements of many biological characteristics is approximated by a symmetrical bell curve. We will consider the significance of this type of distribution next.

Statistical Methods are Used to Evaluate a Frequency Distribution Quantitatively

Statistical tools can be used to analyze a normal distribution in a number of ways. One measure that you are probably familiar with is a parameter called the *mean*. The mean is the sum of all the values in the group divided by the number of individuals in the group. It is computed using the following formula:

$$\bar{X} = \frac{\Sigma x}{N}$$

Where

$\bar{X}$ is the mean

Σx is the sum of all values in the group

N is the number of individual in the group

For example, suppose a bushel of corn cobs had the following lengths (rounded to the nearest centimeter): 15, 14, 13, 14, 15, 16, 16, 17, 15, and 15. Then

$$\bar{X} = \frac{15+14+13+14+15+16+16+17+15+15}{10}$$

$$= 15 \text{ cm}$$

In genetics, we are often interested in the amount of phenotypic variation that exists in a group. Without variation, selective breeding is not possible, and evolution cannot favour one type over another. A common way to evaluate variation within a population is a statistic called the variance. The variance is the sum of the squared deviations from the mean divided by the degrees of freedom (df equals N — 1)

$$V_x = \frac{\Sigma(X-\bar{X})^2}{N-1}$$

where

V_x is the variance

$X-\bar{X}$ is the difference between each value and the mean

N equal the number of observations

For example, if we use the value given previously for corn cob lengths, the variance in the group is calculated as follows:

$$\begin{aligned}\Sigma(X-\bar{X})^2 &= (15-15)^2 + (14-15)^2 + (13-15)^2 + (14-15)^2 + (15-15)^2 + (16-15)^2 + (16-15)^2 + (17-15)^2 + (15-15)^2 + (15-15)^2 \\ &= 0 + 1 + 4 + 1 + 0 + 1 + 1 + 4 + 0 + 0 \\ &= 12 \text{ cm}^2\end{aligned}$$

$$V_x = \frac{\Sigma(X - \bar{X})^2}{N-1}$$

$$= \frac{12\ \text{cm}^2}{9}$$

$$= 1.33\ \text{cm}^2$$

Since the variance is computed from squared deviations, it is a statistic that may be difficult to understand intuitively. For example, weight can be measured in grams; the corresponding variance would be measured in square grams. Even so, variances are centrally important iii the analysis of quantitative traits, because they are additive under certain conditions. Later in this chapter, we will examine how this property is useful in predicting the outcome of genetic crosses.

To gain an intuitive grasp for variation, we can take the square root of the variance. This statistic is called the *standard deviation* (S.D.). Again, using the same values for corn cob length, the standard deviation is:

$$\text{S.D.} = \sqrt{V_x} = \sqrt{1.33\ \text{cm}^2}$$

$$= 1.15\ \text{cm}$$

If the values in a population follow a normal distribution, then it is easy to appreciate the amount of variation by considering the standard deviation. Approximately 68% of all individuals have values within one standard deviation from the mean, either in the positive or negative direction. About 95% are within two standard deviations, and 99.7% are within three standard deviations. When a quantitative characteristic follows a normal distribution, less than 0.3% of the individuals will have values that are more or less than three standard deviations away from the mean of the population. In our corn cob example, three standard deviations equals 3.45 cm. Therefore, we would expect that fewer than 0.3% of the corn cobs would be less than 11.55 cm or greater than 18.45 cm, assuming corn cob length follows a normal distribution.

Some Statistical Methods Compare two Variables with Each Other

In many biological problems, it is useful to compare two different variables. For example, we may wish to compare the occurrence of two different phenotypic traits. Do obese animals have larger hearts? Are brown eyes more likely to occur in people with dark skin pigmentation? A second type of comparison is between traits and environmental factors. Does insecticide resistance occur more frequently in areas that have been exposed to insecticides? Is heavy body weight more prevalent in colder climates? Finally, a third type of comparison is between traits and genetic relationships. Do tall parents tend to produce tall offspring? Do smart women have smart brothers?

To gain insight into such questions, a statistic known as the *correlation* is often applied. To calculate this statistic, we first need to determine the *covariance*, which describes the degree of variation between two variables within a group. The covariance is similar to the variance, except that we multiply together the deviations of two different variables rather than squaring the deviations from a single factor:

$$\mathrm{CoV}_{(X,Y)} = \frac{\Sigma[(X - \bar{X})(Y - \bar{Y})]}{N - 1}$$

where,

X are the values for one variable and X is the mean value in the group

Y are values for another variable and Y is the mean value in that group

N is the total number of pairs of observations

As an example, let's consider the weight of cattle at five years of age between parents and offspring. A farmer might be interested in this relationship to deter mine if genetic factors play a role in the weight of cattle. The data here describe the five-year weights for ten different pairs of cows and their female offspring:

Mother's Weight (kg)	*Offspring's Weight (kg)*	$X - \bar{X}$	$Y - \bar{Y}$	$(X - \bar{X})(Y - \bar{Y})$
570	568	-26	-30	780
572	560	-24	-38	912
599	642	3	44	132
602	580	6	-18	-108
631	586	35	-12	-420
603	642	7	44	308
599	632	3	34	102
625	580	29	-18	-522
584	605	-12	7	-84
575	585	-21	-13	273
$\bar{X} = 596$	$\bar{Y} = 598$			$\Sigma = 1373$

$$\text{CoV}_{(X,Y)} = \frac{\Sigma\left[(X-\bar{X})(Y-\bar{Y})\right]}{N-1}$$
$$= \frac{1373}{10-1}$$
$$= 152.6$$

After we have calculated the covariance, we can evaluate the strength of the association between the two variables by calculating a *correlation coefficient* (r):

$$r_{(X,Y)} = \frac{\text{CoV}_{(X,Y)}}{\text{SD}_X \text{SD}_Y}$$

This value, which ranges between +1 and —1, indicates how two factors vary in relation to each other. A value of +1 is a perfect correlation. It means that two factors vary in a completely predictable way relative to each other; as one factor increases, the other will increase with it. A value of zero indicates there is no detectable way in which the two factors vary relative to each other; the values of the two factors are not related. Finally, an inverse correlation, in which the correlation coefficient is negative, indicates that the two factors tend to vary in opposite ways to each other; as one factor increases, the other will decrease.

Let's use the data of five-year weights for mother and offspring to calculate a correlation coefficient:

$$r_{(X,Y)} = \frac{152.6}{(21.1)(30.5)}$$
$$= 0.237$$

The result is a positive correlation between the five-year weights of mother and offspring. In other words, the positive correlation value suggests that heavy mothers tend to have heavy offspring and lighter mothers, lighter offspring.

However, after a correlation coefficient has been calculated, one must evaluate whether the *r* value represents a true association between the two variables, or whether it could be simply due to chance. To accomplish this, we can test the hypothesis that there is no real correlation (i.e., the null hypothesis): The *r* value differs from zero only as a matter of random sampling error. This is the same approach followed in the chi square analysis. Like the chi square value, the significance of the correlation coefficient is related directly to sample size and the degrees of freedom (*df*). In testing the significance of

Table 9.2. Values of r at the 5% and 1% significance levels

Degrees of Freedom (df)	5%	1%	*Degrees of Freedom (df)*	5%	1%
1	.997	1.000	24	.388	.496
2	.950	.990	25	.381	.487
3	.878	959	26	.374	.478
4	.811	.917	27	.367	.470
5	.754	.874	28	.361	.463
6	.707	.834	29	.355	.456
7	.666	.798	30	.349	.449
8	.632	.765	35	.325	.418
9	.602	.735	40	.304	.393
10	.576	.708	45	.288	.372
11	.553	.684	50	.273	.354
12	.532	.661	60	.250	.325
13	.514	.641	70	.232	.302
14	.497	.623	80	.217	.283
15	.482	.606	90	.205	.267
16	.468	.590	100	.195	.254
17	.456	.575	125	.174	.228
18	.444	.561	150	.159	.208
19	.433	.549	200	.138	.181
20	.423	.537	300	.113	.148
21	.413	.526	400	.098	.128
22	.404	.515	500	.088	.115
23	.396	.505	1000	.062	.081

correlation coefficients, *df* equals N — 2, which is one less than the degrees of freedom of variance (i.e., df for variance equals N — 1). Table 9.2 shows the relationship between the *r* values and degrees of freedom at the 5% and 1% significance levels.

This approach, though, is only valid if several assumptions are met. First, the values of X and Y in the study must have been obtained by an unbiased sampling of the entire population. In addition, this approach assumes that the scores of X and Y follow a normal distribution and that the relationship between X and Y is linear. To illustrate the use of Table 9.2, let's consider the correlation we have just calculated for five-year weights of mother and offspring. In this case, we obtained a value of 0.237 for *r*, and the value of N was 10. Under these conditions, *df* equals 8. According to Table 9.2, it is fairly likely that this value could have occurred as a matter of random sampling error. Therefore, we cannot conclude that the positive correlation is due to a true association between the weights of mother and offspring (i.e., we cannot reject the null hypothesis).

In an actual experiment, however, a researcher would examine many more pairs of mothers and offspring, perhaps 500 to 1000. If a correlation of 0.237 was observed for N = 1000, the value would be significant at the 1% level. We would therefore reject the null hypothesis that weights are not associated with each other. Instead, we would conclude that there is a real association between the weights of mothers and their offspring. Infect, these kinds of experiments have been done for cattle weights, and the correlations between parents and offspring have been found to be significant.

Whenever a statistically significant correlation is obtained, one must be very cautious in interpreting its meaning. An *r* value that is statistically significant need not imply a cause and effect relationship. When parents and offspring display a significant correlation for a trait, one should not jump to the conclusion that genetics is the underlying cause of the positive association. In many cases, parents and offspring share similar environments, so that the positive association might be rooted in environmental factors. In general, correlations are quite useful in identifying positive or negative associations between two variables. Even so, this statistic by itself cannot prove that the association is due to a cause and effect.

Polygenetic Inheritance

In the preceding section, we saw that quantitative traits tend to show a continuum of variation, which can be analyzed with various statistical tools. At the beginning of the 1900s, there was great debate concerning the inheritance of quantitative traits. The biometric school, founded by Galton and Pearson, argued that these types of traits are not controlled by single discrete genes that affect phenotypes in a predictable way. To some extent, the biometric school favoured a blending theory of inheritance, which had been proposed earlier. Alternatively, the followers of Mendel, led by William Bateson in England and William Castle in the United States, held firmly to the idea that traits are governed by genes, which are inherited as discrete units. As we know now, Bateson and Castle were correct. However, as we will see in this section, the difficulty of studying quantitative traits lies in the fact that these traits are influenced by multiple genes and substantial environmental factors. Most quantitative traits are polygenic and exhibit a continuum of phenotypic variation. The term *polygenic inheritance* refers to the transmission of traits that are governed by two or more genes. The locations on chromosomes where these genes reside are called *quantitative trait loci* (QTLs).

Just a few years ago, it was extremely difficult for geneticists to determine the inheritance patterns for genes underlying polygenic traits, particularly those determined by three or more genes having multiple alleles for each gene. Recently, however, molecular genetic tools have enhanced greatly our ability to find regions in the genome where QTLs are likely to reside. This has been a particularly exciting advance in the field of quantitative genetics. In some cases, the identification of QTLs may allow the improvement of quantitative traits in agriculturally important species.

Polygenic Inheritance Creates Overlaps between Genotypes and Phenotypes

The first experiment demonstrating that continuous variation is related to polygenic inheritance was conducted by the Swedish geneticist Herman Nilsson Ehle in 1909. He studied the inheritance of red pigment in the hull of wheat, *Triticum aestivum*. When true-breeding plants with white hulls were crossed to a variety with red hulls, the F_1 generation had an intermediate colour. When the F_1 generation was allowed to self-fertilize, there was a great variation in redness in the F_2 generation, ranging from white, light pink, and pink to intermediate red, medium red, basic red, and dark red. An unsuspecting observer might conclude that this F_2 generation displayed a continuous variation in hull colour. However, Nilsson-Ehle carefully categorized the colours of the hulls and discovered that they fell into a 1: 6 :15 : 20 : 15 : 6: 1 ratio. He concluded that this species is diploid for three different genes that control hull colour, each gene existing in a red or white allelic form. He hypothesized that these three loci must contribute additively to the colour of the hull. The results of Nilsson-Ehle make perfect sense, because we now know that strains of *T. aestivum* are actually hexaploid and therefore have six copies of each gene.

Nilsson-Ehle categorized wheat hull colours into several discrete genotypic categories. However, for many polygenically inherited quantitative traits, this is difficult or impossible. In general, as the number of genes controlling a trait increases, and the influence of the environment increases, the categorization of phenotypes into discrete

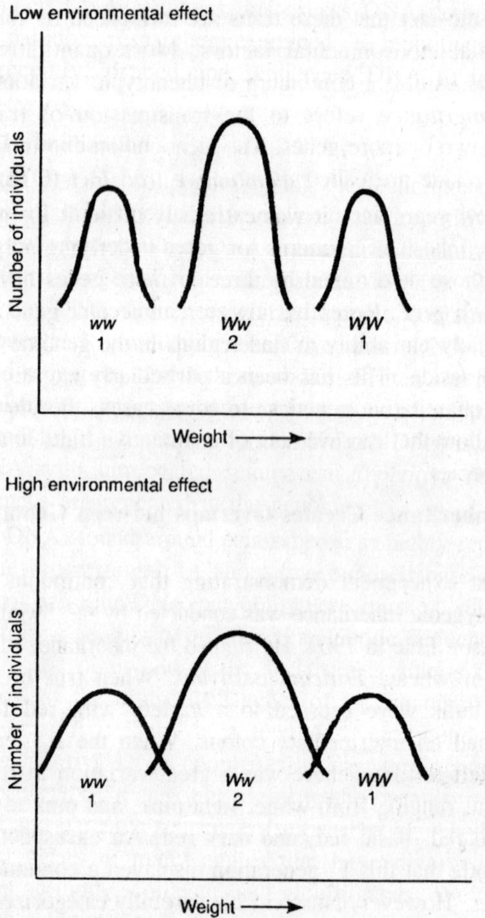

Fig. 9.1. Situations in which seed weight is controlled by one gene, existing in light (w) and heavy (W) alleles.

genotypic classes becomes increasingly difficult if not impossible. In example, the environment (sunlight, soil conditions, and so forth) may affect the phenotypic outcome of a trait in plants (namely, seed weight). Part (a) considers a situation where seed weight is controlled by one gene existing in light (w) and heavy (W) alleles. A heterozygous plant (Ww) is allowed to self-hybridize. When the weight is only slightly influenced by variation in the environment, as seen on the left, the heavy, light, and intermediate seeds fall into separate, well-defined categories. When the environmental variation has a greater impact on seed weight, as shown on the right, there is more phenotypic variation

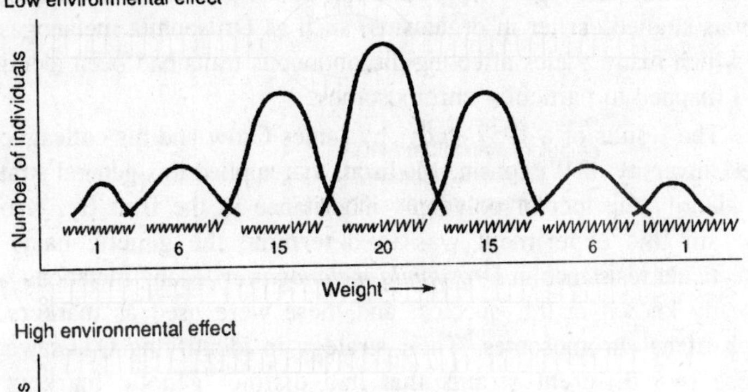

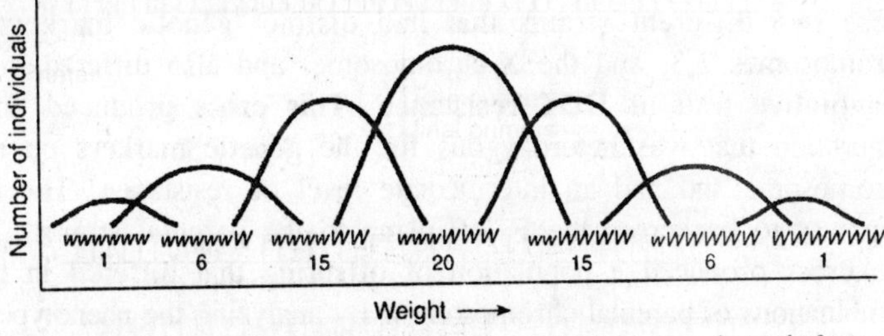

Fig. 9.2. Situations in which seed weight is governed by three genes instead of one, each existing in light and heavy alleles.

in seed weight within each genotypic class. Even so, it is still possible to classify most individuals into the three main categories.

By comparison, a situation where seed weight is governed by three genes instead of one, each existing in light and heavy alleles. When the environmental variation is low and/or plays a minor role in the outcome of this trait, the expected 1: 6 : 15 : 20: 15 : 6: 1 ratio is observed. As shown in the upper illustration in part (b), nearly all of the individuals fall within a phenotypic category that corresponds to their genotypes. When the environment has a more significant effect on phenotype, as shown in the lower illustration, the situation becomes more ambiguous. For example, individuals with five W alleles and one w allele have phenotypes that overlap with individuals having six W alleles or four W alleles and two w alleles. Therefore, it becomes difficult to categorize each phenotype into a unique genotypic class. Instead, the trait displays a continuum ranging from light to heavy seed weight.

When the overlap between phenotypic classes is so great that separating them is not possible, another way to identify the number of genes affecting the polygenic inheritance of a quantitative trait is to look for linkage between these genes and genes affecting discontinuous traits. While this type of approach can be conducted on any organism, it was studied earlier in organisms, such as Drosophila melanogaster, in which many alleles affecting discontinuous traits had been identified and mapped to particular chromosomes.

The results of a 1957 study, by James Crow and his colleagues at the University of Wisconsin, Madison, that applied this general strategy to identify the loci of polygenic inheritance in the fruit fly. Crow's goal in this experiment was to determine the genetic basis for insecticide resistance in *Drosophila melanogaster*. Many mutations were already known in this species, and these were used as markers for each of its chromosomes. Their strategy in identifying QTLs was to cross two different strains that had distinct genetic markers on chromosomes 2,3, and the X-chromosome, and also differed in the quantitative trait of DDT resistance. This cross produced an F_1 generation that was heterozygous for the genetic markers on each chromosome and had an intermediate level of resistance. The next step was to backcross the F_1 offspring to the parental strains. This backcross produced a population of offspring that differed in their combinations of parental chromosomes. By analyzing the phenotypes of the offspring with regard to known chromosomal markers, the researchers could determine whether the chromosomes were inherited from the DDT-resistant or DDT-sensitive strain. These same offspring were analyzed with regard to their ability to survive in the presence of DDT. As a matter of chance, some offspring contained all of the chromosomes from one original parental strain or the other, but most offspring contained a few chromosomes from one parental strain and the rest from the other.

The data in that each copy of the X-chromosome, and chromosomes 2 and 3, confer a significant amount of insecticide resistance. When a fly contains all of its X-chromosomes, and chromosomes 2 and 3 from the insecticide resistant parental strain, it has maximal insecticide resistance. Even when only a single chromosome is derived from the insecticide-sensitive strain, the resistance to DDT is less than maximal. These results are consistent with the hypothesis that insecticide resistance is a polygenic trait involving genes that reside on the X-chromosome and on chromosomes 2 and 3.

Quantitative Trait Loci (QTLs) can now be Mapped by RFLP Analysis

In the past few years, restriction fragment length polymorphisms and other molecular markers have been identified in the genomes of many different organisms. These markers have been used to construct genetic maps of several species' genomes. Once a genome map is obtained, it becomes much easier to analyze the genetic contributions of complex traits. In addition to model organisms such as *Drosophila*, *Arabidopsis*, *Caenorhabditis elegans*, and mice, detailed molecular maps have been obtained for species of agricultural importance. These include crops such as corn, rice, and tomatoes, as well as live stock such as cattle, pigs, and sheep.

When a genetic map is available, geneticists can determine the location of a gene within a genome using RFLP mapping. In 1989, Eric Lander and David Botstein extended this technique to identify QTLs that govern a quantitative trait. The basis of QTL detection is the association between genetically determined phenotypes (e.g., quantitative traits) and molecular markers such as RFLPs. For example, in a species of grain, high yield might be associated with five RFLPs that map to different regions of the genome. These results would indicate that phenotypic variation in yield is influenced by five different loci.

The general strategy for RFLP mapping begins by mating organisms that are very dissimilar in their RFLP markers and also different for quantitative traits. For example, in 1988, Andrew Paterson at Cornell University and his colleagues examined quantitative trait inheritance in the tomato. They studied a domestic strain of tomato and a South American green-fruited variety. These two strains differed in their RFLPs, and they also exhibited dramatic differences in three agriculturally important characteristics: fruit mass, soluble solids content (in the fruit), and fruit pH. The researchers crossed the two strains together, and then they backcrossed the offspring to the domestic tomato. A total of 237 plants were then examined with regard to 70 known RFLP markers. In addition, between 5 and 20 tomatoes from each plant were analyzed with regard to fruit mass, soluble solids content, and fruit pH. Using this approach, they were able to map the genes contributing variation in these traits to particular intervals along the tomato chromosomes. They identified six loci causing variation in fruit mass, four affecting soluble solids content, and five with effects on fruit pH.

This method of analyzing quantitative traits via RFLP mapping has provided an important tool in the analysis and manipulation of quantitative traits. In addition, other types of genetic markers such as sequence-tagged sites, which are identified by PCR methods, can be used in mapping studies of QTLs. These types of studies are advancing our understanding of how quantitative traits are influenced by genes. In some cases, a quantitative trait may be influenced by many genes, with only a few having major importance in determining the phenotypic outcome. Once the principal genes have been identified, researchers or breeders can determine if two or more different genes act additively to affect a quantitative trait. Also, the mapping of QTLs can allow geneticists to identify alleles within the same gene and determine how they affect the trait in a homozygous or heterozygous state. In the future, it is clear that RFLP mapping will enhance greatly our basic understanding of quantitative traits and may provide exciting applications in the field of agriculture.

Heritability

As we have just seen, recent approaches in molecular mapping have enabled re searchers to identify the genes that contribute to a quantitative trait. The other key factor that affects the phenotypic outcomes of quantitative traits is the environment. All traits of biological organisms are influenced by genetics and the environment, and this is particularly pertinent in the study of quantitative traits.

A geneticist, however, can never actually determine the relative amount of a quantitative trait that is controlled by genetics, or the amount that is governed by the environment. When you think about it, a researcher cannot raise an organism without an environment and discover the amount of a trait that is governed by genetics. Likewise, it is impossible to rear an organism without its genes and conclude how much the environment contributes to the outcome of a trait. Instead, the focus of our analysis of quantitative traits is to explain how variation, both genetic and environmental, will affect the phenotypic results.

The term *heritability* refers to the amount of phenotypic variation within a group of individuals that is due to genetic factors. If all of the phenotypic variation in a group were due to genetic variation, the heritability would have a value of 1. If all the variation were due to environmental effects, the heritability would equal 0. For most groups of organisms, the heritability for a given trait lies between these two extremes. For example, both genes and diet affect the size that an individual will attain. Some individuals will inherit genes that tend to make them large, and a proper diet will also promote larger size. Other individuals will inherit genes that make them small, and an inadequate diet may contribute further to small size. Taken together, both genetics and the environment influence the phenotypic results.

In the study of quantitative traits, a primary goal is to determine how much of the phenotypic variation arises from genetic factors, and how much comes from environmental factors. In this section, we will examine how geneticists analyze the genetic and environmental components that affect quantitative traits. As we will see, this approach has been applied successfully in agricultural breeding strategies to produce domesticated species with desirable characteristics.

Phenotypic Variance is Due to the Additive Effects of Genetic Variance and Environmental Variance

Earlier in this chapter, we examined the amount of phenotypic variation within a group by calculating the variance. In studying quantitative trait variation, the first step is to partition this variation into components that are attributable to different causes. If we assume that genetic and environmental factors are the only two components that determine a trait, and if genetic and environmental factors are independent of each other, then the total variance for a trait in a group of individuals is

$$V_T = V_G + V_E$$

where

V_T is the total variance. It reflects the amount of variation that is measured at the phenotypic level.

V_G is the relative amount of variance due to genetic factors.

V_E is the relative amount of variance due to environmental factors.

The partitioning of variance into genetic and environmental components allows us to estimate their relative importance in influencing the variation within a group. If V_G is very high and V_E is very low, genetics plays a greater role in promoting variation within a group. Alternatively, if V_G is low and V_E is high, environmental causes underlie much of the phenotypic variation. As will be de scribed later in this chapter, a livestock breeder might want to apply selective breeding if V_G for an important (quantitative) trait is high. In this way, the characteristics of the herd may be improved. Alternatively, if V_G is negligible, it would make more sense to investigate (and manipulate) the environmental causes of phenotypic variation.

With experimental animals, one possible way to determine V_G and V_E is by comparing the variation in traits between genetically identical and genetically disparate groups. For example, researchers have developed genetically homogeneous strains of mice. After many generations of brother—sister matings, these strains have become monomorphic for all of their genes. Within such a strain of mice, V_G

equals zero. Therefore, all phenotypic variation is due to V_E. When studying quantitative traits such as weight, an experimenter might want to know the genetic and environmental variance for a different, genetically heterogeneous group of mice. To do so, the genetically homogeneous and heterogeneous mice could be raised under the same environmental conditions, and their weights measured. The phenotypic variance for weight could then be calculated as described earlier. Let's suppose we obtained the following results:

V_T = 0.30 sq oz for the group of genetically homogeneous mice

V_T = 0.52 sq oz for the group of genetically heterogeneous mice

In the case of the homogeneous mice, $V_T = V_E$, because V_G equals zero. Therefore, V_E equals 0.30 sq oz. To estimate V_G for the heterogeneous group of mice, we assume that V_E (i.e., the environmentally produced variance) is the same for them as it is for the homogeneous mice, since the two groups were raised in identical environments. This assumption allows us to calculate the genetic variance for the heterogeneous mice:

$$V_T = V_G + V_E$$
$$0.52 = V_G + 0.30$$
$$V_G = 0.22 \text{ sq oz}$$

This result tells us that some of the phenotypic variance in the genetically heterogeneous group is due to the environment (namely, 0.30 sq oz) and some (0.22 sq oz) is due to genetic variation in alleles that affect the weight.

Heritability is the Relative Amount of Phenotypic Variation that is due to Genetic Factors

Another way to view variance is to focus our attention on the genetic contribution to phenotypic variation. Heritability is the proportion of the phenotypic variance that is attributable to genetic factors. If we assume again that environment and genetics are the only two components, then

$$H_B^2 = V_G / V_T$$

where

H_B^2 is the heritability in the broad sense

V_G is the variance due to genetics

V_T is the total phenotypic variance

The heritability defined here, H_B^2 is called the *true heritability in the broad sense*. It takes into account all genetic factors that may affect the phenotype.

As we have seen throughout this text, genes can affect phenotypes in various ways. As described earlier in this chapter, the Nilsson-Ehle experiment showed that the alleles determining hull colour in wheat affect the phenotype in an additive way. Alternatively, alleles affecting other traits may show a dominant—recessive relationship. In this case, the alleles are not strictly additive, because the heterozygote has a phenotype closer to, or perhaps the same as, the homozygote containing two copies of the dominant allele. In addition, another complicating factor is epistasis. The alleles for one gene may influence the phenotypic expression of the alleles of another gene. To account for these differences, geneticists usually subdivide VG into these three different genetic factors:

$$V_G = V_A + V_D + V_I$$

where

V_A is the variance due to additive alleles

V_D is the variance due to alleles that follow a dominant/recessive pattern of inheritance

V_I is the variance due to genes that interact in an epistatic manner

In analyzing quantitative traits, geneticists often focus on V_A and neglect the contributions of V_D and V_I. This is done for scientific as well as practical reasons. For many quantitative traits, the additive effects of alleles often dominate the phenotypic outcome in genetic crosses. In addition, when the alleles behave additively, we can predict the outcomes of crosses based on the quantitative characteristics of the parents. The heritability of a trait due to the additive effects of alleles is called the *narrow sense heritability*:

$$h_N^2 = V_A / V_T$$

The narrow sense heritability (h_N^2 may be an inaccurate measure of the true heritability if V_D and V_I are not small values. However, for many quantitative traits, geneticists have found that the value of V_A is very large compared with V_D and V_I. In such cases, the true heritability in the broad sense and the narrow sense heritability are similar to each other.

There are several ways to estimate narrow sense heritability. A common strategy is to measure a quantitative trait among groups of genetically related individuals. For example, agriculturally important traits, such as egg weight in poultry, can be analyzed in this way. To calculate the heritability, one would determine the ob served egg weights between individuals whose genetic relationships are known, such as a mother and her female offspring. These data could then be used to compute a correlation between the parent and offspring using the methods described earlier in this chapter. The narrow sense heritability is then calculated as

$$h_N^2 = r_{obs} / r_{exp}$$

where r_{obs} is the observed phenotypic correlation between related individuals

r_{exp} is the expected correlation based on the known genetic relation ship

In our example, r_{obs} is the observed phenotypic correlation between parent and offspring. In research studies, the observed phenotypic correlation for egg weights between mothers and daughters has been found to be about 0.25 (although this will vary among strains). The expected correlation, r_{exp}, is based on the known genetic relationship. A parent and child share 50% of their genetic material, so, that r_{exp} equals 0.50. So,

$$h_N^2 = r_{obs} / r_{exp}$$

$$= 0.25 / 0.50 = 0.50$$

Note: For siblings, $r_{exp} = 0.50$; for identical twins, $r_{exp} = 1.0$; and for an uncle niece relationship, $r_{exp} = 0.25$.

According to this calculation, about 50% of the phenotypic variation in egg weight is due to genetic factors; the other half is due to the environment.

When calculating heritabilities from correlation coefficients, keep in mind that this computation assumes that genetics and the environment are independent variables. This is not always the case. The environments of parents and offspring are often more similar to each other than they are to those of unrelated individuals. There are several ways to avoid this confounding factor. First, in human studies, one may analyze the heritabilities from correlations between adopted children and their biological parents. Alternatively, one can examine a variety of relation ships (uncle—niece, identical twins versus fraternal twins, etc.) and see if the heritability values are roughly the same in all cases. This approach was applied in the study to be described next.

Holt Found that the Heritability of Dermal Ridge Count in Human Fingerprints is very High

Fingerprints are inherited as a quantitative trait. It has been long known that identical twins have fingerprints that are very similar, whereas fraternal twins show considerably less agreement. Galton was the first researcher to study fingerprint patterns, but this trait became more amenable to genetic studies when Kristine Bonnevie at the University of Kristiania in Norway developed a method in 1924 for counting the number of ridges within a human fingerprint.

Human fingerprints can be categorized as having an arch, a loop, or a whorl (or a combination of these patterns). The primary difference among these patterns is the number of triple junctions, each known as a *triradius*. At a triradius, a ridge emanates in three directions. An arch has zero triradius, a loop has one, and a whorl has two. In Bonnevie's method of counting, a line is drawn from a triradius to the center of the fingerprint. The ridges that touch this line are then counted. (Note: The triradius ridge itself is not counted, and the last ridge is not counted if it forms the center of the fingerprint.) With this method, one can obtain a ridge count for all ten fingers. Bonnevie conducted a study on a small population and found that ridge count correlations were relatively high in genetically related individuals.

In the experiment described here, published in 1961, Sarah Holt at the University College of London, who was also interested in the inheritance of this quantitative trait, carried out a more exhaustive study of ridge counts in a British population. In groups of 825 males or 825 females, the ridge count varied from 0 to 300, with mean values of approximately 145 for males and 127 for females:

Hypothesis

Dermal ridge count has a genetic component. The goal of this experiment is to determine the contribution of genetics in the variation of dermal ridge counts.

Testing the hypothesis

Starting material. A group of human subjects from Great Britain.

1. Take a person's finger and blot it onto an ink pad.
2. Roll the person's finger onto a recording surface to obtain a print.
3. With a low-power binocular microscope, count the number of ridges using the Bonnevie method described previously.
4. Calculate the correlation coefficients between different pairs of individuals as described earlier in this chapter.

Interpreting the data

When we look at the data shown here, it becomes apparent that genetics plays the major role in explaining the variation in this trait. Genetically unrelated individuals (namely, parent-parent relationships) have a negligible correlation for this trait. By comparison, individuals who are related genetically have a substantially higher correlation.

When we divide the observed correlation coefficient by the expected correlation coefficient based on the known genetic relationships, the average heritability value is 0.97, which is very close to 1.0. These values indicate that nearly all of the variation in fingerprint pattern is genetic variation. Significantly, fraternal and identical twins have substantially different correlation coefficients, even though we expect that they have been raised in very similar environments.

These results support the idea that genetics is playing the major role in promoting variation and that the results are not biased heavily by environmental similarities that may be associated with genetically related individuals. From an experimental viewpoint, the results show us how the determination of correlation coefficients between related and unrelated individuals can provide insight regarding the relative contributions of genetics and environment to the variation of a quantitative trait.

Heritability Values are only Relevant to Particular Groups Raised in a Particular Environment

Table 9.3. describes some heritability values that have been calculated for particular populations. Unfortunately, heritability is a widely misunderstood concept. Heritability describes the amount of phenotypic variation due to genetic factors for a particular population raised in a particular environment. The words "variation," "particular population," and "particular environment" cannot be overemphasized. For example, in one population of cattle the heritability for milk production may be 0.35, while in another group (with less genetic variation) the heritability may be 0.1. Second, if a group displays a heritability of 1.0 for a particular trait, this does not mean that the environment is unimportant in affecting the outcome of the trait. A heritability value of 1.0 only means that the amount of variation within this group is due to genetics. Perhaps, the group has been raised in a relatively homogeneous environment, so that the environment has not caused a significant amount of variation. Nevertheless, the environment may be quite important. It just is not causing much variation within this particular group.

As a hypothetical example, let's suppose that we take a species of rodent and raise a group on a poor diet; we find their weights range from 1.5 to 2.5 pounds, with a mean weight of 2 pounds. We allow them to mate, and then raise their off spring on a healthy diet of rodent chow. The weights of the offspring range from 2.5 to 3.5 pounds, with a mean weight of 3 pounds.

Table 9.3. Examples of heritabilities for quantitative traits

Trait	*Heritability value*
Humans	
Stature	0.65
Cattle	
Body weight	0.65
Butterfat, %	0.40
Mi]k yield	0.35
Mice	
Tail length	0.40
Body weight	0.35
Litter size	0.20
Poultry	
Body weight	0.55
Egg weight	0.50
Egg production	0.10

In this hypothetical experiment, we might find a positive correlation in which the small parents tended to produce small offspring, and the large parents, large offspring. The correlation of weights between parent and offspring might be, say, 0.5. In this case, the heritability for weight would be calculated as r_{obs}/r_{exp}, which equals 0.5/0.5, or 1.0. The value of 1.0 means that all of the variation within the groups is due to genetics. The offspring vary from 2.5 to 3.5 pounds because of genetic variation, and also the parents range from 1.5 to 2.5 because of genetics. However, as we see here, environment has played an important role. Presumably, the mean weight of the offspring is higher because of their better diet.

This example is meant to emphasize the point that heritability only tells us the relative contributions of genetics and environment in influencing phenotypic *variation* in a *specific group* under a *specific set of conditions*. Heritability may not indicate the relative importance of these two factors in determining the outcomes of traits.

Selective Breeding of Species can alter Quantitative Traits Dramatically

The term *selective breeding* refers to programs and procedures designed to mod ilk phenotypes in species of economically important plants and animals. This phenomenon, also called *artificial selection*, is related to natural selection. In fact, in forming his theory of natural selection, Charles Darwin was influenced by his observations of selective breeding by pigeon fanciers and other breeders. The primary difference between artificial and natural selection is how the parents are chosen. Natural selection is due to natural variation in reproductive success; in artificial selection, the breeder chooses individuals who possess traits that are desirable from a human perspective.

For centuries, humans have been practicing selective breeding to obtain domestic species with interesting or agriculturally useful characteristics. The common breeds of dogs and cats have been obtained by selective breeding strategies. As shown here, it is very striking how selective breeding can modify the quantitative traits in a species. When comparing a Chihuahua with a Saint Bernard, the magnitude of the differences is fairly amazing. They hardly look like members of the same species.

Likewise, most of the food we eat is obtained from species that have been modified profoundly by selective breeding strategies. This includes products such as grains, fruit, vegetables, meat, milk, and juices. As seen here, certain quantitative traits in the domestic strains differ considerably from those of the original wild species.

The concept that underlies selective breeding is variation. Within a group of individuals, there may be allelic variation that affects the outcomes of quantitative traits. The fundamental strategy of the selective breeder is to choose parents who will pass on advantageous alleles to their offspring. In general, this means that the breeder will choose parents with desirable phenotypic characteristics. For example, if a breeder wants large cattle, he/she will choose the largest members of the herd as parents for the next generation. Presumably, these large cattle will transmit alleles to their offspring that confer large size. The breeder will often choose genetically related individuals (e.g., brothers and sisters) as the parental stock. The practice of mating between genetically related individuals is known as *inbreeding*.

A common outcome when selective breeding is conducted for a quantitative trait. This experiment began with 163 ears of corn with an oil content ranging from 4 to 6%. In each of 80 succeeding generations, corn plants were divided into two separate groups. In one group, members with the highest oil content were chosen as parents of the next generation. In the other group, members with the lowest oil content were chosen. After many generations, the oil content in the first group rose to over 18%; in the other group, it dropped to less than 1%. These results show that selective breeding can modify quantitative traits in a very directed manner.

Similar results have been obtained for many other quantitative traits. An experiment in 1941, by K. Mather of the John Innes Horticultural Institution, in which flies were selected on the basis of their bristle number. The starting group had an average of 40 bristles for females and 35 bristles for males. After eight generations, the group selected for high bristle number had an average of 46 bristles for females and 40 for males, while the group selected for low bristle number had an average of 36 bristles for females and 30 for males.

Quantitative traits are often at an intermediate value in unselected populations. Therefore, artificial selection can be used to increase or decrease the magnitude of the trait. Oil content can go up or down, and bristle number can increase or decrease. Nevertheless, there is a limit to how far this can continue. Presumably, the starting population possesses a large of amount of genetic variation, which contributes to the diversity in phenotypes. In other words, at the beginning of these

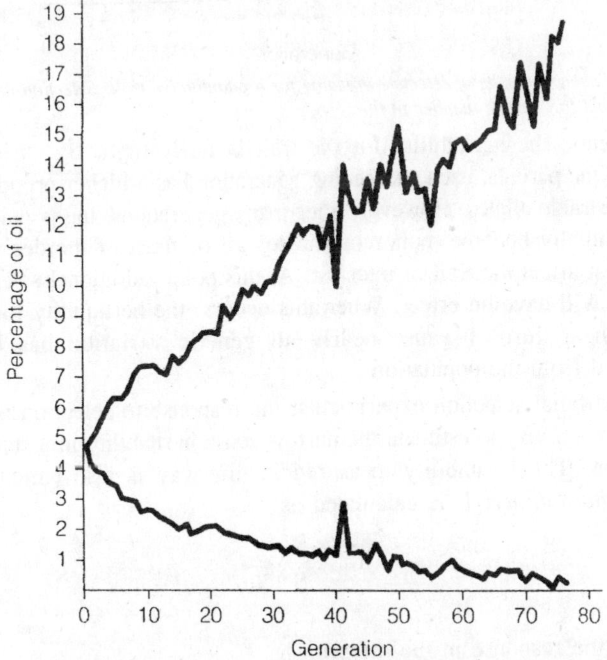

Fig. 9.3. Common results of selective breeding for a quantitative trait: Selection for high and low oil content in corn.

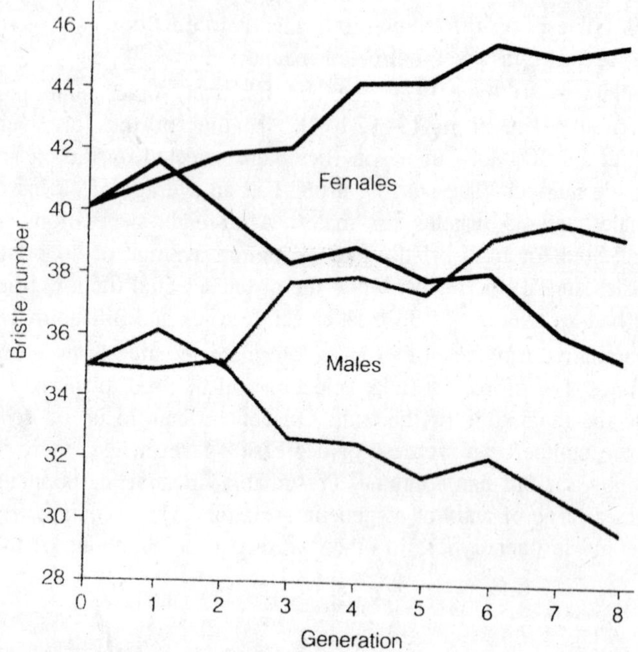

Fig. 9.4. Common results of selective breeding for a quantitative trait: Selection for high and low bristle number in flies.

experiments, the heritability for the trait is fairly high. By carefully choosing the parents, each succeeding generation has a higher proportion of the desirable alleles. However, after many generations, the population will eventually become monomorphic for all or most of the desirable alleles that affect the trait of inter est. At this point, additional selective breeding will have no effect. When this occurs, the heritability for the trait is near zero, because nearly all genetic variation has been eliminated from the population.

In artificial selection experiments, the response to selection is the most common way to estimate the narrow sense heritability in a starting population. The heritability measured in this way is also called the *realized heritability*. It is calculated as

$$h_N^2 = \frac{R}{S}$$

where,

R is the response in the offspring

S is the selection differential in the parents

Here,

$$R = \bar{X}_O - \bar{X}$$
$$S = \bar{X}_P - \bar{X}$$

where,

$\bar{X}$ is the mean of the starting population

$\bar{X}_O$ is the mean of the offspring

$\bar{X}_P$ is the mean of the parents

So,

$$h_N^2 = \frac{\bar{X}_O - \bar{X}}{\bar{X}_P - \bar{X}}$$

An example, let's suppose we began with a population of fruit flies in which the average bristle number for both genders was 37.5. The parents chosen from this population had an average bristle number of 40. The offspring of the next generation had a average bristle number of 38.7. With these values, the realized heritability is

$$h_N^2 = \frac{38.7 - 37.5}{40 - 37.5}$$
$$= \frac{1.2}{2.5} = 0.48$$

This result tells us that about 48% of the phenotypic variation is due to the additive effects of alleles.

An important aspect of narrow sense heritabilities is their ability to predict the outcome of selective breeding. Solved Problem 1 at the end of this chapter illustrates this idea.

Heterosis may be Explained by Dominance or Overdominance

As we have just seen, selective breeding can be used to alter the phenotypes of domestic species in a highly directed way. An unfortunate consequence of inbreeding, however, is that it tends to decrease the overall genetic variation within a group and may unknowingly promote homozygosity for deleterious alleles. In agriculture, it is widely observed that when two different inbred strains are crossed to each other, the resulting offspring are more vigorous (e.g., larger, longer-lived) than either of the inbred parental strains. This phenomenon is called *heterosis* or *hybrid vigour*. In modern agricultural breeding practices, many strains of plants and animals are hybrids produced by crossing two different inbred lines. In fact, much of the success in agricultural breeding programs is founded in heterosis. In rice, for example, hybrid strains have a 15-20% yield advantage over the best conventional inbred varieties under similar cultivation conditions. The genetic basis for heterosis has been debated for more than 80 years, and the controversy is still not resolved. There are two major hypotheses to explain heterosis:

Dominance hypothesis to explain heterosis

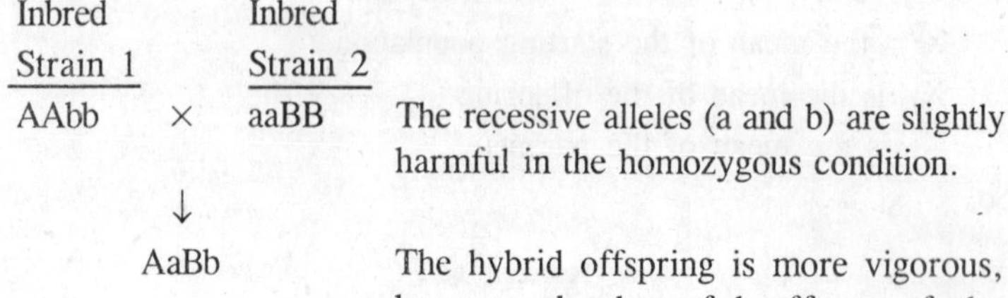

Inbred Strain 1		Inbred Strain 2	
AAbb	×	aaBB	The recessive alleles (a and b) are slightly harmful in the homozygous condition.
	↓		
	AaBb		The hybrid offspring is more vigorous, because the harmful effects of the recessive alleles are masked by the dominant alleles.

Overdominance hypothesis to explain heterosis

Inbred Strain 1		Inbred Strain 2	
A-1A-l	×	A-2A-2	Neither the A—1 nor A—2 allele is recessive.
	↓		
	A—JA—2		The hybrid offspring is more vigorous, because the heterozygous combination of alleles exhibits over dominance. This means that the A—1A—2 heterozygote is more vigorous than either the A—1 A—1 or A—2A—2 homozygote.

In 1908, the *Dominance Hypothesis* was proposed by Charles Davenport at Cold Spring Harbor. He suggested that highly inbred strains have become homozygous for one or more recessive genes that are somewhat deleterious (but not lethal). This undesirable homozygosity can happen randomly as result of inbreeding or genetic drift. Because the homozygosity occurs by chance, two different inbred strains are likely to be homozygously recessive for different genes. Therefore, when they are crossed to each other, the resulting hybrids are heterozygous and do not suffer the consequences of homozygosity for deleterious recessive alleles. In other words, the dominance of the beneficial alleles explains the observed heterosis.

In 1908, a second hypothesis, known as the *Over dominance Hypothesis*, was proposed by George Shull at Cold Spring Harbor and Edward East at the Cormecticut Agricultural Experimental Station.

Over dominance occurs when the heterozygote is more vigorous than either corresponding homozygote. According to this idea, heterosis occurs because the resulting hybrids are heterozygous for one or more genes that display over dominance. A classic case of overdominance involves the β-globin allele for sickle cell anemia. In areas where malaria is prevalent, the heterozygote has a higher survival rate than both the wild-type homozygote and the homozygote affected with sickle cell anemia.

As mentioned earlier in this chapter, the recent advent of RFLP linkage mapping has made it possible to identify the QTLs that underlie heterosis. In corn, Charles Stuber at the U.S. Department of Agriculture in Raleigh, North Carolina, and his colleagues have found that heterozygotes of most QTLs for grain yield had higher production than the corresponding homozygotes. These results support the Over dominance Hypothesis. The researchers noted, however, that over dominance is very difficult to distinguish from *pseudo-overdominance*, a phenomenon initially suggested by James Crow in 1952. Pseudo-over dominance is really the same as dominance, except that the two (or more) genes involved are very closely linked. In our last example, *a* and *B* may be closely linked in one strain while *A* and *b* are closely linked in another strain. The hybrid is really heterozygous (*AaBb*) for two different genes, but this may be difficult to discern in mapping experiments because the genes are so close together. Therefore, without very fine mapping, which is currently difficult to do for QTLs, it is hard to distinguish between over dominance and pseudo-over dominance. By comparison, Steven Tanksley at Cornell University, working with colleagues in China, found that heterosis in rice is due to dominance rather than over dominance. Over the next few years, it will be interesting to see if heterosis among agriculturally important species is usually explained by dominance, over dominance, or a combination of the two.

Concluding Remark

Quantitative genetics is the study of traits that vary in a continuous, quantitative way in populations. Such *quantitative traits* can be anatomical, physiological, or behavioural, and some diseases exhibit characteristics and inheritance patterns analogous to those of quantitative traits.

Within populations, quantitative traits commonly exhibit a continuum of phenotypic variation that may follow a *normal distribution*. Statistical methods can be used to analyze such a distribution. The mean describes the average value among the population; the *standard deviation* (S.D.) provides a measure of the amount of variation in the group. The variance is a measure of the squared deviations from the mean. Variances are useful statistics, because they are additive. Also, some statistical methods compare two variables with each other. The *correlation coefficient* evaluates the strength of association between two variables. In genetics, it is often used to see if there are phenotypic correlations between genetically related individuals.

Most quantitative traits are *polygenic*. The chromosomal locations of genes that influence quantitative traits are called *quantitative trait loci* (QTLs). The reasons most quantitative traits show a continuum are that each trait is governed by several genes existing in two or more alleles, and the environment contributes a substantial amount of variation to the phenotypic outcome.

All traits of biological organisms are influenced by genetics and the environment, particularly quantitative traits. Phenotypic variance is due to the additive effects of genetic variance and environmental variance. *Heritability* is the fraction of the phenotypic variance that is attributable to genetic factors; it describes the amount of phenotypic variation that is due to genetic factors for a particular population raised in a particular environment. *Heritability in the broad sense* takes into account all of the genetic factors that could affect the phenotype. By comparison, the heritability of a trait due to the additive effects of alleles is called the *narrow sense heritability*. For many quantitative traits, the broad sense and narrow sense heritabilties are similar.

Geneticists measure the quantitative traits of an individual and describe them numerically. In a population, experimentally obtained numerical values from each individual are used to calculate statistics such as the mean, standard deviation, variance, and correlation.

More than one method can be used to determine the number of genes or QTLs that influence a quantitative trait. For example, QTLs can be identified via their linkage to genes on chromosomes. More recently, RFLPs are being used as molecular markers to determine the number of QTLs for a given quantitative trait. In some cases, this approach enables plant and animal breeders to understand the genetic nature of quantitative traits and to develop useful breeding strategies.

A central issue for quantitative geneticists is the relative contributions of genetics and the environment that underlie the phenotypic variation seen in quantitative traits. In this chapter, we have considered two experimental methods to measure the heritability of quantitative traits. One method is to compare the correlation coefficients among related and unrelated individuals.

A second method is selective breeding, which is frequently used to improve quantitative traits in agriculturally important species. Quantitative traits are often at an intermediate value in unselected populations. Therefore, artificial selection can be used to increase or decrease the magnitude of the trait. The response to selection can be used as a way to estimate the heritability in the starting population. The heritability measured in this way is called the realized heritability.

Selective breeding sometimes involves the repeated breeding of related individuals to produce strains that are highly inbred. An unfortunate consequence of inbreeding is that it decreases the overall genetic variation within a group and may inadvertently promote homozygosity for deleterious alleles. When two different in bred strains are crossed to each other, the resulting offspring are often more vigorous than either of the inbred parental strains. This phenomenon, called heterosis or hybrid vigor, may be due to the dominant effects of masking harmful recessive alleles or to over dominance of alleles in the heterozygous state.

10

DEVELOPMENTAL GENETICS

Multicellular organisms begin their lives with a fairly simple organization (namely, a fertilized egg) and then proceed step by step to a much more complex arrangement. As this occurs, cells divide, migrate, and change their characteristics as they become highly specialized units within a multicellular individual. In an adult, each cell plays its own particular role for the good of the entire individual. In animals, for example, muscle cells allow an organism to move, while intestinal cells facilitate the absorption of nutrients. This division of labour among the various cells and organs of the individual works collectively to promote its survival.

Developmental genetics, currently one of the hottest fields in molecular biology, is concerned with the roles genes play in orchestrating the changes that occur during development. In this chapter, we will examine how the sequential actions of genes provide a program for the development of an organism from a fertilized egg to an adult.

The last couple of decades has seen staggering advances in our understanding of developmental genetics at the molecular level. Scientists have chosen a few experimental organisms, such as the fruit fly, nematode, frog, mouse, and *Arabidopsis*, and worked toward the identification and characterization of the genes required for running their developmental programs. In certain organisms, notably the fruit fly, most of the genes that play a critical role in the early stages of development have been identified. Researchers are now exploring how the proteins encoded by these genes control the course of development. In this chapter, we will consider several examples in which geneticists understand how the actions of genes govern the developmental process.

Invertebrate Development

We will begin our discussion of multicellular development by considering two model organisms, *Drosophila melanogaster* and the nematode *Caenorhabditis elegans*, that have been pivotal in our understanding of developmental genetics. As we have seen throughout this text, *Drosophila* has been a favourite subject of geneticists since 1910, when Thomas Hunt Morgan isolated his first white-eyed mutant. It has been used to determine many of the fundamental principles of genetics, including the chromosome theory of inheritance, the random mutation theory, and linkage mapping, to mention a few.

In developmental biology, *Drosophila* is also useful, for a variety of reasons. First, researchers have identified many mutant strains with altered developmental pathways. The techniques for generating and analyzing mutants in this organism are more advanced than in any other animal. Second, at the larval stage, *Drosophila* is large enough to conduct transplantation experiments, yet small enough to determine where particular genes are expressed at critical stages of development.

By comparison, *C. elegans* is used by developmental geneticists because of its simplicity. The adult organism is a small transparent worm composed of only about 1000 somatic cells. Starting with the fertilized egg, the pattern of cell division and the fate of each cell within the embryo are completely known.

In this section, we will begin by describing the general features of *Drosophila* development. We will then focus our attention on embryonic development (embryogenesis), because it is during this stage of development that the overall body plan is determined. We will see how the expression of particular genes and the localization of gene products within the embryo influences the developmental process. We will then briefly consider development in *C. elegans*. In this organism, we will examine how the timing of gene expression plays a key role in determining the developmental fate of particular cells.

Early Stages of Embryonic Development Determine the Pattern of Structures in the Adult Organism

Multicellular development follows a body plan or pattern. The term pattern refers to the spatial arrangement of different regions of the body At the cellular level, the body pattern is due to the arrangement of cells and their specialization.

The progressive growth of a fertilized egg into an adult organism involves four types of cellular events: cell division, cell movement, cell differentiation, and cell death. It is the coordination of these four events that leads to the formation of a body with a particular pattern. As we will see, the temporal expression of genes and the localization of gene products at precise regions in the fertilized egg and early embryo are the critical phenomena that underlie this coordination.

The oocyte is the most critical cell in determining the pattern of development in the adult organism. It is an elongated cell with preestablished axes and a well-defined cytoplasmic organization. After fertilization takes place, the zygote goes through a series of nuclear divisions that are not accompanied by cytoplasmic division. Initially, the resulting nuclei are scattered throughout the yolk, but eventually they migrate to the periphery. This is the syncytial blastoderm stage.

After the nuclei have lined up along the cell membrane, individual cells are formed as portions of the cell membrane envelop each nucleus; this creates a structure called a *cellular blastoderm*. This structure is composed of a sheet of cells on the outside with yolk in the center. In this arrangement, the cells are distributed asymmetrically. At the posterior end are a group of cells called the pole cells. These are the primordial germ cells that eventually will give rise to gametes in the adult organism.

After blastoderm formation is complete, some dramatic changes occur during *gastrulation*. This stage involves a great deal of *cell migration*, which produces three cell layers known as the ectoderm, mesoderm, and endoderm. In general, the ectoderm remains on the outside of the gastrula, the endoderm is on the inside, and the mesoderm is wedged in the middle.

As this process occurs, the embryo begins to be subdivided into morphologically detectable units. Initially, shallow grooves divide the embryo into 14 *parasegments*. However, this is a transient condition. A short time later, these grooves disappear, and new boundaries are formed that divide the embryo into morphologically discrete *segments*. The segmented pattern of a *Drosophila* embryo at about 10 hours postfertilization. Later in this section, we will explore how the coordination of gene expression underlies the formation of these parasegments and segments.

At the end of *embryogenesis*, a larva will hatch from the egg and begin feeding on its own. In *Drosophila*, there are three larval stages, separated by molts. During molting, the larva sheds its cuticle, a hardened extracellular shell that is secreted by the epidermis.

After the third larval stage, *Drosophila* proceeds through a process known as *metamorphosis*. Groups of cells called *imaginal disks* were produced earlier in development. During metamorphosis, these imaginal disks grow and differentiate into the structures found in the adult fly (e.g., head, wings, legs, abdomen).

In metazoa (i.e., animals that are more complex than unicellular protozoa), the final result of development commonly is an adult body organized along three axes: the *dorsoventral axis*, the *anteroposterior axis*, and the *right-left axis*. An additional axis, used mostly for designating limb parts, is the *proximodistal axis*.

Although many interesting events occur during the three larval stages and the adult stage, we will focus most of our attention on the genetic events that occur during embryonic development. Even before hatching, the embryo develops the basic body plan that will be found in the adult organism. In other words, during the early stages of development, the embryo is divided into segments that correspond to the segments of the larva and adult. Therefore, an understanding of how these segments form in the embryo is critical to our understanding of pattern formation.

Study of *Drosophila* Mutants with Disrupted Development Patterns has Identified Genes that Control Development

Mutations that alter the course of *Drosophila* development have contributed greatly to our understanding of the normal process of mutations in a complex of genes called the *bithorax* complex. This mutant fly has four wings instead of two; the halteres (a balancing organ that resembles a miniature wing), which are found on the third thoracic segment, are changed into wings, normally found on the second thoracic segment. The term bithorax refers to the observation that the characteristics of the second thoracic segment are duplicated.

Edward Lewis at the California Institute of Technology, a pioneer in the genetic study of development, became interested in the bithorax phenotype and began investigating it in 1946. He discovered that the mutant chromosomal region actually contained a complex of three genes involved in specifying developmental pathways in the fly. A gene that plays a central role in specifying the final outcome of a body region is called a *homeotic gene*. We will discuss particular examples of homeotic genes later in this chapter.

During the 1960s and 1970s, interest in the relationship between genetics and embryology blossomed as biologists began to appreciate the role of genetics at the molecular and cellular levels. It soon became clear that the genome of multicellular organisms contains groups of genes that initiate a program of development involving networks of gene regulation. By identifying mutant alleles that disrupt development, geneticists have begun to unravel the pattern of gene expression that underlies the normal pattern of multicellular development.

Early in development, a category of genes known as *segmentation genes* plays a role in the formation of body segments. The expression of segmentation genes in specific regions of the embryo causes it to become segmented into the pattern. In the 1970s, while working at the European Molecular Biology Laboratory in Germany, Christiane Nusslein-Vothard and Eric Wieschaus undertook a systematic search for *Drosophila* mutants with disrupted segmentation patterns. It was their pioneering effort that led to the identification of most of the genes required for the embryo to develop into a segmented pattern.

Figure 10.1 illustrates a few of the interesting phenotypic effects that Nusslein Volhard and Wieschaus observed when a particular segmentation gene is defective. The gray boxes indicate the regions that are missing in the resulting larvae. The segments adjacent to the deleted regions exhibit a mirror-image duplication. They identified three classes of segmentation genes: *gap genes*, *pair-rule genes*, and *segment-polarity genes*. When a mutation inactivates a gap gene, a contiguous section of the embryo is missing . In other words, there is a gap of several segments. For example, when the *Kruppel* gene is defective, about eight segments are missing from the embryo. By comparison, a defect in a pair-rule gene causes alternating parasegments to be deleted For example, when the *even-skipped* gene is defective, the even-numbered para segments are missing. Finally, segment-polarity mutations cause portions of segments to be missing either an anterior or a posterior region. When this gene is defective, the anterior portion of each segment is missing from the larva.

Overall, the phenotypic effects of mutant segmentation genes provide geneticists with important clues regarding the roles of these genes in the developmental process of segmentation. Later in this section, we will examine when and where the segmentation genes are expressed, and how their pattern of expression leads to the segmentation of the embryo.

Generation of a Body Pattern Depends on the Positional Information that each Cell Receives during Development

The identification of mutant alleles that disrupt the developmental process has permitted great insight into the genes controlling pattern formation. Before we discuss how these genes function, let's consider a central concept in developmental biology known as *positional*

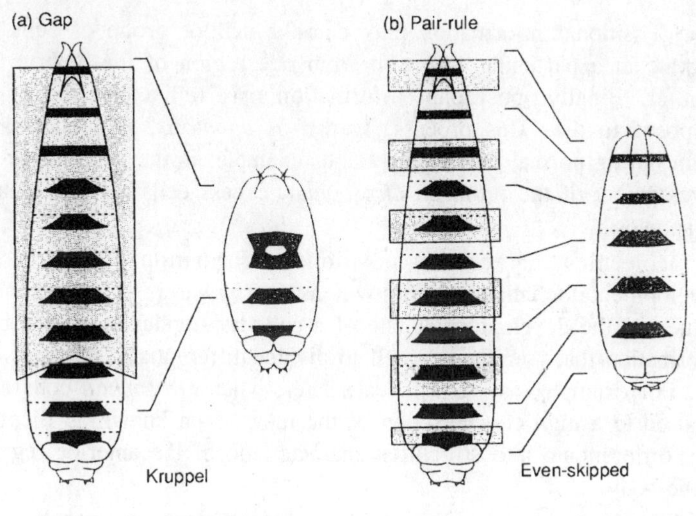

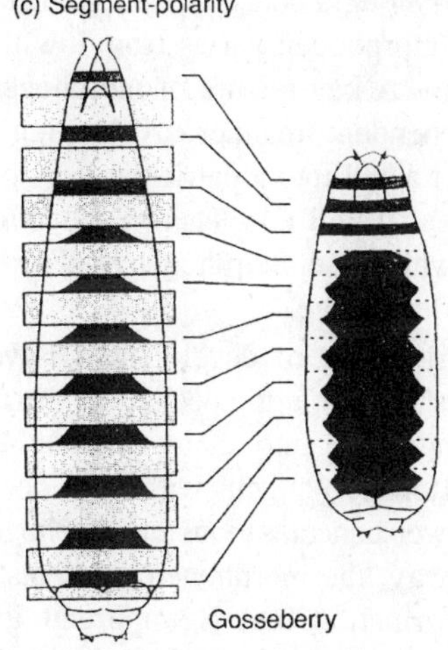

Fig. 10.1. Phenotypic effects in Drosophila larvae that have mutations in segmentation genes. Effects shown are caused by defects in (a) gap genes, (b) pair-rule genes, and (c) segment-polarity genes.

information. For an organism to develop into a segmented pattern with unique morphological and cellular features, each cell of the body must somehow know its position relative to the other cells. A cell may respond to positional information in various ways. For example, positional information may stimulate a cell to divide into two daughter cells. Or it may cause a cell to differentiate into a particular cell type. Positional information may cause a cell or group of cells to migrate in a particular direction from one region of the embryo to another. Finally, positional information may tell a cell that it is supposed to die. This process, known as *apoptosis*, is a necessary event during normal development. For example, in the later stages of development of the retina in *Drosophila*, excess cells are eliminated by apoptosis.

Molecules that convey positional information and promote developmental changes are known as *morphogens*. A morphogen influences the developmental fate of a cell. It provides the positional information that stimulates a cell to divide, differentiate, migrate, or die. For example, as we will learn later, when *Drosophila* cells are exposed to a high concentration of the morphogen known as Bicoid, they differentiate into structures characteristic of the anterior region of the body.

Within an oocyte and during embryonic development, morphogens typically are distributed along a concentration gradient. In other words, the concentration of a morphogen varies from low to high in different regions of the organism. A key feature of morphogens is that they act in a concentration-dependent manner. At a high concentration, a morphogen will restrict a cell into a particular developmental pathway; at a lower concentration, it will not. There is often a critical *threshold concentration* above which the morphogen will exert its effects but below which it is ineffective.

During the earliest stages of development, several morphogenic gradients are preestablished within the oocyte. At the cellular blastoderm stage, the zygote subdivides into many smaller cells. Due to the preestablished gradient of morphogens within the oocyte, these smaller cells have higher or lower concentrations of morphogens, depending on their location. In this way, the morphogen gradients in the oocyte can provide positional information that is important in establishing the general polarity of an embryo along two main axes: the anteroposterior axis and the dorsoventral axis. This topic will be described in greater detail later in this chapter.

A morphogen gradient can also be established by cell secretion and transport. A certain cell or group of cells may synthesize and secrete a morphogen at a specific stage of development. After secretion, the morphogen may be transported to neighbouring cells. The concentration of the morphogen is usually highest near the cells that secrete it . The morphogen may then influence the developmental fate of the cells that are exposed to it. When a cell or group of cells governs the developmental fate of neighbouring cells by producing a morphogen, this process is known as *induction*.

In addition to morphogens, positional information is conveyed by *cell adhesion*. Each cell makes its own collection of surface receptors; these receptors cause it to adhere to other cells and/or to the extracellular matrix (ECM), which consists primarily of carbohydrates and fibrous proteins. Such receptors are known as *cell adhesion molecules* (CAMs). A cell may gain positional information via the combination of contacts it makes with other cells or with the ECM.

The phenomenon of cell adhesion, and its role in multicellular development, was first recognized by H. V. Wilson in 1907 while working at the University of North Carolina in Chapel Hill. He took multicellular sponges and disaggregated them into individual cells. Remarkably, the cells actively migrated until they adhered to one another to form a new sponge, complete with the chambers and canals that characterize a sponge's internal structure! When sponge cells from different species were mixed, they sorted themselves properly, adhering only to cells of the same species. Overall, these results indicate that cell adhesion plays an important role in governing the position that a cell will adopt during development.

The general phenomena that underlie positional information during embryonic development. With these ideas in mind, we can begin to examine how specific genes in *Drosophila* encode morphogens that convey positional information. In this organism, the establishment of the body axes and division of the fly into segments involves the participation of a few dozen genes. These genes are often given interesting names based on the phenotypic effects when they are mutant. It is beyond the scope of this text to examine how all of these genes exert their effects during embryonic development. Instead, we will consider a few examples that illustrate how the expression of a particular gene and the localization of its gene product have a defined effect on the pattern of development.

Gene Products of Maternal Effect Genes are Deposited Asymmetrically into the Oocyte and Establish the Anteroposterior and Dorsoventral Axes at a Very Early Stage of Development

The first stage in *Drosophila* embryonic pattern development is establishment of the body axes. This occurs before the embryo becomes segmented. In fact, the morphogens necessary to establish these axes are distributed prior to fertilization. During oogenesis, certain gene

Table 10.1. Examples of *Drosophila* genes that play a role in pattern development.

Description	*Examples*
Some genes play a role in determining the axes of development. Also, certain genes govern the formation of extreme terminal (anterior and posterior) regions.	Anterior: *bicoid*, *exuperantia*, *hunchback*, *swallow* Posterior: *nanos*, *cappuccino*, *oskar*, *pumilio*, *spire*, *staufen*, *tudor*, *vasa* Terminal: *torso*, *torsolike* Dorsoventral: *toll*, *cactus*, *dorsal easter*, *gurken*, *nudel*, *pelle*, *pipe*, *snake*, *spatzle*
Some gene play a role in promoting the subdivision of the embryo into segments. These are called segmentation genes. As described later, there are three types of segmentation genes, known as gap genes, pair-rule genes, and segment-polarity genes.	Gap genes: *empty spiracles*, *huckebein hunchback*, *knirps*, *Kruppel*, *tailless* Pair-rule genes: *even-skipped*, *hairy*, *runt*, *fushi tarazu*, *odd-paired*, *odd-skipped*, *paired*, *sloppy paired* Segment-paired genes: *engrailed*, *hedgehog*, *wingless*, *gooseberry*
Some genes play a role in determining the fate of particular segments. These are known as homeotic genes. *Drosophila* has two clusters of homeotic genes, known as the *antennapedia* complex and the bithorax complex.	Antennapedia complex: *labial*, *proboscipedia*, *deformed*, *sex combs reduced*, *antennapedia* Bithorax complex: *ultrabithorax*, *abdominal A*, *abdominal B*

products, which are important in early developmental stages, are deposited asymmetrically within the egg. Later, after the egg has been fertilized and development begins, these gene products will establish independent developmental programs that govern the formation of the four major body regions of the embryo. These are the anterior, posterior, terminal, and dorsoventral regions.

In each of these regions, the products of several different genes ensure that proper development will occur. A few gene products act as key morphogens, or receptors for morphogens, that initiate changes in embryonic development. As shown here, these gene products are deposited asymmetrically in the egg. For example, the product of the *bicoid* gene is necessary to initiate development of the anterior structures of the organism. During oogenesis, the mRNA for bicoid accumulates in the anterior region of the oocyte. In contrast, the mRNA from the *nanos* gene accumulates in the posterior end. Later in development,

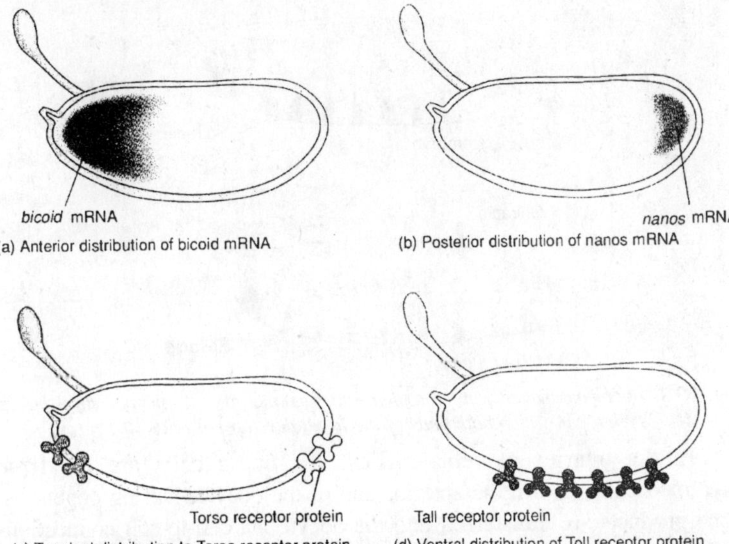

Fig. 10.2. The establishment of the axes of polarity in the Drosophila embryo.

the nanos mRNA will be translated into protein, which functions to influence posterior development. Nanos is required for the formation of the abdomen. In addition to *bicoid* and *nanos*, the development of the structures at the extreme anterior and posterior ends of the embryo are regulated in part by a receptor protein called *Torso*. This receptor, which is activated only at the anterior and posterior ends of the egg, is necessary for the formation of the head and abdomen. The fourth major system is the dorsoventral system. A receptor protein known as *Toll* is activated along the ventral midline of the egg and initiates the establishment of the dorsoventral axis.

Let's now take a closer look at the molecular mechanism of one of these modulators, namely *bicoid*. The *bicoid* gene got its name because a larva defective in this gene develops with two posterior ends. This allele exhibits a maternal effect pattern of inheritance. A female fly that is phenotypically normal (because its mother was heterozygous for the normal *bicoid* allele), but genotypically homozygous for an inactive *bicoid* allele (because it inherited the in active allele from its mother and father), will produce 100% affected offspring even when mated to a male that is homozygous for the normal *bicoid* allele. In other words, the genotype of the mother determines the phenotype of the offspring. This occurs because the *bicoid* gene product is provided to the oocyte via the nurse cells.

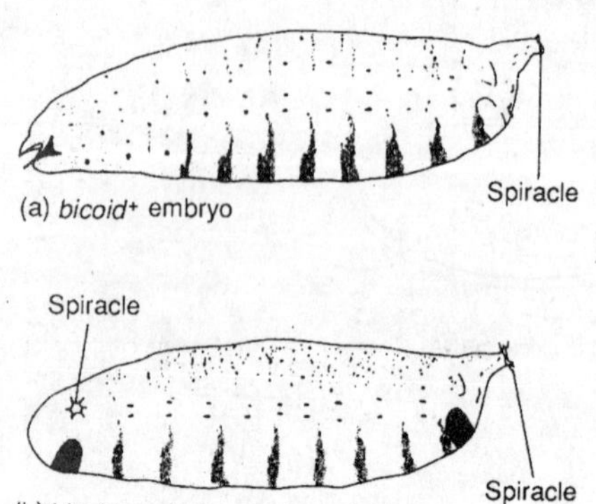

Fig. 10.3. The bicoid mutation in Drosophila. (a) A normal, bicoid$^+$ embryo. (b) A bicoid$^-$ embryo, in which both ends of the larva develop posterior structures.

In the ovaries of female flies, the nurse cells are localized asymmetrically toward the anterior end of the oocyte. During oogenesis, gene products are transferred into the oocyte via cell-to-cell connections called cytoplasmic bridges. Maternally encoded gene products enter one side of the oocyte. This side will eventually become the anterior end of the embryo. The *bicoid* gene is actively transcribed in the nurse cells, and *bicoid* mRNA is transported into the anterior end of the oocyte. The 3' end of bicoid mRNA contains a signal that is recognized by binding proteins thought necessary for the transport of this mRNA into the oocyte. After it enters the oocyte, the bicoid mRNA is trapped near the anterior side.

In situ hybridization experiment in which a *Drosophila* egg was examined via a probe complementary to the bicoid mRNA. As seen here, the *bicoid* mRNA is highly concentrated near the anterior pole of the egg cell. When the *bicoid* mRNA subsequently is translated, a gradient of Bicoid protein is established. After fertilization occurs, the Bicoid protein functions as a transcription factor. A remarkable

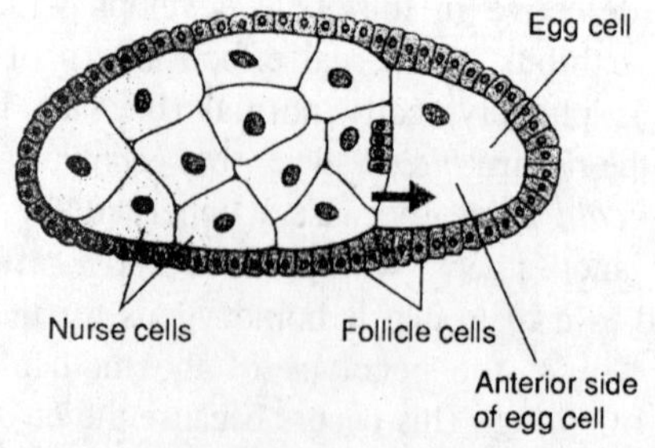

Fig. 10.4. Asymmetrical localization of gene products during oogenesis in Drosophila.

feature of this protein is that its ability to influence gene expression is tuned exquisitely to its concentration. Depending on the distribution of the Bicoid protein, this transcription factor will only activate genes in certain regions of the embryo. For example, Bicoid stimulates a gene called *hunchback* in the anterior half of the embryo, but it does not activate the *hunchback* gene in the posterior half.

Gap, Pair-rule, and Segment-polarity Genes act Sequentially to Divide the *Drosophila* Embryo into Segments

After the anteroposterior, dorsoventral, and terminal regions of the embryo have been established by maternal effect genes, the next developmental process is to organize the embryo transiently into parasegments and then permanently into segments. This pattern of positional information will be maintained, or remembered throughout the rest of development. In other words, each of the segments in the embryo will give rise to unique morphological features in the adult. For example, T2 will become a thoracic segment with a pair of legs and a pair of wings, and A8 will become a segment of the abdomen.

As shown, the boundaries of the segments are out of register with the boundaries of the parasegments. An appreciation of this feature is critical to our understanding of segmentation. From the viewpoint of genes, we will see that the parasegments are the locations where gene expression is controlled spatially. The anterior compartment of each segment overlaps with the posterior region of a parasegment; the posterior compartment of a segment overlaps with an anterior region of the next parasegment. The pattern of gene expression that occurs in the anterior region of one parasegment and the posterior region of an adjacent parasegment results in the formation of a segment.

Now that we have a general understanding of the way the *Drosophila* embryo is subdivided, we can examine how particular genes cause it to become segmented into this pattern. As mentioned, the genes that play a role in the formation of body segments are called *segmentation genes*; there are three classes of segmentation genes: gap genes, pair-rule genes, and segment-polarity genes. The expression and activation patterns of these genes in specific regions of the embryo cause it to become segmented.

A partial, simplified scheme of the genetic hierarchy that leads to a segmented pattern in the *Drosophila embryo*.. This figure presents the general sequence of events that occurs during the early stages of embryonic development. As described in this figure, the following steps occur:

1. Maternal effect gene products, such as *bicoid* mRNA, are deposited asymmetrically into the oocyte. These gene products form a gradient that will later influence the formation of axes, such as the anteroposterior axis.
2. After fertilization, maternal effect gene products activate *zygotic genes*. In contrast to maternal effect genes, which are expressed during oogenesis, zygotic genes are expressed after fertilization. The first set of zygotic genes to be activated are the gap genes. Gap genes are activated as broad bands within particular regions of the embryo. These bands do not correspond to parasegments or segments within the embryo.
3. The gap genes and maternal effect genes then activate the pair-rule genes. Note that these pair-rule genes are expressed in particular parasegments. *Ftz* is expressed in the odd-numbered parasegments, and *even-skipped* is expressed in the even-numbered segments.
4. Once the pair-rule genes are activated in an alternating banding arrangement, their gene products then regulate the segment-polarity genes. As shown in the segment-polarity gene *engrailed* is expressed in the anterior region of each parasegment. Another segment polarity gene, *wingless*, is expressed in the posterior region. Later in development, the anterior end of one parasegment and the posterior end of another parasegment will develop into a segment with particular morphological characteristics.

Expression of Homeotic Genes Controls the Phenotypic Characteristics of Segments

We have considered how the *Drosophila* embryo becomes organized along axes and then into a segmented body pattern. Now we will examine how each segment develops its unique morphological features. Geneticists often use the term *cell fate* to describe the ultimate morphological features that a cell or group of cells will adopt. For example, the fate of the cells in segment T2 in the *Drosophila* embryo is to develop into a thoracic segment containing two legs and two wings. In *Drosophila*, the cells in each segment of the body have their fate determined at a very early stage of embryonic development, long before the morphological features become apparent.

Our understanding of developmental fate has been aided greatly by the identification of mutant genes that alter cell fates. In animals, the first mutant of this type was described by the German entomologist G.Kraatz in 1876. He observed a sawfly (*Climbex axillaris*) in which part of an antenna was replaced with a foot. During the late 19th century, the English zoologist Wffliam Bateson collected many of these types of observations and published them in 1894 in a book entitled Materials far the Study of Variation Treated with Especial Regard to Discontinuity in the Origin of Species. In this book, Bateson coined the term *homeotic* to describe mutant alleles in which one body part is replaced by another.

As mentioned earlier in this chapter, Edward Lewis at Caltech began to study strains of *Drosophila* having homeotic mutations. This work, which began in 1946, was the first systematic study of homeotic genes. Each homeotic gene controls the fate of a particular region of the body. *Drosophila* contains two clusters of homeotic genes called the *bithorax* complex and the *antennapedia* complex. The antennapedia complex contains five genes, designated *lab*, *pb*, *dfd*, *scr*, and *antp*. The bithorax complex has three genes, *ubx*, *abdA*, and *abdB*. Both of these complexes are located on chromosome 3, but a large segment of DNA separates them. As noted, the order of these genes along chromosome 3 correlates with the anteroposterior axis of the body. For example, *lab* is expressed in the anterior segment and governs the formation of mouth structures. The *antp* gene is expressed strongly in the thoracic region during embryonic development and controls the formation of thoracic structures. Transcription of the *abdB* gene occurs in the posterior region of the embryo; this gene controls the formation of the posteriormost abdominal segments.

The role of homeotic genes in determining the identity of particular segments has been revealed by mutations that alter their function. Antennapedia mutation is a *gain of function mutation* in the *antp* gene that causes it to be expressed in an additional place in the embryo. In this case, the antp gene is expressed abnormally in the anterior segment that normally gives rise to the antennae. In other words, there has been a gain of *antp* function in this segment. The abnormal expression of *antp* in this region causes the antennae to be converted into legs!

Investigators have also studied many loss of function alleles in homeotic genes. When a particular homeotic gene is defective, the region that it normally governs will be controlled by the homeotic gene that acts in the adjacent anterior region. For example, the *ubx* gene normally functions within parasegments 5 and 6. If this gene is missing, this section of the fly becomes converted to the structures that are found normally in parasegment 4. The homeotic genes are part of the genetic hierarchy that produces the morphological characteristics of the fly. They are regulated in a very complex way. Their expression is controlled by gap genes and pair-rule genes, and they are also regulated by interactions among themselves. In addition, a group of genes known as the *polycomb* genes represses the expression of homeotic genes in regions where they should not act. Overall, the concerted actions of many gene products cause the homeotic genes to be expressed only in the appropriate region of the embryo.

Since they are part of a genetic hierarchy, it is not too surprising that homeotic genes encode transcription factors. The coding sequence of homeotic genes contains a 180 base pair consensus sequence, known as a *homeobox*. This sequence was first discovered in the *antp* and *ubx* genes, and it has since been found in all *Drosophila* homeotic genes and in some other genes affecting pattern development, such as bicoid. The protein domain encoded by the homeobox is called a *homeodomain*. The arrangement of α-helices within the homeodomain promotes the binding of the protein to the major groove of DNA. In this way, homeotic proteins can bind to DNA in a sequence-specific manner. In addition to DNA-binding ability, homeotic proteins also contain a transcriptional activation domain that functions to activate the genes to which the homeodomain can bind.

The transcription factors encoded by homeotic genes activate the next category of genes, collectively known as *realizator genes*. These genes produce the morphological characteristics of each segment. Much current research attempts to identify realizator genes and determine how their expression in particular regions of the embryo leads to morphological changes in the embryo, larva, and adult.

In some cases, realizator genes also encode transcription factors. Presumably, such realizator genes control the expression of other sets of genes that will alter the morphological characteristics of cells. In other cases, realizator genes encode proteins involved in cell-to-cell signaling pathways. The activation of these signaling pathways is thought to play a key role in the ability of cells and groups of cells to adopt their correct morphologies. It is expected that research during the next few decades will shed considerable light on the pathways by which realizator genes control morphological changes in the fruit fly.

Developmental Fate of each Cell in the Nematode *Caenorhabditis elegans* is Known

We now turn our attention to another invertebrate, *C. elegans*, that has been the subject of numerous studies in developmental genetics. As does *Drosophila*, this worm begins its development as a fertilized egg. The embryo develops within the egg shell and hatches when it reaches a size of 558 cells. After hatching, it continues to grow and mature as it passes through four successive moults. It takes about three days for a fertilized egg to develop into an adult worm.

With regard to gender, *C. elegans* can be a male (and only produce sperm) or a hermaphrodite (capable of producing sperm and egg cells). An adult male is composed of 1031 somatic cells and produces about 1000 sperm. A hermaphrodite consists of 959 somatic cells and produces about 2000 gametes (both sperm and eggs).

A remarkable feature of this organism is that the pattern of cellular development is extremely invariant from worm to worm. In the early 1960s, Sydney Brenner pioneered the effort to study the pattern of cell division in *C. elegans*. To do so, a researcher can identify a particular cell at an embryonic stage, follow that cell as it divides, and observe where its descendant cells will be located in the adult.

Because *C. elegans* is transparent and composed of relatively few cells, researchers can follow cell division step by step, beginning with a fertilized egg and ending with an adult worm. An illustration that depicts how cell division proceeds is called a *fate map*. It describes the fate of any cell's descendants. At the first cell division, the egg divides to produce two cells, called AB and P_1. AB then divides into two cells, AB a and AB p; and P_1 divides into two cells, EMS and P_2. As noted in the EMS cell then divides into two cells, called MS and E. The cellular descendants of the E cell give rise to the worm's intestine. In other words, the fate of the E cell's descendants is to develop into intestinal cells. This diagram also illustrates the concept of a *cell lineage*. This term refers to a series of cells that are derived from each other by cell division. For example, the EMS cell, E cell, and the intestinal cells are all part of the same cell lineage.

Having a cellular fate map for an organism is an important experimental advantage. It allows researchers to investigate how gene expression in any cell, at any stage of development, may affect the outcome of a cell's fate. In the experiment described next, we will see how the timing of gene expression is an important parameter in the fate of a cell's descendants.

Horvitz and Ambros Found that Heterochronic Mutations Disrupt the Timing of Developmental Changes in *C. elegans*

Our discussion of *Drosophila* development focused on how the spatial expression and localization of gene products can lead to a particular pattern of embryonic development. Another important issue in development is timing. The cells of a multicellular organism must know when to divide and when to differentiate into a particular cell type. If the timing of these processes is not coordinated, certain tissues will develop too early or too late, disrupting the developmental process.

In *C. elegans*, the timing of developmental events can be examined carefully at the cellular level. As mentioned, the fate of each cell has been determined. Using a microscope (with Nomarski differential interference contrast optics), a researcher can focus on a particular cell within this transparent worm and watch it divide into two cells, then four cells, and so forth. Therefore, a scientist can judge whether a cell is behaving as it should during the developmental process.

To identify genes that play a role in the timing of cell fates, researchers have searched for mutant alleles that disrupt the normal timing process. In a collaboration in the late 1970s, H. Robert Horvitz at the Massachusetts Institute of Technology and John Sulston at the MRC Laboratory of Molecular Biology in England set out to identify mutant alleles in *C. elegans* that disrupt cell fates or the timing of cell fates.

Prior to this work, they did not know what phenotypic effects to expect from a mutation that altered the fate of cells within a cell lineage. Using a microscope, they screened thousands of worms for altered morphologies that might indicate an abnormality in development. During this screening process, one of the phenotypic abnormalities they found was an *egg-laying defective* phenotype. They reasoned that since the egg-laying system depended on a large number of cell types (vulval cells, muscle cells, and nerve cells) an abnormality in any of the cell lineages leading to these cell types might cause an inability to lay eggs.

In *C. elegans*, an egg-laying defective phenotype is easy to identify, because the hermaphrodite will be able to fertilize its own eggs but will be unable to lay them. When this occurs, the eggs actually hatch within the hermaphrodite's body. This leads to the death of the hermaphrodite as it becomes filled with hatching worms. This egg-laying defective phenotype, in which the hermaphrodite becomes filled with its own offspring, is called a "bag of worms":

In their initial study, published in 1980, the egg-laying defective phenotype yielded several mutant strains that were defective in particular cell lineages. In 1984, in the experiment described here, Victor Ambros and Horvitz took this same approach and were able to identify genes that play a key role in the timing of cell fate.

Hypothesis

Mutations that cause an egg-laying defective phenotype may affect the timing of cell lineages.

Testing the hypothesis

Starting material: Prior to this work, many laboratories had screened thousands of *C. elegans* worms and identified many different mutant strains that were egg-laying defective. (Note: There are many different genes that when mutated may cause an egg-laying defective phenotype. Only some of them are expected to be genes that alter the timing of cell fate within a particular cell lineage.)

Experimental level

1. Obtain a large number of *C. elegans* strains that have an egg-laying defective phenotype. The wild-type strain was also studied as a control.
2. Right after hatching, observe the fate of particular cells via microscopy. This involves long hours of viewing specific cells within a worm and watching to see if they divide at the appropriate time. Typically, the viewer looks at the cell nuclei (which are relatively easy to see in this transparent worm) and keeps track of when they divide. A researcher can watch and time the division of the cell nuclei as a way to monitor cell division. In this example, a researcher began watching a cell called the *T cell*, and monitored its division pattern, and the pattern of subsequent daughter cells, during the first and second larval stages. These patterns were examined in both wild-type and egg-laying defective worms.

These data show the division pattern of one particular cell lineage. The top of this lineage begins with a cell called the T cell. It is part of the cell lineage that in Judes the AB p cell The cells at the left of each lineage are located anteriorly in the worm; those on the right are located posteriorly. Neurons are labeled in blue, epidermal cells in red.

Interpreting the data

As shown in the data, the wild-type strain shows a particular pattern of cell division for the T cell lineage, which occurs at specific times during the L1 and L2 larval stages. In the normal strain, the T cell divides during the L1 larval stage to produce a T.a and T.p cell. The T.a cell also divides during L1 to produce a T.aa and T.ap cell. The T.p cell divides during L1 to produce Tpa and T.pp. These cells also divide during L1, eventually producing five neurons and one cell that is programmed to die (designated with an X). During the L2 larval stage, the Tap cell resumes division to produce four cells: three epidermal cells and one neuron.

The other T cell lineages are from worms that carry mutations in a gene called *lin*-14. The allele designated n536 has caused the reiteration of the normal events of L1 during the L2 larval stage. In L2, the only cell of this lineage that is supposed to divide is Tap. In worms carrying the n536 allele, however, this cell behaves as if it were a T cell, rather than a Tap cell. It produces a group of cells that are identical to what a T cell produces during the L1 stage. In the L3 stage, the cells in the n536 strain behave as if they were in L2. Besides the egg laying defect, the phenotypic outcome of this irregularity in the timing of cell fates is a worm that has a few more cells and goes through five or six larval stages instead of the normal four.

A more severe allele that causes multiple reiterations is the n355 allele. This strain continues to reiterate the normal events of L1 during the L2, L3, and L4 stages. In contrast, the n540 allele has an opposite effect on the T cell lineage. During the L1 larval stage, the T cell behaves as if it were a Tap cell in the L2 stage. In this case, it skips the divisions and cell fates of the L1 and proceeds directly to cell fates that occur during the L2 stage.

The types of mutations described here are called *heterochronic mutations*. This term refers to the fact that the timing of fates for particular cell lineages is not synchronized with the development of the rest of the organism. More recent molecular data have shown that this is due to an irregular pattern of gene expression. In wild-type worms, the Lin-14 protein accumulates during the L1 stage and promotes the T cell division pattern shown for the wild-type. During L2, the Lin- i4 protein diminishes to negligible levels. The n536 and n355 alleles are examples of gain of function mutations. In these alleles, the Lin-14 protein persists during later larval stages. By comparison, the n540 allele is a loss of function mutation. This allele causes Lin-14 to be inactive during L1, so that it cannot promote the normal L1 pattern of cell division and cell fate.

Overall, the results described in this experiment are consistent with the idea that the precise timing of lin-14 expression during development is necessary to correctly control the fates of particular cells in *C. elegans*. Mutations that alter the expression of lin-14 lead to phenotypic abnormality, namely the inability to lay eggs. These results illustrate the importance of the correct timing of developmental events.

Vertebrate Development

Embryologists have studied the morphological features of development in many vertebrate species. Historically, amphibians and birds have been studied extensively, because their eggs are rather large and easy to manipulate. For example, the early developmental stages of the chicken and frog (*Xenopus laevis*) have been described in great detail. In more recent times, the successes obtained in *Drosophila* have shown the great power of genetic analyses in elucidating the underlying molecular mechanisms that govern biological development. With this knowledge, many researchers are attempting to understand the genetic pathways that govern the development of the more complex body structure found in vertebrate organisms.

Several vertebrate species have been the subject of genetic studies of development. These include the mouse, the chicken, *Xenopus*, and the small aquarium zebrafish (*Brachydanio rerio*). In this section, we will discuss primarily the genes that are important in mammalian development, particularly those that have been characterized in the mouse. Among mammals, the most extensive genetic analyses have been performed on the mouse. As we will see, several genes affecting its develop mental pathways have been cloned and characterized. In this section, we will examine how these genes affect the course of mouse development.

Researchers have Identified Homeotic Genes in Vertebrates

In most vertebrates, which have long generation times and produce relatively few offspring, it is not practical to screen large numbers of embryos or offspring in search of mutant phenotypes with developmental defects. As an alternative, the most successful way of identifying genes that affect vertebrate development has been the use of molecular techniques to identify vertebrate genes similar to those that control development in simpler organisms such as *Drosophila*.

Species that are related evolutionarily to each other often contain genes with similar DNA sequences. When two or more genes have similar sequences because they are derived from the same ancestral gene, they are called *homologous genes*. Since they have similar sequences, a DNA strand from one gene will hybridize to a complementary strand of a homologue. The general approach of using cloned *Drosophila* genes as probes to identify homologous vertebrate genes has been quite successful. Using this method, researchers have found complexes of homeotic genes in many vertebrate species that bear striking similarities to those in the fruit fly. In the mouse, these groups of adjacent homeotic genes are called *Hox complexes*. As shown, the mouse has four *Hox complexes*, designated *HoxA* (on chromosome 6), *HoxB* (on chromosome 11), *HoxC* (on chromosome is), and *HoxD* (on chromosome 2). There are a total of 38 genes in the four complexes. There are 13 different gene types within the four *Hox complexes*, although each complex contains less than all 13 types of genes. Among the first 6 types of genes, 5 of them are homologous to genes found in the antennapedia complex of *Drosophila*. Among the last 7, 3 are homologous to the genes of the bithorax complex.

Like the *bithorax* and *antennapedia* complexes in *Drosophila*, the arrangement of *Hox genes* along the mouse chromosome reflects their pattern of expression from The anterior to the posterior end. This phenomenon shows the results of the expression pattern for a group of *HoxB* genes in a mouse embryo. Overall, these results are consistent with the idea that the *Hox genes* play a role in determining the fates of segments along anteroposterior axis.

Currently, researchers are trying to understand the functional roles of the genes within the *Hox* complexes in vertebrate development. In the fly, great advances in developmental genetics have been made by studying mutant alleles in genes that control development. In mice, however, there are few natural mutations affecting development. This has made it difficult to understand the role that genetics plays in the development of the mouse and other vertebrate organisms.

To circumvent this problem, geneticists are taking an approach known as *reverse genetics*. In this strategy, researchers first identify the wild-type gene using cloning methods. In this case, the *Hox* genes in vertebrates have been cloned using *Drosophila* genes as probes. The next step is to create a mutant version of a *Hox gene* in vitro. This mutant allele is then reintroduced into a mouse. When the function of the wild-type gene is thereby eliminated, this is called a *gene knockout*. In this way, researchers can determine how the mutant allele affects the phenotype of the mouse.

The term reverse genetics reflects that the experimental steps occur in an order opposite to that in the conventional approach used in *Drosophila*. In the fly, the mutant alleles were identified by their phenotype first, and then they were cloned. In the mouse, the genes were cloned first, the mutations were made in vitro, and then these were introduced into the mouse to observe their phenotypic effects.

During the 1990s, several laboratories have used a reverse genetics approach to understand how the *Hox* genes affect vertebrate development. In *Drosophila*, loss of function alleles for homeotic genes usually show an anterior transformation. This means that the segment where the defective homeotic gene is expressed now exhibits characteristics that resemble the adjacent anterior segment. Similarly, certain gene knockouts (e.g., HoxA-2, B-4, and C-8) also show anterior transformations within particular regions of the mouse. However, knockouts of other Hox genes (e.g., A-5 and A-11) have posterior transformations, and knockouts of A-3 and A-1 exhibit abnormalities in morphology but no clear homeotic transformations. Overall, the current picture indicates that the *Hox genes* in vertebrates play an important role in homeotic transformations. Nevertheless, additional research will be necessary to understand the individual roles that each of the 38 *Hox* genes play during embryonic development.

Genes that Encode Transcription Factors also Play a Key Role in Cell Differentiation

Throughout most of this chapter, we have focused our attention on patterns of gene expression that occur during the very early stages of development. These genes control the basic body plan of the organism. At later stages of development, cells reach their predetermined destinations and eventually become *differentiated*. This means that a cell's morphology and function have changed, usually permanently, into a highly specialized cell type. For example, an undifferentiated mesodermal cell may differentiate into a specialized muscle cell, or an ectodermal cell may differentiate into a nerve cell.

At the molecular level, the profound morphological differences between muscle cells and nerve cells arise from gene regulation. Though nerve and muscle cells contain the same genetic material (i.e., the same set of genes), they regulate the expression of their genes in very different ways. Certain genes that are transcriptionally active in muscle cells are completely inactive in nerve cells, and vice versa. Therefore, nerve and muscle cells express different proteins, which affect the morphological and physiological characteristics of the respective cells in distinct ways. In this manner, differential gene regulation underlies cell differentiation.

We learned earlier that in *Drosophila* a hierarchy of gene regulation is responsible for establishing the body pattern. Maternal effect genes control the expression of gap genes, which control the expression of pair-rule genes, and so forth. A similar type of hierarchy is thought to underlie cell differentiation. Researchers have identified specific genes, realizator genes, the expression of which causes cells to differentiate into particular cell types. These genes cause undifferentiated cells to differentiate into their proper cell fates.

In 1987, Harold Weintraub and his colleagues at the Hutchinson Cancer Research Center in Seattle identified a gene, which they called *myoD*. This gene plays a key role in skeletal muscle cell differentiation. Experimentally, when the cloned *myoD* gene was expressed in fibroblast cells in a laboratory, the fibroblasts differentiated into skeletal muscle cells. This result was particularly remarkable, since fibroblasts normally differentiate into osteoblasts (bone cells), chrondrocytes (cartilage cells), adipocytes (fat cells), and smooth muscle cells, but they never differentiate, *in vivo*, into skeletal or cardiac muscle cells.

Since this initial discovery, researchers have found that *myoD* belongs to a small group of genes that play a role in initiating muscle development. Besides *myoD*, these include myogenin myf-5$^+$ and MRF-4$^+$. All four of these genes encode transcription factors that contain a *basic* domain and a *helix-loop-helix* domain (bHLH). The basic domain is responsible for DNA binding and the activation of skeletal muscle cell—specific genes. The HLH domain is necessary for dimer formation between transcription factor proteins. Because of their common structural features and their role in muscle differentiation, myoD, myogenin, myf-5, and MRF-4 are called myogenic *bHLH proteins*. They are found in all vertebrates, and have been identified in several invertebrates, such as *Drosophila* and *C. elegans*. In all cases, the myogenic bHLH genes are activated during skeletal muscle cell development.

In addition to myogenic bHLH genes, there are other bHLH genes that are not muscle cell specific. These other bHLH genes encode transcription factors known as *E proteins*. E proteins are synthesized in a wide variety of tissues. They are also important in cell differentiation.

At the molecular level, two key features enable myogenic bHLH proteins to promote muscle cell differentiation. First, the basic domain binds specifically to a muscle cell—specific enhancer sequence; this sequence is adjacent to genes that are expressed only in muscle cells. Therefore, when bHLH proteins are activated, they can bind to these enhancers and activate the expression of many different muscle cell—specific genes. In this way, myogenic bHLH proteins act as master switches that activate the expression of many muscle-specific genes.

When the encoded proteins are synthesized, they change the characteristics of an undifferentiated cell into those of a highly specialized skeletal muscle cell.

Another important aspect of myogenic bHLH proteins is that their activity is regulated by dimerization. When a heterodimer forms between a myogenic bHLH protein and an E protein, which also contains a bHLH domain, the heterodimer binds to the DNA and activates gene expression. However, when a heterodimer forms between a myogenic bHLH protein and a protein called Id (for inhibitor of differentiation), the heterodimer cannot bind to DNA, because the Id protein lacks a basic domain. The Id protein is produced during early stages of development and prevents myogenic bHLH proteins from promoting muscle differentiation too soon. At later stages of development, the amount of Id protein falls, and myogenic bHLH proteins can then combine with E proteins to induce muscle differentiation.

Plant Development

In developmental plant biology, the model organism for genetic analysis is *Arabidopsis thaliana*. Unlike most plants, which have long generation times and large genomes, *Arabidopsis* has a generation time of about two months and a genome size of 7×10^7 bp, which is similar to *Drosophila* and *C. elegans*. A flowering Arabidopsis plant is small enough to be grown in the laboratory, and it produces a large number of seeds. Like *Drosophila*, *Arabidopsis* can be subjected to mutagens such as X-rays to generate mutations that alter developmental processes. The small genome size of this organism makes it relatively easy to map these mutant alleles and eventually clone the relevant genes.

The morphological patterns of growth are markedly different between plants and animals. As described previously, animal embryos become organized along anteroposterior, dorsoventral, and lateral axes, and then they subdivide into segments. By comparison, the form of higher plants has two key features. The first is the root—shoot axis. Most plant growth occurs via cell division near the tips of the shoots and the bottoms of the roots.

Second, this growth occurs in a well-defined radial pattern. For example, early in *Arabidopsis* growth, a rosette of leaves is produced from leaf buds that emanate in a spiral pattern directly from the main shoot. Later, the shoot generates branches that will also produce leaf buds as they grow. Overall, the radial pattern in which a plant shoot gives off the buds that give rise to branches, leaves, and flowers is an important mechanism that determines much of the general morphology of the plant.

At the cellular level too, plant development differs markedly from animal development. For example, cell migration does not occur during plant development. In addition, the development of a plant does not rely on morphogens that are deposited asymmetrically in the oocyte. In plants, an entirely new individual can be regenerated from most types of somatic cells. In other words, most plant cells are *totipotent*, meaning that they have the ability to produce an entire individual. By comparison, animal development invariably relies on the organization within an oocyte as a starting point for development.

In spite of these apparent differences, the underlying molecular mechanisms of pattern development in plants still share some similarities with those in animals. In this section, we will consider a few examples in which genes encoding transcription factors play a key role in plant development.

Plant Growth occurs from Meristems that are Formed during Embryonic Development

In animal development, the earliest developmental stages serve to subdivide the embryo into segments that eventually will give rise to adult structures. The growth of the embryo into the adult is simply an expand of the embryo body pattern. However, plants do not grow like this. Instead, the parts of adult plants are formed from groups of dividing cells known as *apical meristems*. A meristem is an organized group of actively dividing cells. The *shoot apical meristem* grows upward and gives rise to the shoot structures, while *root apical meristems* grow downward to produce the roots. As they grow, a meristem produces off-shoots of proliferating cells. On the shoot, for example, these offshoots or buds give rise to structures such as leaves and flowers.

The organization of a plant into separate root and shoot meristems occurs during the early embryonic stages. After fertilization, the first cellular division is asymmetrical and produces a smaller cell, called the terminal cell, and a larger basal cell. The terminal cell will give rise to most of the embryo, and it will later develop into the shoot of the plant. The basal cell will give rise to the root, along with extraembryonic tissue that is required for seed formation. At the heart stage, which is composed of only about 100 cells, the basic organization of the plant has been established. The shoot apical meristem will arise from a group of cells located between the cotyledons. The root apical meristem is located at the opposite side of the embryo.

In *Arabidopsis*, geneticists are also identifying genes that play a role during development. G. Jurgens and his colleagues at the University of Munchen in Germany have identified a category of genes, known as the *apical-basal-patterning genes*, that are important in early stages of development. For example, the *gurke* gene is necessary for apical development. When it is defective, the embryo lacks apical structures. The fackel gene influences the formation of the central region of the embryo. When it is defective, this region of the embryo is missing. The *monopterous* gene is necessary for the formation of basal structures. Finally, certain genes, like *gnom*, are necessary for the formation of the terminal regions of the embryo. When this gene is defective, these regions of the embryo do not form.

Plant Homeotic Genes Control Flower Development

Although the term homeotic was coined by William Bateson to describe homeotic mutations in animals, the first known homeotic genes were described in plants. In ancient Greece and Rome, for example, double flowers in which stamens were replaced by petals were noted. In current research, geneticists have been studying these types of mutations to better understand developmental pathways in plants. Many homeotic mutations affecting flower development have been identified in *Arabidopsis* and also in the snapdragon (*Antirrhinum majus*).

Examples in *Arabidopsis* are shown. Part (a) shows a normal *Arabidopsis* flower. It is composed of four concentric whorls of structures. The outer whorl contains four sepals, which protect the flower bud before it opens. The second whorl is composed of four petals, and the third whorl contains six stamens. The stamens are the structures that make the male gametophyte, pollen. Finally, the innermost whorl contains two carpels, which are fused together. The carpel produces the female gametophyte. The homeotic mutants have undergone transformations of particular whorls. For example, the sepals have been transformed into carpels, the petals into stamens.

By analyzing the effects of many different homeotic mutations in *Arabidopsis*, Elliot Meyerowitz at the California Institute of Technology and his colleagues have proposed the ABC model for flower development. In this model, three classes of genes, called A, B, and C, govern the formation of sepals, petals, stamens, and carpels. In the outer most whorl (whorl 1), the gene A products are made. This promotes sepal formation. In whorl 2, both gene A and gene B products are made, which promotes petal formation. In whorl 3, the expression of genes B and C causes stamens to be made, and in whorl 4, only gene C is expressed, which promotes carpel formation.

Now let's take a look at what happens in certain homeotic mutants. In this model, genes A and C repress each other's expression, and gene B functions independently. In a mutant defective in gene A expression, gene C will also be expressed in whorls 1 and 2. This produces a carpel—stamen—stamen—carpel arrangement. When gene B is defective, a flower cannot make petals or stamens. Therefore, a gene B defect yields a flower with a sepal—sepal—carpel—carpel arrangement. Finally, when gene C is defective, gene A is expressed in all four whorls. This results in a sepal—petal—petal—sepal pattern.

Overall, it appears that the types of genes promote cell differentiation that leads to either sepal, petal, stamen, or carpel structures. But what happens if all three types of genes are defective? These results indicate that the leaf structure is the default pathway and that the A, B, and C genes cause development to deviate from a leaf structure in order to make something else. In this regard, the sepals, petals, stamens, and carpels can be viewed as modified leaves.

In *Arabidopsis*, there are two types of gene A (*apetala 1* and *apetala* 2), two types of gene B (*apetala 3* and *pistillata*), and one type of gene C (*agamous*). All of these plant homeotic genes encode transcription factor proteins. The proteins contain a DNA-binding domain and a dimerization domain. However, the Arabidopsis homeotic genes do not contain a sequence similar to the homeobox found in animal homeotic genes. Instead, most of them (except for apetala2) contain a *MADS box* (an acronym of the first four plant genes of this type that were identified: MCM1, AG, DEF, and SRF). In the transcription factor proteins, the MADS domain promotes the binding of the transcription factor to specific DNA sequences.

Like the *Drosophila* homeotic genes, plant homeotic genes are thought to be part of a hierarchy of gene regulation. Genes that are expressed within the flower bud primordium produce proteins that activate the expression of these homeotic genes. Once they are transcriptionally activated, the homeotic genes then regulate the expression of realizator genes, the products of which promote the formation of sepals, petals, stamens, or carpels.

Concluding Remark

In this chapter, we have examined the role genetics plays in the *development* of multicellular organisms. Each organism has its own developmental program, driven by a hierarchy of genes that encode transcription factors and other types of regulatory proteins. In *Drosophila*, the developmental program begins in the oocyte. It is here that a few key gene products act as morphogens that are asymmetrically located or activated. This provides *positional information* that initiates development of the body plan or pattern in this organism. The maternal effect gene products serve to organize the *anteroposterior* and *dorsoventral* axes soon after fertilization. Once these axes have been established, the next step is to divide the embryo into *segments*.

Three categories of *segmentation* genes, known as *gap genes*, *pair-rule genes*, and *segment-polarity genes*, act sequentially to promote segmentation. The pair-rule genes control the identity of each *parasegment*, and the segment-polarity genes divide each parasegment into anterior and posterior regions. A segment, which is a morphological feature, is composed of the anterior region of one parasegment and the posterior region of an adjacent parasegment. The role of *homeotic genes* is to dictate the morphological characteristics of each segment. To do so, they activate *realizator genes*, which express proteins that determine the morphological characteristics of cells.

In *C. elegans*, the availability of a *fate map* has allowed geneticists to identify *heterochronic mutations* that alter the timing of *cell fate*. These mutations lead to abnormalities in morphology. This phenomenon illustrates the importance of coordinated cell division in determining cell fate during development.

In vertebrates, much less is known about the programs that promote development. Nevertheless, they appear to bear many similarities to the programs in simpler invertebrates. For example, the *Hox complex* in mice contains groups of genes that are homologous to homeotic genes in *Drosophila*. Furthermore, experiments with *gene knockouts* of particular *Hox* genes suggest that they act as homeotic genes in vertebrates.

A well-studied feature of vertebrate development is cell *differentiation*. Researchers have identified genes involved in the differentiation of undifferentiated cells into highly specialized cell types. For example, the *myogenic bHLH proteins* play a critical role in the differentiation of skeletal muscle cells.

Morphologically, plant development differs markedly from animal development. Plants are organized along a root—shoot axis and growth occurs along *apical meristems*. Plant geneticists have identified some of the genes that play a role in the general organization of the plant embryo. Geneticists have also identified many homeotic genes that dictate the morphology of particular structures later in development. As was described in this chapter, three classes of homeotic genes, genes A, B, and C, are involved in the formation of the four whorls that make up a flower. These results show that plant development is dictated by a genetic program that bears some similarities to those found in animals.

In this chapter, we have seen that a genetic approach (namely, the identification of mutant alleles) has been essential in our understanding of development. Researchers, particularly those studying invertebrates, have identified many different mutant alleles that alter key steps in the developmental process. In *Drosophila*, for example, these include mutations in maternal effect genes, segmentation genes, and homeotic genes. The molecular characterization of these genes has shown that many of them encode transcription factors that influence the expression of other genes. The study of developmental mutations has enabled scientists to piece together a genetic hierarchy that underlies many developmental programs.

The characterization of developmental mutants has also been correlated with the spatial localization of gene products using techniques such as in situ hybridization. This enables researchers to determine when and where a gene product is made during oogenesis or embryogenesis. This approach has shown that some mutations that alter development are gain of function alleles, which cause a gene to be expressed in the wrong place or at the wrong time. Other mutations are loss of function alleles, which cause a defect in gene expression. In *C. elegans*, the availability of a fate map has also allowed researchers to examine in detail how mutations (e.g., gain of function and loss of function alleles) can affect the timing of developmental steps.

The key experimental advantage of studying invertebrates is the ability to identify mutations that alter developmental steps. This approach is not as easy in vertebrates, but a reverse genetic approach is proving to be successful. Using invertebrate genes as probes, researchers have identified vertebrate genes, such as the Hox genes in mice, that are homologous to the homeotic genes of invertebrates. To understand their role, the wild-type Hox genes have been mutated in vitro and reintroduced into mice. The results of this method are consistent with the idea that the Hox genes function as the mouse homologues of the *Drosophila* homeotic genes. However, further research will be needed to understand the individual roles of these 38 genes.

In plants, a genetic approach is also proving effective in elucidating the development process. Early in development, a category of genes, known as the apical-basal-patterning genes, appear necessary for the formation of the apical and basal regions of the embryo. These genes were identified by investigating the effects of mutations on the developing plant embryo. Similarly, mutations in plant homeotic genes have been discovered by their ability to abnormally transform one plant structure into another (e.g., sepals into petals). A comparison of single, double, and triple homeotic mutations has provided the framework for the ABC model of flower development.

11

POPULATION GENETICS

The central issue in *population genetics* is genetic variation. Population geneticists want to know the extent of genetic variation within populations, why it exists, and how it changes over the course of many generations. Population genetics emerged as a branch of genetics in the 1920s and 1930s. Its mathematical foundations were developed by theoreticians who extended the principles of Mendel and Darwin by deriving formulae to explain the occurrence of genotypes within populations. These foundations can be attributed largely to three individuals: Sir Ronald Fisher, at the Rothamsted Experimental Station at Harpenden, England; Sewall Wright at Cold Spring Harbor; and J.B.S.Haldane at Cambridge University. As we will see, support for their mathematical theories was provided by several researchers who analyzed the genetic composition of natural and experimental populations. More recently, population geneticists have used laboratory techniques to probe genetic variation at the molecular level. In addition, the staggering improvement in computer technology has aided population geneticists in the analysis of their genetic theories and data.

GENES IN POPULATIONS

Population genetics may seem like a significant departure from other topics in this text, but it is a direct extension of our understanding of Mendel's laws of inheritance, molecular genetics, and the ideas of Darwin. The focus is shifted away from the individual and toward the population of which the individual is a member. Conceptually, all the genes in a population make up the *gene pool*, defined as the totality of all genes within a particular population. In this regard, each member of the population is viewed as receiving its genes from the gene pool.

Furthermore, if an individual reproduces, it contributes to the gene pool of the next generation. Population geneticists study the genetic variation within the gene pool, and how this variation changes from one generation to the next. In this introductory section, we will examine some of the general features of populations and gene pools.

Population is a Group of Interbreeding Individuals who Share a Gene Pool

In genetics, the term population has a very specific meaning. A *population* is a group of individuals of the same species that can interbreed with one another. Many species occupy a wide geographic range and arc divided into discrete populations. For example, distinct populations of a given species may be located on different continents.

A large population usually is composed of smaller groups called *subpopulations*, *local populations*, or *demes*. The members of a subpopulation are far likelier to breed among themselves than with other members of the general population. Sub-populations are often separated from each other by moderate geographic barriers. As shown here, two populations of Douglas fir (*Pseudotsuga menziesii*) are separated by a wide river bottom, where the fir trees are unlikely to grow. The groups of trees on opposite sides of this river bottom constitute local populations. Interbreeding is much more apt to occur among members of each local population than between members of neighbouring populations. On relatively rare occasions, however, pollen can be blown across the river bottom, which allows interbreeding between these two local populations.

Populations typically are dynamic units that change from one generation to the next. A population may change its size, geographic location, and genetic composition. With regard to size, natural populations commonly go through cycles of "feast or famine during which the population swells or shrinks. In addition, natural predators or disease may periodically decrease the size of a population to significantly lower levels; the population later may rebound to its original size. Populations or individuals within populations may migrate to a new site and establish a distinct population in this location. This new geographic location may differ in environment from the original site. As population sizes and locations change, their genetic composition generally changes as well. As will be described later in this chapter, population geneticists have developed mathematical theories that predict how the gene pool will change in response to fluctuations in size, migration, and new environments.

Some Genes are Monomorphic, and Others are Polymorphic

In population genetics, the term *polymorphism* (meaning many forms) refers to the observation that many traits display variation within a population. Historically, polymorphism first referred to the variation in traits that are observable with the naked eye. Polymorphisms in colour and pattern have long attracted the attention of population geneticists. These include studies of melanism in the peppered moth and of variation in snail colour, which are discussed later in this chapter. All of the individuals shown in this figure are from the same species, but they differ in alleles that affect colour and pattern.

At the DNA level, polymorphism is due to two or more alleles that influence the phenotype of the individual who inherits them. In other words, it is due to genetic variation. Geneticists also use the term polymorphic to describe the variation in genes that govern a polymorphic trait. A gene that commonly exists as two or more alleles in a population is described as polymorphic. By comparison, a *monomorphic* gene exists predominantly as a single allele in a population. By convention, when a single allele is found in at least 99% of all instances of a gene, the gene is considered monomorphic. To be judged polymorphic, a gene must have one or more additional alleles that make up at least 1% of the alleles in the population.

During the 1960s, molecular techniques became available that could assess variation in alleles that influence enzyme structure. In 1966, John Hubby and Richard Lewontin, at the University of Chicago, studied allelic variation in populations of the fruit fly *Drosophila pseudoobscura*. This species of fruit fly was chosen because (unlike *D. melanogaster*) *D. pseudoobscura* does not share its habitat with humans, and so it is considered truly wild. In addition, it has a wide geographic distribution in North and Central America. Hubby and Lewontin studied many structural genes that were known to encode enzymes. At the molecular level, genetic variation in a structural gene may result in the production of an enzyme with slight differences in its amino acid sequence compared with the wild-type enzyme. Based on previous work, Hubby and Lewontin knew that these differences in amino acid sequence can be detected as changes in mobility of the enzymes during gel electrophoresis. Two enzymes with alterations in their gel mobilities due to small differences in their amino acid sequences are called *allozymes*. In other words, they are alleles of the same *enzyme*.

In their study, Hubby and Lewontin looked for genetic variation in 18 different enzymes in five different fruit fly populations. On average, 30% of the enzymes were found as two or more allozymes. This means that the genes encoding these enzymes had slightly different DNA sequences; these differences resulted in alleles that encoded slightly different amino acid sequences. This study tends to underestimate genetic variability, since some amino acid substitutions do not alter protein mobility during gel electrophoresis. Nevertheless, the important conclusion from this work was that the genetic variability within these populations is quite high. Around the time of this work, Harry Harris of University College in London also found that approximately 30% of human genes encoding enzymes are polymorphic. More recently, many population geneticists have investigated genetic variation using DNA sequencing methods.

In most natural populations, a substantial percentage of genes are polymorphic.. However, in small populations that are near extinction, genetic variation is expected to be low, since the gene pool is derived from a small number of individuals. An extreme example is the African cheetah population, which has a genetic variation near zero. Later in this chapter, we will examine how small population size contributed to this problem. By comparison, approximately 30% of human genes are polymorphic. Since humans have approximately 100,000 different genes, this means that roughly 30,000 genes can be found in two or more different alleles.

Keep in mind that polymorphism refers to the diversity of genes in a population. Within a single individual, genetic variation will be less, because an individual may be homozygous for polymorphic genes. For example, the ABO blood type is determined by three alleles, designated I^A, I^B, and *i*. A person with type *O* blood is homozygous, ii, and therefore has less genetic variation than the population as a whole. Inhuman populations, about 30% of all genes are polymorphic, but within any individual less than 10% of all genes are heterozygous.

Population Genetics is Concerned with Allele and Genotype Frequencies

As we have seen, population geneticists want to understand the prevalence of polymorphic genes within populations. Much of their work involves evaluating the prevalence of genes in a quantitative way. Two fundamental calculations are central to population genetics: *allele frequencies* and *genotype frequencies*. The allele and genotype frequencies are defined as

$$\text{allele frequency} = \frac{\text{number of copies of an allele in a population}}{\text{total number of all alleles for that gene in a population}}$$

$$\text{genotype frequency} = \frac{\text{No. of individual with particular genotype in population}}{\text{total number of individuals in a population}}$$

Though these two frequencies are related, a clear distinction between them must be kept in mind. As an example, let's consider a population of 100 pea plants with the following genotypes:

64 tall plants with the genotype *TT*

32 tall plants with the genotype *Tt*

4 dwarf plants with the genotype *tt*

When calculating an allele frequency, homozygous individuals have two copies of an allele, whereas heterozygote only have one. For example, in tallying the *t* allele, each of the 32 heterozygotes has one copy of the *t* allele, and each dwarf plant has two copies. The allele frequency for *t* equals

$$t = \frac{32 + (2)(4)}{(2)(64) + (2)(32) + (2)(4)}$$

$$= \frac{40}{200} = 0.2, \text{ or } 20\%$$

This result tells us that the allele frequency of *t* is 20%. In other words, 20% of the alleles for this gene in the population are the t allele.

Let's now calculate the genotype frequency of tt (dwarf) plants:

$$tt = \frac{4}{64 + 32 + 4}$$

$$= \frac{4}{100} = 0.04, \text{ or } 4\%$$

We see that 4% of the individuals in this population are dwarf plants.

Allele and genotype frequencies are always less than or equal to 1 (i.e., less than or equal to 100%). If a gene is monomorphic, the allele frequency for the single allele will equal a value of 1.0. For polymorphic genes, if we add up the frequencies for all of the alleles in the population, we should obtain a value of 1.0. In our pea plant example, the allele frequency of t equals 0.2. The frequency of the other allele, *T*, equals 0.8. If we add the two together, we obtain a value of 0.2 + 0.8 = 1.0. In the next section, we will learn how allele and genotype frequencies within populations are related.

HARDY-WEINBERG EQUILIBRIUM

Now that we have a general understanding of genes in populations, we can begin to relate these concepts to mathematical expressions in order to examine whether allele and genotype frequencies will change over the course of many generations. In 1908, Godfrey Harold Hardy, professor of mathematics at Cambridge University; and Wilhelm Weinberg, a physician in Stuttgart, Germany, independently derived a simple mathematical expression that predicted stability of allele and genotype frequencies from one generation to the next. This expression, known as the *Hardy—Weinberg equation*, relates allele and genotype frequencies within a population. It is also called an equilibrium, because the allele and genotype frequencies do not change over the course of many generations. This relationship established a framework on which to understand genetic stability.

In subsequent decades, as the field of population genetics developed, it became apparent that genetic stability is not always present in natural populations. Furthermore, genetic change underlies the theory of evolution. Therefore, in the 1920s and 1930s, population geneticists turned their efforts largely toward understanding how genetic stability is altered. In this section, we will examine the Hardy—Weinberg equilibrium and the conditions that must be met for it to be valid. At the end of this chapter, we will learn how natural populations usually violate the Hardy—Weinberg equilibrium and thereby cause allele frequencies to change.

Hardy—Weinberg Equation can be Used to Calculate Genotype Frequencies Based on Allele Frequencies

The Hardy—Weinberg equation is a simple mathematical expression that relates genotype and allele frequencies. Let's first examine the mathematical components of the Hardy—Weinberg equation, and then we will look at the conditions necessary an equilibrium to be achieved.

Let's begin by considering a situation in which a gene is polymorphic and exists as two different alleles, *A* and *a*. If the allele frequency of *A* is denoted by the variable p, and the allele frequency of *a* by q, then:

$$p + q = 1$$

For example, if $p = 0.8$, then q must be 0.2. In other words, if the allele frequency *A* equals 80%, the remaining 20% of alleles must be *a*, because together they equal 100%.

The Hardy—Weinberg equation states that:

$$p^2 + 2pq + q^2 = 1 \text{ (Hardly-Weinbreg equation)}$$

If this equation is applied to a gene that exists in alleles designated A and a, then

p^2 equals the genotype frequency of *AA*

$2pq$ equals the genotype frequency of *Aa*

q^2 equals the genotype frequency of *aa*

If $p = 0.8$ and $q = 0.2$, then

$$AA = p^2 = (0.8)^2 = 0.64$$

$$Aa = 2pq = 2(0.8)(0.2) = 0.32$$

$$aa = q^2 = (0.2)^2 = 0.04$$

In other words, if the allele frequency of *A* is 80% and the allele frequency of *a* is the genotype frequency of *AA* is 64%,*Aa* 32%, and *aa* 4%.

To see the relationship between allele frequencies and genotypes, compares the Hardy—Weinberg equation with the Punnett square approach. As seen here, the Hardy—Weinberg equation results from the way gametes combine randomly with each other to produce offspring. In a population, the frequency of a gamete carrying a particular allele is equal to the allele frequency in that population. For example, the frequency of a gamete carrying the *A* allele equals 0.8.

We can use the product rule to determine the frequency of genotypes. For example, the frequency of producing an *AA* homozygote is $0.8 \times 0.8 = 0.64$, or 64%. Likewise, the probability of inheriting the a allele is $0.2 \times 0.2 = 0.04$, or 4%. In our Punnett square, there are two different ways to produce heterozygotes. An offspring could inherit the A allele from its father and a from its mother, or A from its mother and a from its father. Therefore, the frequency of heterozygotes is $pq + pq$, which equals $2pq$; in our example, this is $2(0.8)\ (0.2) = 0.32$, or 32%.

The Hardy—Weinberg equation predicts an equilibrium (i.e., unchanging allele and genotype frequencies) if a certain set of conditions are met in a population. These are as follows:

1. The population is so large that allele frequencies do not change due to random sampling effects.
2. The members of the population mate with each other without regard to their phenotypes and genotypes.
3. There is no migration between different populations.
4. There is no survival or reproductive advantage for any of the genotypes. In other words, no natural selection occurs.
5. No new mutations occur within the population.

According to the equilibrium, the Hardy—Weinberg equation provides a quantitative relationship between allele and genotype

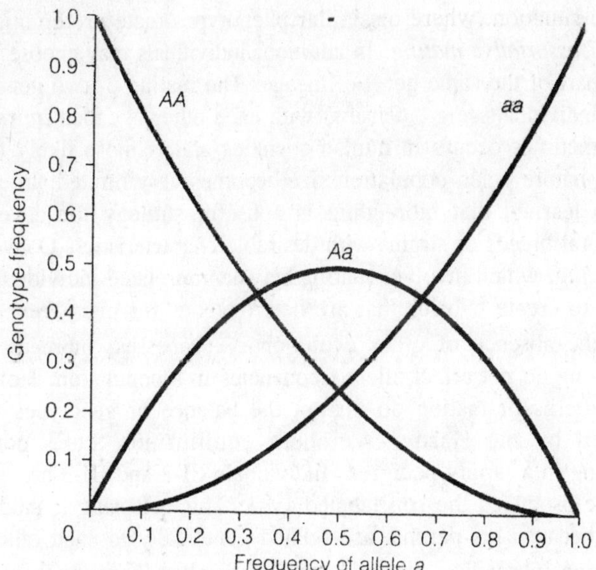

Fig. 11.1. The relationship between allele frequencies and genotype frequencies according to Hardy-Weinberg equilibrium.

frequencies in a population. As expected, when the allele frequency of *A* is very low, the *aa* genotype predominates; when the *A* allele frequency is high, the *AA* homozygote is most frequent in the population. When the allele frequencies of A and a are intermediate in value, the heterozygote predominates.

In reality, no population satisfies the Hardy—Weinberg equilibrium completely. Nevertheless, in large natural populations with little migration and negligible natural selection, the Hardy—Weinberg equilibrium may be nearly approximated. In addition, as discussed later in this chapter, deviations from the Hardy—Weinberg equilibrium help us understand how populations are changing from one generation to the next.

Nonrandom Mating may Occur in Natural and Human Populations

As mentioned earlier, one of the conditions required to establish the Hardy—Weinberg equilibrium is random mating. This means that individuals choose their mates irrespective of their genotypes and phenotypes. In many cases, particularly human populations, this condition is violated frequently.

When two individuals are more likely to mate due to similar phenotypic characteristics, this is known as *assortative mating*. The opposite situation, where dissimilar phenotypes mate preferentially, is called *disassortative mating*. In addition, individuals may choose a mate who is part of the same genetic lineage. The mating of two genetically related individuals (e.g., cousins) with each other is called *inbreeding*. This sometimes occurs in human societies and is more likely to take place in nature when population size becomes very limited. In earlier, we also learned that inbreeding is a useful strategy for developing agricultural breeds or strains with desirable characteristics. Conversely, *outbreeding*, which involves mating between unrelated individuals, can be used to create hybrids that are beterozygous for many genes.

In the absence of other evolutionary processes, inbreeding and outbreeding do not affect allele frequencies in a population. However, these patterns of mating do disrupt the balance of genotypes that is predicted by the Hardy—Weinberg equilibrium. Let's consider inbreeding in a family pedigree. Individuals III-2 and III-3 are cousins and have produced the son labeled IV-1. This offspring is said to be *inbred*, because his parents are related genetically to each other.

During inbreeding, the gene pool is smaller, because the parents are related genetically. In the 1910s to 1930s, Wright and Fisher developed methods to quantify the degree of inbreeding. A *coefficient of inbreeding* (F) can be computed by analyzing the degree of relatedness within a pedigree. As an example, let's find the coefficient of inbreeding for individual IV-1. To begin this problem, we must first identify all of the common ancestors that this individual has. A *common ancestor* is anyone who is an ancestor to both of an individual's parents.

Our next step is to determine the inbreeding paths. An *inbreeding path* for an individual is the shortest path through the pedigree that includes both parents and the common ancestor. In a pedigree, there

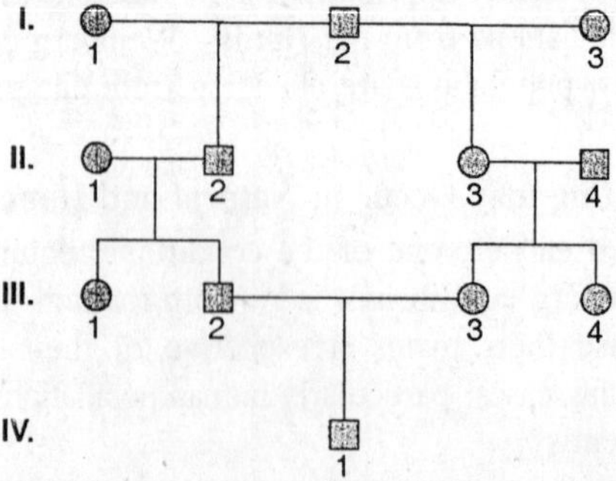

Fig. 11.2. A human pedigree containing inbreeding. Individual IV-1 is the result of a consanguineous mating.

is an inbreeding path for each common ancestor. The length of each inbreeding path then is calculated by adding together all of the individuals in the path except the individual of interest. In this case, there is only one path, since there is only one common ancestor. To add the members of the path, we begin with individual IV-1, but we do not count him. We then move to his father (III-2), to his grandfather (II-2), to I-2, his great-grandfather (the common ancestor), and back down to his other grandmother (II-3), and finally to his mother (III-3). This path has five members in it. Finally, to calculate the in breeding coefficient, we use the following formula:

$$F = \Sigma(1/2)^n (1 + F_A)$$

where,

- F is the inbreeding coefficient of the individual of interest
- n is the number of individuals in the inbreeding path, excluding the inbred offspring
- F_A is the inbreeding coefficient of the common ancestor
- Σ indicates that we add together $(1/2)^n (1 + F_A)$ for each inbreeding path

In this case, there is only one common ancestor. Also, we do not know anything about the heritage of the common ancestor, and so we assume that F_A is zero.

$$F = \Sigma(1/2)^n (1 + 0)$$
$$= (1/2)^5 = 1/32 = 3.125\%$$

Our inbreeding coefficient, 3.125%, tells us the probability that a gene in the inbred individual (IV- 1) is homozygous due to its inheritance from a common ancestor (I-2). In this case, therefore, each gene in individual IV-1 has a 3.125% chance of being homozygous because he has inherited the same allele twice from his great-grandfather (I-2), once through each parent. For example, let's suppose that the common ancestor (I-2) is heterozygous for the allele that causes cystic fibrosis. There is a 3.125% probability that the inbred individual (IV-1) is homozygous for this gene because he has inherited both copies from his great-grandfather. He has a 1.56% probability of inheriting the normal allele twice and a 1.56% probability of inheriting the mutant allele twice. The coefficient of inbreeding is also called the *fixation coefficient*. That is why the inbreeding coefficient is denoted by the letter F (for Fixation). It is the probability that an allele will be fixed in the homozygous condition. The term fixation signifies that the homozygous individuals can pass only one type of allele to their offspring.

In other pedigrees, an individual may have two or more common ancestors. In this case, the inbreeding coefficient F is calculated as the sum of the inbreeding paths.

Inbreeding can have both positive and negative consequences in a population. From an agricultural viewpoint, it results in a higher proportion of homozygotes, which may exhibit a desirable trait. For example, an animal breeder may use in breeding to produce animals that are larger because they have become homozygous for alleles promoting larger size. On the negative side, many genetic diseases are inherited in a recessive manner. For these rare recessive disorders, inbreeding increases the likelihood that an individual will be homozygous and therefore afflicted with the disease.

Factors that Change Allele Frequencies in Population

The Hardy—Weinberg equilibrium and the high prevalence of polymorphisms within natural populations seem, at first glance, to contradict each other. On the one hand, the Hardy—Weinberg equilibrium predicts that allele frequencies will not change from one generation to the next. On the other hand, research has shown that many genes are polymorphic, indicating that two or more alleles can attain a significant percentage within a population. This leads to a central question in population genetics: How do genetic polymorphisms originate, and why are they maintained in natural populations?

For very rare, detrimental alleles, the explanation is straightforward. Random mutations are far likelier to produce deleterious alleles than beneficial ones, because the intricate functioning of proteins will almost certainly be disrupted rather than improved by a haphazard change. Therefore, mutation tends to favour the accumulation of deleterious alleles in a population. However, as we will learn in this section, natural selection promotes the elimination of harmful alleles in a population. So, in a population, the low rate of mutation produces detrimental alleles, and the process of natural selection works to eliminate them. The opposing actions of these two processes maintain detrimental alleles at very low frequencies.

However, many genetic polymorphisms exist in which two or more alleles are found at substantial frequencies in a population. In these cases, the allele percent ages commonly observed in genetic polymorphisms cannot be explained by new mutations; the allele frequencies are too high, the rate of new mutations too low. For example, let's use our previous example of a polymorphic gene existing in alleles *A* and *a* at respective frequencies of 0.8 and 0.2. One possibility is that this current population is derived from an ancestral population that was originally monomorphic, containing only the *A* allele in its gene pool. Over time, new mutations could have converted the *A* allele into the *a* allele in members of the population. However, the present-day frequency of 0.2 for the a allele is unlikely to be merely the result of the gradual accumulation of new mutations of *A* to *a*, because the rate of such mutations is exceedingly low. Therefore, though new mutations provide a source of allelic variation, they do not occur fast enough to generate the allele frequencies observed in many polymorphisms. After that new mutation occurred, other evolutionary forces must have increased the frequency of the a allele from near zero to a frequency of 0.2. But what are these other evolutionary forces, and how do they work?

We can divide evolutionary processes into two categories: neutral and adaptive. A major debate in population genetics centers on which of these two types of processes accounts for most of the genetic variation in natural populations. *Neutral processes* (or neutral forces) alter allele frequencies in a random manner. In other words, neutral processes change allele frequencies without any regard to the survival of the individual. They can occur as a matter of chance. The two key neutral processes are migration and genetic drift.

By comparison, *adaptive forces* are processes in which alleles that confer greater survival or reproductive success increase in frequency. Natural selection produces individuals that are well adapted to their environment. Those individuals who happen to possess these beneficial alleles are more likely to survive, reproduce, and contribute to the gene pool of the next generation. In this section, we will ex amine the contributions of neutral and adaptive processes in creating and maintaining genetic variation.

Mutations Provide the Source of Genetic Variation

Mutations can involve changes in gene sequences, chromosome structure, and/or chromosome number. They are random events that can occur spontaneously at a low rate, or can be caused by mutagens at a higher rate. In this chapter, we will be concerned primarily with the effects of gene mutations on the frequency of new alleles within a population.

In 1926, the Russian geneticist Sergei Tshetverikov was the first to suggest that mutational variability provides the raw material for evolution, but does not constitute evolution itself. In other words, mutation can provide new alleles to a population but does not act as a major force in dictating the final balance of allele frequencies. Tshetverikov concluded that populations in nature absorb mutations "like a sponge" and retain them in a heterozygous condition, thereby providing a source of variability for future change.

Table 11.1. Processes that alter frequencies

Mutation	Mutation introduces new alleles into populations, but at a very low rate. New mutations may be beneficial, neutral, or deleterious. For alleles to rise to a significant percentage in a population, it is necessary for other evolutionary processes (namely, genetic drift, migration, and/or natural selection) to operate.
Neutral Forces	
Random genetic drift	A change in allele frequencies due to random survival and drift reproduction. In other words, allele frequencies may change as a matter of chance from one generation to the next. This occurs to a greater degree in smaller populations.
Migration	Migration can occur between two populations having different allele frequencies. The introduction of migrants into the recipient population may change the latter's allele frequencies.
Adaptive Forces	
Natural selection	Natural selection is the enhanced survival or reproduction of individuals having genotypes that confer beneficial phenotypes. Likewise, natural selection is also the diminished survival and reproduction of members with unfavourable phenotypes.

In population genetics, it is most useful to consider new mutations as they affect the survival and reproductive potential of the individual who inherits them. In this regard, a new mutation may be beneficial, neutral, or deleterious. For structural genes that encode proteins, the effects of new mutations will depend on their impact on protein function. Neutral and deleterious mutations are far likelier to occur than beneficial mutations. For example, alleles can be altered in many different ways that render an encoded protein defective. Deletions, frameshift mutations, missense mutations, and nonsense mutations all may cause a gene to express a protein that is non functional or less functional than the wild-type protein. Neutral mutations can also occur in several different ways. For example, a neutral mutation can change the wobble base without affecting the amino acid sequence of the encoded protein, or it can be a missense mutation that has no effect on protein function. Such mutations are point mutations at specific sites within the coding sequence. Neutral mutations can also occur within non coding sequences of genes, such as introns. By comparison, beneficial mutations are relatively uncommon. To be advantageous, a new mutation could alter the amino acid sequence of a protein to yield a better-functioning product. While such mutations do occur, they are expected to be very rare for a population in a stable environment.

The *mutation rate* is the probability that a gene will be altered by a new mutation. It is expressed as the number of new mutations in a given gene per generation. It is commonly thought to be in the range of 1 in 100,000 to 1 in 1,000,000, or 10^{-5} to 10^{-6} per generation. However, studies of mutation rates typically follow the change of a normal (functional) gene to a deleterious (nonfunctional) allele. The mutation rate producing beneficial alleles is expected to be substantially less.

It is clear that new mutations provide genetic variability. Population geneticists also want to know how the mutation rate affects the allele frequencies in a population over time. To appreciate this idea, let's take the simple case where a gene exists in a functional allele, *A*; the allele frequency of *A* is denoted by the variable *p*. A deleterious mutation can convert the *A* allele into a nonfunctional allele, *a*. The allele frequency of *a* is designated by *q*. The conversion of the A allele into the *a* allele by mutation will occur at a rate that we call *u*. If we assume that the rate of the reverse mutation (*a* to *A*) is negligible, the increase in the frequency of the a allele after one generation will be

$$\Delta q = up$$

For example lets consider the following condition

p = 0.8(i.e frequency of A is 80%)

q = 0.2(i.e frequency of A is 80%)

u = 10^{-5} (i.e the mutation rate of converting A to a)

$\Delta q = (10^{-5})(0.8) = (0.00001)(0.8) = 0.000008$

Therefore in the next generation

$$q_{n+1} + 1 = 0.2 + 0.000008 = 0.200008$$

$$p_{n+1} + 1 = 0.8 - 0.000008 = 0.799992$$

As we can see from this calculation, new mutations do not significantly alter the allele frequencies in a single generation.

We can use the following equation to calculate the change in allele frequency after any number of generations:

$$(1-u)^t = \frac{p_t}{p_0}$$

where,

u is the mutation rate of the conversion of A to a

t is the number of generations

p_0 is the allele frequency of A in the starting generation

p_t is the allele frequency of A after t generations

As an example, let's suppose that the allele frequency of *A* is 0.8, $u = 10^{-5}$ and we want to know what the allele frequency will be after 1000 generations ($t = 1000$). Plugging these values into the preceding equation and solving for P_t,

$$(1-0.00001)^{100} = \frac{p_t}{0.8}$$

$$p_t = 0.792$$

Therefore, after 1000 generations the frequency of A has dropped only from 0.8 to 0.792. Again, these results point to how slowly the mutation rate effects changes in allele frequencies. In natural populations, the rate of new mutation is rarely a significant catalyst in shaping allele frequencies. Instead, other processes such as migration, genetic drift, and natural selection have far greater effects on allele frequencies. We will examine how these other factors work in the remainder of this chapter.

In Small Populations, Allele Frequencies can be Altered Random Genetic Drift

During the 1930s, Sewall Wright played a large role in developing the concept of *random genetic drift*, which refers to random changes in allele frequencies due to ampling error. In other words, allele frequencies may drift from generation to generation as a matter of chance. This situation is similar to flipping a coin. If we lipped a coin 10 times, we might get 4 heads and 6 tails; the frequency of heads is 0.4. If we do it again, we might get 5 heads and 5 tails for a frequency of 0.5. The frequency of heads is drifting due to random sampling error.

In populations, genetic drift favours either the loss or the fixation of an allele. The rate at which an allele is lost or becomes fixed at 100% depends on the population size. At the beginning of this hypothetical simulation, all of these populations have identical allele frequencies: A = 0.5 and a –0.5. In the five small populations, this allele frequency fluctuates substantially from generation to generation. Eventually, in these small populations, one of the alleles is eliminated and the other is fixed at 100%. At this point, an allele has become monomorphic and cannot fluctuate any further. By comparison, the allele frequencies in the large population fluctuate much less, since random sampling error is expected to have a smaller impact on allele frequencies. Nevertheless, genetic drift will lead to homozygosity even in large populations, but this will take many more generations to occur.

An important controversy in population genetics centers on whether most genetic variation is due to the accumulation of neutral mutations by genetic drift, or to mutations with a survival advantage that are promoted by natural selection. The DNA sequencing of genes has revealed that both processes have been important in the evolution of species.

Now let's ask two questions:

1. How many new mutations do we expect in a natural population?

2. How likely is it that a new mutation will be either fixed in or eliminated from a population due to random genetic drift?

With regard to the first question, the average number of new mutations depends on the mutation rate (u) and the number of individuals in a population (N). If each individual has two copies of the gene of interest, then the expected number of new mutations in this gene is:

$$\text{expected number of new mutations} = 2Nu$$

From this, we can see that a new mutation is more likely to occur in a large population than in a small one. This makes sense, since the larger population has more copies of the gene to be mutated. If a new mutation does occur, it may be eliminated or fixed in a population due to random genetic drift. If a mutation is neutral, the probability of fixation of a newly arising allele is:

$$\text{probability of fixation} = 1/2N$$ (assuming equal numbers of males and females contribute to the next generation)

In other words, the probability of fixation is the same as the allele frequency in the population. For example, if $N = 20$, the probability of fixation equals $1/(2 \times 20)$, or 2.5%. Conversely, a new allele may be lost from the population:

$$\text{probability of elimination} = 1 - \text{probability of fixation}$$
$$= 1 - 1/2N$$

As you may have noticed, the value of N has opposing effects with regard to mutations and their eventual fixation in a population. When N is very large, new mutations are much more likely to occur. Each new mutation, however, has a greater chance of being eliminated from the population due to random genetic drift. On the other hand, when N is small, the probability of new mutations is also small, but if they occur, the likelihood of fixation is relatively large.

Now that we have an appreciation for the phenomenon of genetic drift, we can ask a third question:

3. If fixation does occur, how many generations is it likely to take?

Again, this will depend on the number of individuals in the population:

$$\bar{t} \approx 4N$$

where

$\bar{t}$ equals the average number of generations to achieve fixation

N equals the number of individuals in the population, assuming that males and females contribute equally to each succeeding generation

As you may have expected, allele fixation will take much longer in large populations. If there are 1 million breeding members in a population, it will take 4 million generations, perhaps an insurmountable period of time, to reach fixation. In small group of 100 individuals, however, fixation will only take 400 generations. The drifting of neutral alleles among different populations and species provides a way to measure the rate of evolution and can be used to determine evolutionary relationships.

Our preceding discussion of random genetic drift has emphasized two important points. First, genetic drift ultimately operates in a directional manner with regard to allele frequency. It leads to either allele fixation or elimination. Second, its impact is more significant in smaller populations. In nature, there are several interesting ways that small population size and genetic drift affect the genetic composition of a species. For example, some species occupy wide ranges in which small papulation become geographically isolated from the rest of the species. The allele frequencies within these small populations are more susceptible to genetic drift. Since this is a random process, small isolated populations tend to become more genetically disparate in relation to other populations.

A second example of genetic drift is called the *bottleneck effect*. In nature, a population can be reduced dramatically in size by events such as earthquakes, floods, drought, and the human destruction of habitat. These types of events may randomly eliminate most of the members of the population without regard to genetic composition. The period of the bottleneck, when the population size is very small, may be influenced by genetic drift. First, the original surviving members may have allele frequencies that differ from those of the original population. In addition, allele frequencies are expected to drift substantially during the generations when the population size is small. In extreme cases, alleles may even be eliminated. Eventually, the bottlenecked population may regain its original size. However, the new population will have less genetic variation than the original one. As mentioned earlier in this chapter, the African cheetah population

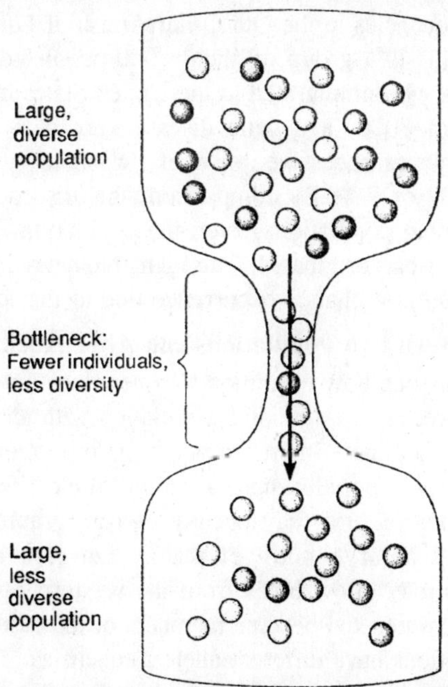

Fig. 11.3. A representation of the bottleneck effect.

has lost nearly all of its genetic variation. This is due to a bottleneck effect. An analysis by population geneticists has suggested that a severe bottleneck occurred approximately 10,000—12,000 years ago, reducing the population size to near extinction. The population eventually rebounded, but the bottleneck reduced the genetic variation to very low levels.

A third interesting case of genetic drift is the *founder effect*. This refers to the phenomenon in which a small group of individuals separates from a larger population and establishes a colony in a new location. For example, a few individuals may migrate away from a large continental population and become the founders of an island population. The founder effect has two important consequences. First, the founding population is expected to have less genetic variation than the original population from which it was derived. Second, as a matter of chance, the allele frequencies in the founding population may differ markedly from those of the original population.

Population geneticists have studied many examples where isolated populations have been derived via colonization of members of another population. In the 1960s, Victor McKusick of Johns Hopkins University studied allele frequencies in the Old Order Amish of Lancaster County, Pennsylvania. This is a group of about 8000 people, descended from just three couples who immigrated to the United States in 1770. Among this population of 8000, a genetic disease known as the Ellis—van Creveld syndrome (a recessive form of dwarfism) was found at a frequency of 0.07 or 7%. By comparison, this disorder is extremely rare in other human populations, even the population from which the founding members had originated. The high frequency in the Lancaster County population is a chance occurrence due to the founder effect

Migrations between two Populations can Alter Allele Frequencies

We have just seen how migration to a new location by a relatively small group can result in a founding population with an altered genetic composition due to genetic drift. In addition, migration between two different established populations can alter allele frequencies. For example, a species of bird may occupy two geographic regions that are separated by a large body of water. On rare occasions, the prevailing winds may allow birds from the western population to fly over this body of water and become members of the eastern population. If the two populations have different allele frequencies, and if migration occurs in sufficient numbers, this may alter the allele frequencies in the eastern population.

After migration has occurred, the new (eastern) population is called a *conglomerate*. To calculate the allele frequencies in the conglomerate, we need two kinds of information. First, we must know the original allele frequencies in the donor and recipient populations. Second, we must know the proportion of the conglomer ate population that is due to migrants. With these data, we can calculate the change in allele frequency in the conglomerate population using the following equation:

$$\Delta p_C = m(p_D - p_R)$$

where,

Δp_C is the change in allele frequency in the conglomerate population

p_D is the allele frequency in the donor population

p_R is the allele frequency in the original recipient population

m is the proportion of migrants that make up the conglomerate population:

$$m = \frac{\text{number of donor individuals in the conglomerate population}}{\text{total number of individuals in the conglomerate population}}$$

As an example, let's suppose the allele frequency of A is 0.7 in the donor population and 0.3 in the recipient population. A group of 20 individuals migrate and join the recipient population, which originally had 80 members. Thus,

$$m = \frac{20}{20 + 80}$$
$$= 0.2$$
$$\Delta p_c = m(P_D - P_R)$$
$$= 0.2(0.7 - 0.3)$$
$$= 0.08$$

We can now calculate the allele frequency in the conglomerate:

$$p_C = p_R + \Delta p_C$$
$$= 0.3 + 0.08 = 0.38$$

Therefore, in the conglomerate population, the allele frequency of *A* has changed from 0.3 (its value before migration) to 0.38. This increase in allele frequency arises from the higher allele frequency for A in the donor population. Since population geneticists are concerned with allele frequencies rather than the migration of individuals, this phenomenon is called *gene flow*. It occurs whenever individuals migrate between populations having different allele frequencies. In our previous example, we considered the consequences of a unidirectional migration from a donor to a recipient population. In nature, it is common for individuals to migrate in both directions. This bidirectional migration has two important consequences. Depending on its rate, migration tends to reduce differences in allele frequencies between neighbouring populations. In fact, population geneticists can analyze allele frequencies in two different populations to evaluate the rate of migration between them. Populations that frequently mix their gene pools via migration tend to have similar allele frequencies, whereas isolated populations are expected to be more disparate. In addition, migration can enhance genetic diversity within a population. As discussed earlier in this chapter, new mutations are relatively rare events. Therefore, a particular mutation may only arise in one population. Migration may then introduce this new allele into neighbouring populations.

Natural Selection Favours the Survival of the Fittest

In the 1850s, Charles Darwin and Alfred Russell Wallace independently proposed the theory of *natural selection*. According to this idea, there is a "struggle for existence." The conditions found in nature result in the selective survival and reproduction of individuals whose characteristics make them well adapted to their environment. These surviving individuals are more likely to re produce and contribute offspring to the next generation. More recently, population geneticists have realized that natural selection can be related not only to differential survival, but also to mating efficiency and fertility.

A modern description of natural selection can relate our knowledge of molecular genetics to the phenotypes of individuals:

1. Within a population, there is genetic variation arising from differences in DNA sequences. Distinct alleles may encode proteins of differing function.
2. Some alleles may encode proteins that enhance an individual's survival or reproductive capability as compared with other members of the population. For example, an allele may produce a protein that is more efficient at a higher temperature, conferring on the individual a greater probability of survival in a hot climate.
3. Individuals with beneficial alleles are more likely to survive and contribute to the gene pool of the next generation.
4. Over the course of many generations, allele frequencies of many different genes may change through natural selection, thereby significantly altering the characteristics of a species. The net result of natural selection is to make a population better adapted to its environment and/or more successful at reproduction.

Natural selection acts on phenotypes (which are derived from an individual's genotype). With regard to quantitative traits, there are three ways that natural selection may operate:

1. *Directional selection* favours the survival of one extreme of a phenotypic distribution that is better adapted to an environmental condition. For example, in a dimly lit forest, directional selection would favour snails with darker shells, because they would be better camouflaged and less susceptible to predation.
2. *Disruptive* (or *diversifying*) *selection* favors the survival of two (or more) different phenotypic classes of individuals. For example, disruptive selection favours the survival of dark-shelled and light-shelled snails that occupy a heterogeneous environment. Disruptive selection is likely to occur in species that occupy diverse environments so that some members of the species will survive in each type of environment.
3. *Stabilizing selection* favours the survival of individuals with intermediate phenotypes. For example, moderately coloured snails may survive better in an environment that has an intermediate level of light, such as a tall grassland.

During the 1920s and 1930s, Haldane, Fisher, and Wright developed mathematical relationships to explain the theory of natural selection. As our knowledge of natural selection increased, it became apparent that there are many different ways for natural selection to operate. In this text, we will consider a few examples of natural selection involving a single gene that exists in two alleles. In reality, however, natural selection acts on populations of individuals, in which many genes are polymorphic and each individual contains thousands or tens of thousands of different genes.

To begin our quantitative discussion of natural selection, we must examine the concept of *Darwinian fitness*. Fitness is the relative likelihood that a phenotype will survive and contribute to the gene pool of the next generation as compared with another phenotype. Although this property often correlates with physical fitness, the two ideas should not be confused. Darwinian fitness is a measure of reproductive superiority. An extremely fertile phenotype may have a higher Darwinian fitness than a less fertile phenotype that appears more physically fit. To consider Darwinian fitness in a quantitative way, let's use our example of two genes existing in the *A* and *a* alleles. We can assign fitness values to each of the three genotypic classes according to their relative reproductive potential. For example, let's suppose that in a population the average reproductive success of the three genotypes are as follows:

AA produce 5 offspring

Aa produce 4 offspring

aa produce 1 offspring

By convention, the genotype with the highest reproductive ability is given a fitness value of 1.0. Fitness values are denoted by the variable *W*. The fitness values of the other genotypes are assigned values relative to this 1.0 value:

$$\text{Fitness of AA: } W_{AA} = 1.0$$

$$\text{Fitness of Aa: } W_{Aa} = 4/5 = 0.8$$

$$\text{Fitness of aa: } W_{aa} = 1/5 = 0.2$$

Keep in mind that the differences in reproductive achievement among these three genotypes could stem from different reasons. A very common reason is that the most fit phenotype is more likely to survive. Of course, an individual must survive into adulthood to reproduce. A second possibility is that the most fit phenotype is more likely to mate. For example, a bird with brightly coloured feathers may have an easier time attracting a mate than a bird with duller plumage. Finally, a third possibility is that the fittest phenotype maybe more fertile. It may produce a higher number of gametes, or gametes that are more successful at fertilization.

An opposite parameter that population geneticists also use in their calculations is the *selection coefficient* (s). This measures the degree to which a genotype is selected against:

$$s = 1 - W$$

By convention, of the existing genotypes, the one with the highest fitness has an s value of zero. Genotypes at a selective disadvantage have s values that are greater an zero but less than or equal to 1.0. An extreme case is a recessive lethal allele. It could have an s value of 1.0 in the homozygote, while the s value in the heterozygote could be zero. Natural selection alters allele frequencies in a step-by-step, generation-per generation way. To appreciate how this occurs, let's take a look at how fitness affects the Hardy—Weinberg equilibrium and allele frequencies. Again, let's suppose that a gene exists in two alleles, A and a. The three fitness values are

$$W_{AA} = 1.0$$

$$W_{Aa} = 0.8$$

$$W_{aa} = 0.2$$

In the next generation, we expect that the Hardy—Weinberg equilibrium will be modified in the following way:

Frequency of AA: p^2W_{AA}

Frequency of Aa: $2pqW_{Aa}$

Frequency of aa: q^2W_{aa}

In a population that is changing due to natural selection, these three terms will not add up to 1.0, as they would in the Hardy—Weinberg equilibrium. Instead, the three terms sum to a value known as the *mean fitness of the population*:

$$p^2W_{AA} + 2pqW_{Aa} + q^2W_{aa} = \bar{W}$$

Dividing both sides of the equation by the mean fitness of the population,

$$\frac{p^2W_{AA}}{\bar{W}} + \frac{2pqW_{Aa}}{\bar{W}} + \frac{q^2W_{aa}}{\bar{W}} = 1$$

Using this equation, we can calculate the expected genotype and allele frequencies after one generation of natural selection:

Frequency of AA genotype: $\frac{p^2W_{AA}}{\bar{W}}$

Frequency of Aa genotype: $\frac{2pqW_{Aa}}{\bar{W}}$

Frequency of aa genotype: $\frac{q^2W_{aa}}{\bar{W}}$

Allele frequency of A: $P_A = \frac{p^2W_{AA}}{\bar{W}} + \frac{pqW_{Aa}}{\bar{W}}$

allele frequency of a: $q_a = \frac{q^2W_{aa}}{\bar{W}} + \frac{pqW_{Aa}}{\bar{W}}$

As an example, let's suppose that the starting allele frequencies are A = 0.5 and 0.5, and use fitness values of 1.0, 0.8, and 0.2 for the three genotypes, AA, Aa, aa, respectively. We begin by calculating the mean fitness of the population:

$$p^2W_{AA} + 2pqW_{Aa} + q^2W_{aa} = \bar{W}$$

$$\bar{W} = (0.5)2(1) + 2(0.5)(0.5)(0.8) + (0.5)2(0.2)$$

$$= 0.25 + 0.4 + 0.05 = 0.7$$

After one generation of selection, we get,

Frequency of AA genotype: $\frac{p^2 W_{AA}}{\bar{W}} = \frac{(0.5)^2(1)}{0.7} = 0.36$

Frequency of Aa genotype: $\frac{2pqW_{Aa}}{\bar{W}} = \frac{2(0.5)(0.5)(0.8)}{0.7} = 0.57$

Frequency of aa genotype: $\frac{q^2 W_{aa}}{\bar{W}} = \frac{(0.5)^2(0.2)}{0.7} = 0.07$

Allele frequency of A: $p_A = \frac{p^2 W_{AA}}{\bar{W}} + \frac{pqW_{Aa}}{\bar{W}}$

$$= \frac{(0.5)^2(1)}{0.7} + \frac{(0.5)(0.5)(0.8)}{0.7} = 0.64$$

Allele frequency of a: $q_a = \frac{q^2 W_{aa}}{\bar{W}} + \frac{pqW_{Aa}}{\bar{W}}$

$$= \frac{(0.5)^2(0.2)}{0.7} + \frac{(0.5)(0.5)(0.8)}{0.7} = 0.36$$

After one generation, the allele frequency of A has increased from 0.5 to 0.64, while the frequency of a has decreased from 0.5 to 0.36. This is because the AA genotype has the highest fitness, while the Aa and aa genotypes have progressively lower fitness values. Another interesting feature of natural selection is that it raises the mean fitness of the population. If we assume that the individual fitness values are constant, the mean fitness of this next generation is

$$\bar{W} = p^2 W_{AA} + 2pqW_{Aa} + q^2 W_{aa}$$
$$= (0.64)^2(1) + 2(0.64)(0.36)(0.8) + (0.36)^2(0.2)$$
$$= 0.80$$

The mean fitness of the population has increased from 0.7 to 0.8. This population is more well adapted to its environment than the previous one. Another way of viewing this calculation is that the subsequent population has a greater reproductive potential than the previous one. We could perform the same types of calculations to find the allele frequencies and mean fitness value in the next generation. If we assume that the individual fitness values remain constant, the frequencies of A and a in the next generation are 0.85 and 0.15, respectively, and the mean fitness increases to 0.931. As we can see, the general trend is to increase A, decrease a, and increase the mean fitness of the population.

In the previous example, we considered the effects of natural selection by beginning with allele frequencies at intermediate levels (namely, A = 0.5 and a = 0.5). What would happen if a new mutation introduced the A allele into a population that was originally monomorphic for the a allele. As before, the AA homozygote has a fitness of 1.0, the Aa heterozygote 0.8, and the recessive homozygote 0.2. Initially, the A allele is at a very low frequency in the population. If it is not lost initially due to genetic drift, its frequency slowly begins to rise and then, at intermediate values, rises much more rapidly. Eventually, this type of natural selection may lead to the fixation of the beneficial allele. However, this new beneficial allele is in a precarious situation when its frequency is very low. As mentioned earlier in this chapter, random genetic drift is likely to eliminate new neutral mutations due to sampling error. Genetic drift can also eradicate beneficial alleles, although this is not as likely, because natural selection favors them.

Balanced Polymorphisms may Exist due to Heterozygote Superiority or Heterogeneous Environments

As we have just seen, selection for an allele that is beneficial in a homozygote may lead to its eventual fixation in a population. Likewise, genetic drift also eliminates or fixes neutral alleles. So how do we explain the high levels of polymorphisms observed for genes in natural populations?

One possibility is that many of these polymorphisms involve neutral alleles not subject to natural selection. The genes become polymorphic due to genetic drift, and there has not been enough time for fixation to occur. (As we have seen, fixation can require vast amounts of time.) In other cases, a polymorphism may have reached an equilibrium where opposing selective forces have reached a balance; the population is not evolving toward fixation or allele elimination. This situation, known as a *balanced polymorphism*, can occur for several different reasons.

In some cases, a balanced polymorphism occurs when the heterozygote is at a selective advantage. The higher fitness of the heterozygote is balanced by the lower fitnesses of both corresponding homozygotes. Let's consider the following case of relative fitness:

$$W_{AA} = 0.7$$
$$W_{Aa} = 1.0$$
$$W_{aa} = 0.4$$

This is an example of *heterozygote advantage*, also called *overdominance*. The selection coefficients are:

$$s_{AA} = 1 - 0.7 = 0.3$$

$$s_{Aa} = 1 - 1.0 = 0$$

$$s_{aa} = 1 - 0.4 = 0.6$$

Under these conditions, the population will reach an equilibrium in which

$$\text{Allele frequency of A} = \frac{s_{aa}}{s_{AA} + s_{aa}}$$

$$= \frac{0.6}{0.3 + 0.6} = 0.67$$

$$\text{Allele frequency of a} = \frac{s_{AA}}{s_{AA} + s_{aa}}$$

$$= \frac{0.3}{0.3 + 0.6} = 0.33$$

Balanced polymorphisms can sometimes explain the high frequency of alleles that are deleterious in a homozygous condition. A classic example is the H allele of the human β-globin gene. A homozygous H^SH^S individual displays sickle cell anemia, a disease that leads to the sickling of the red blood cells. The H^SH^S homozygote has a lower fitness than a homozygote with two normal copies of the globin gene, H^AH^A. However, the heterozygote, H^AH^S, has the highest level of fitness in areas where malaria is endemic. Compared with nor mal, H^AH^A, homozygotes, the heterozygotes have a 10—15% better chance of survival if infected by the malarial parasite, *Plasmodium falciparum*. Therefore, the H^S allele is maintained in populations where malaria is prevalent, even though the allele is detrimental in the homozygous state.

Besides sickle cell anemia, several other gene mutations that cause human disease in the homozygous state are thought to be prevalent because of heterozygote advantage. These include cystic fibrosis, in which the heterozygote is resistant to diarrheal disease; PKU, in which the heterozygous fetus is resistant to abortion caused by a fungal toxin; and Tay Sachs disease, in which the heterozygote is resistant to tuberculosis.

Balanced polymorphism also may occur when a species occupies a region that contains heterogeneous environments. One environment may favor a particular allele, while a neighbouring environment may favour a different allele. This is a photograph of land snails, *Cepaea nemoralis*, that live in woods and open fields. This snail is polymorphic in colour and banding patterns. In 1954, A.J. Cain and P. M. Sheppard of the University of Oxford found that snail colour was correlated with the environment. The highest frequency of brown shell colour was found in snails in the beechwoods, where there are wide expanses of dark soil. Pink snails are most common in the leaf littes of forest floors in deciduous woods. The yellow snails are most abundant in the sunny, unwooded areas of hedgerows and rough herbage. It has been proposed that disruptive selection has favored the survival of these different phenotypes. Migration between the snail populations keeps the polymorphism in balance among these different environments.

Kettlewell Studied Industrial Melanism in the Moth *Biston betularia* as a Modern Example of Natural Selection

The concept of natural selection provides a framework for explaining how allele frequencies are maintained in populations; it also explains how species become adapted to the environments in which they live. As scientists, we tend to examine the ultimate result of natural selection. In other words, we can observe species in their native environments, and in some cases, it seems obvious how certain characteristics provide an organism with a survival advantage. The smell of a skunk and the ability of a cactus to retain water are classic examples. To support the theory of natural selection, population geneticists would like to witness the process in a species over time.

The first observed example of modern natural selection occurred in the decades following the Industrial Revolution (which began in the latter half of the 19th century). As a result of industrialization, large areas of the earth's surface are blanketed by the fallout of smoke particles. In and around industrial areas, this fall out is dramatically more evident than in undeveloped areas. The smoke particles tend to kill vegetative lichens on the trunks and boughs of trees. Rain washes the nollutants down the tree trunks until they are bare and black. By comparison, trcc runks in unpollutcd arcas arc much lighter and are speckled with growing lichens. During the early 1900s, scientists in Europe and North America began to notice the proliferation of melanic moth varieties that are darker in colour than the previously abundant varieties. This moth, *Biston betularia*, is polymorphic. One variety, *carbonaria*, is very dark, whereas the typical variety is much lighter in colour. As seen here, the carbonaria morph is easily seen on a lichen-covered tree trunk, and the typical morph is readily visible on dark trunks. These observations led geneticists to propose that the Indus trial Revolution was selecting for the survival of carbonaria. It was hypothesized that the darkened colour of the carbonaria morph made it less susceptible to predation by birds. Supporting this idea, a field study in England showed that the carbonaria morph is more prevalent in polluted areas of Birmingham, while the typical morph is predominant in unpolluted areas

In the experiment described here, conducted in 1956, H. Bernard Kettlewell, a British biologist, set out to test the hypothesis that the colouration polymorphism in *B. betularia* is due to natural selection involving predation by birds.

Hypothesis

The colouration of *B. betularia* affects the probability that birds will see and eat them. This natural selection leads to the survival of those moths with colouration more closely matching that on which they rest, whether that be in polluted or unpolluted woods.

Testing the hypothesis

Starting the material: Samples of *B. betularia*, both carbonaria and typical, were collected from the wild. In some cases, they were bred in the laboratory, to be released later into the wild.

1. Mark the underside of *carbonaria* and *typical* moths with cellulose paint.
2. Release equal numbers of marked *carbonaria* and *typical* moths in either polluted woods near an industrial area of Birmingham, or in an unpolluted woods in Dorset, England.
3. On several consecutive days, recapture moths that have been marked. This can be done by attracting them to a mercury vapour light. Record the number of recaptured moths.
4. Also, sit in a blind and observe the moths on a tree trunk. Record the numbers and types of sitting moths that are eaten by birds.

Interpreting the data

As shown in the data, the *carbonaria* moths were more likely to be eaten by birds in an unpolluted woods. This observation was corroborated by the lower percentage of *carbonaria* moths that were recaptured in this area. Conversely, the *typical* moths were much more likely to be eaten in a polluted woods. Again, this observation was supported by the lower percentage of recapture for *typical* moths in Birmingham. Taken together, these results supported the hypothesis that *carbonaria* have a greater chance of surviving in polluted woods, *typicals* in unpolluted woods. In other words, the consequences of the Industrial Revolution have resulted in natural selection for the *carbonaria* variety. As stated by Kettlewell, "These experiments showed that birds act as selective agents and that the melanic forms of betularia are at a cryptic advantage in an industrial area such as Birmingham."

Genetic Load is the Negative Consequence of Genetic Variation

As we have just seen, genetic variation in a population can lead to adaptation via natural selection. In general, we often think of genetic variation in a positive light. It allows species to adapt to their environment and provides the raw material for evolution. Furthermore, genetic diversity adds interest and beauty to our perception of biology. On the negative side, however, certain kinds of genetic variation are detrimental to the survival of a population. The term *genetic load* (L) refers to genetic variation that decreases the average fitness of a population as compared with a (theoretical) maximum or optimal value:

$$L = \frac{W_{max} - W}{W_{max}}$$

where,

L equals the genetic load

W equals the average fitness of a population

W_{max} equals the maximal fitness of a population

There are many factors that contribute to the genetic load in populations. Several of these are listed here:

1. *Mutation*: Most new mutations are deleterious. Therefore, the accumulation of new mutations in a population contributes to the genetic load.
2. *Segregation*: As mentioned earlier in this chapter, certain mutations are beneficial in the heterozygous condition but detrimental to homozygotes. Therefore, the segregation of alleles to create homozygotes causes a genetic load in the population.
3. *Recombination*: Sometimes, linked genes occur in optimal combinations. For example, two genes may be found in *A* and *a* alleles and *B* and *b* alleles. Let's suppose that the *AB* and *ab* combinations are beneficial, but the *Ab* and *aB* combinations are less fit. Although natural selection may select for chromosomes containing the *AB* and *ab* pairs of genes, recombination (i.e., crossing over) may create *Ab* or *aB* combinations and thereby decrease the fitness of the population.

4. *Heterogeneous environments*: As mentioned earlier in this chapter (namely, for snail colouration), some species occupy two or more types of environments. This creates polymorphisms that are suited to different environments. Some members of the population will have genotypes that are not well suited to their environment. For example, a darkly colored snail may be located in a sunny environment.
5. *Meiotic drive/gamete selection*: Certain types of alleles and chromosomal rearrangements are transmitted preferentially to gametes even though they may be detrimental to the survival of the organism. For example, some alleles favor gamete viability even though they may not be beneficial in the adult.
6. *Maternal—fetal incompatibility*: In a few cases, the genotype of the mother may interact in an unfavourable way with that of a fetus. A classic example is the Rh^+ factor, a type of red blood cell antigen. If a mother is Rh^- and a fetus is Rh^+ the mother's body will occasionally mount an immunological reaction against the fetus, thereby diminishing its chances for survival.
7. *Finite population*: In a small population, genetic drift causes allele frequencies to deviate from their optimal fitness values.
8. *Migration*: When individuals from one environment migrate to a new environment, they usually do not possess the optimal combination of alleles for survival. Therefore, migration tends to decrease the fitness of a population. While it is beyond the scope of this text, population geneticists have derived mathematical formulae that describe the amount of genetic load that these various factors may contribute to a population.

Concluding Remark

In this chapter, we have surveyed some of the critical issues in population genetics. The primary focus of this field is understanding genetic variation within populations. A substantial proportion (e.g., about 30% in humans) of genes in natural populations are *polymorphic*, existing in two or more alleles. Population geneticists want to understand why this variation in the *gene pool* exists and how it may change.

According to the *Hardy—Weinberg equilibrium*, *allele frequencies* remain stable as long as several conditions are met. These include no new mutations, random mating, large population size, no migration, and no natural selection. While these conditions are never truly met, they can be approximated in large, well-adapted populations. Therefore, the Hardy—Weinberg equilibrium predicts *genotype frequencies*, and genetic stability when conditions are appropriate.

However, we know that many processes in nature promote genetic change. As we will see in the following chapter, these forces lead to the evolution of species. The initial source of genetic variation is mutation. However, the *mutation rate* is too low to account for the allele frequencies observed for most genes in natural populations. Instead, mutation provides the source of variation, and other evolutionary forces act to increase the frequency of some new alleles. In small populations, *neutral forces* such as *random genetic drift* can alter allele frequencies dramatically. Simply due to random sampling error, a new neutral mutation may be eliminated from a population or it may become fixed in it. This can occur via a *founder effect*, when a small population moves to a new site, or a *bottleneck effect*, when a population is reduced in size due to negative environmental events. Allele frequencies can also be altered by migration. When two populations differ in their allele frequencies, migration can alter the existing allele frequencies. This is known as *gene flow*.

Finally, *adaptive forces*, such as *natural selection*, make a species better adapted to its environment. Natural selection favors individuals with the greatest reproductive potential (i.e., the greatest *Darwinian fitness*). These may be individuals who are more likely to survive, more likely to mate, or more fertile. When natural selection favors one homozygote, it selects for the fixation of the favourable allele. However, when the heterozygote has the highest fitness, or when two or more phenotypes in an environmentally diverse region have the highest fitnesses, a *balanced polymorphism* will result.

A central approach in population genetics is to measure variation experimentally, and then examine the mathematical relationships between genetic variation and the occurrence of genotypes and phenotypes within populations. In this chapter, we have focused on how population geneticists explain phenotypic variation via mathematical theories. Experimentally, phenotypic variation can be measured by tallying the organisms within a population that display particular phenotypes. For example, we can count the number of dark snails and light snails that live in a particular location. This provides a measure of the phenotypic variation within this population. We also learned that variation can be studied at the molecular level by identifying allozymes. After variation is measured, population geneticists attempt to develop and apply mathematical relationships that provide insight into the causes and dynamics of the variation. Although geneticists understand that mutation is the source of new variation, their mathematical formulae indicate that the rate of new mutations is too low to explain the experimentally observed variation in natural populations. Instead, other processes, such as genetic drift, migration, and natural selection must alter the frequency of new mutations after mutations have occurred. The mathematical theories of population geneticists tell us that these forces may be directional, leading to the eventual fixation or elimination of an allele, or they may promote a balanced polymorphism within a population. These theories also show us that variation may have negative consequences, known as the *genetic load*.

12

ECOLOGICAL GENETICS

Ecology, the study of the interrelationships between organisms and their environment, has come to have a substantial interface with molecular genetics. In large part, this relationship between fields has grown because differences between a specific protein's characteristics are more likely to be due to alleles of one gene than are differences in the size, shape, or colour of the organism. The ecologist who asks why certain plants are tall in the valleys and short on the mountains may have to analyze a trait that is influenced by many genes and/or only partially heritable. The ecologist who asks why sickle cell anemia is so much more prevalent in some area of Africa than in others is going to be able to look at the single allelic difference between hemoglobin A and hemoglobin S. This molecular ecologist can use the data directly in the population genetics equations. Among molecular ecologists, electrophoresis is very commonly used to identify variant proteins that have undergone changes in net charge density due to mutations from the original wild type form. As mentioned in earlier chapters, hemoglobin S and the normal form, hemoglobin A, can be separated in this way.

Evolutionary biology is an inferential science, seeking to determine past relationships of organisms that are presently in different taxonomic categories. The questions asked by an evolutionary biologist are ultimately genetic: how similar or different are the two organismic groups at the genetic level? The reason for including this topic with ecological genetics is that evolution largely proceeds through natural selection of individuals who are best adapted to their environments, there fore, by definition, fittest. The influence of genetic constitution of creatures in different environments, *ecological genetics*, is one component of the analysis of an evolutionary geneticist, then. This idea is well expressed by the title of a book, The Ecological Theater and the Evolutionary Play, by G. Evelyn Hutchinson.

Ecological Niches

The exact temporal and spatial part of the physical and biological environment occupied by an organism is called its *ecological niche*. It is the ecological niche to which natural selection adapts the organism. An example of such selection can be seen in *Cepaea nemoralis*, the European land snail. The colour of the shell in *C. nemoralis* is largely determined by one gene with several alleles. Three of these alleles are: Y_2 brown, dominant to either of the other alleles; Y_1 pink, dominant to Y but recessive to Y_2; and Y yellow, recessive to both Y_1 and Y_2. In woodlands, the natural snail populations are brown and pink, but in the open green fields, most snails are yellow. It appears likely that these traits are under strong selective pressure, since bird predators search mainly by sight for the snails. A finding, which suggests that the situation is more complex, is that yellow snails heat up more slowly in the open than other colours.

Watt has examined the species of *Colias* butterflies that are found at different altitudes. He found that species common at higher altitudes had a dark melanin patch on the hind-wing, where it was effective in

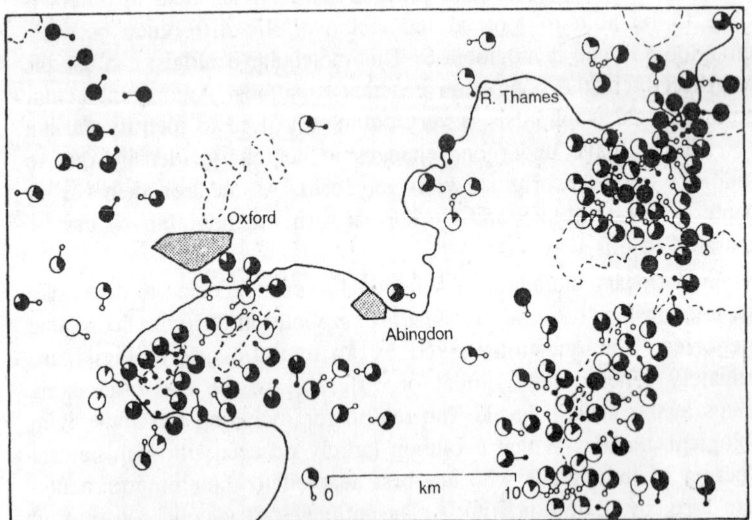

Fig. 12.1. Pattern of distribution of yellow colour allele in the snail Cepaea nemoralis.

enhancing the warming rates of the bodies of the butterflies. The lower altitude forms had much less melanin in this area.

A molecular-level example of the relation of an ecological niche to the genetic status of the population in that niche involves the sickle cell hemoglobin allele. The distribution of the hemoglobin S allele and the distribution of the malaria parasite in Africa and surrounding areas. This correlation suggested a testable hypothesis: that the sickle cell heterozygote (often referred to as having the sickle cell trait) is at an advantage in areas infested with this parasite. Allison succeeded in showing that heterozygous children have malaria less frequently and less severely than children without the hemoglobin S. The advantage of thc heterozygote over the homozygote with hemoglobin A is thus malaria resistance. The heterozygote has an advantage over the homozygote for hemoglobin S in that the anemia is recessive. Note that the heterozygote superiority to hemoglobin A homozygotes applies only where the malaria parasite is a part of the niche.

Bottlenecks in Population Size and Founder Effects

A few individuals from any population might have the opportunity to found a new population. This opportunity could come from migration to a new site (an island, for example) or from near decimation of the original population, leaving only a few survivors to reestablish the population. The new population may be very different from the original population in gene frequencies. This type of dramatic shift in gene frequencies due to a *bottleneck* (temporary small size) in population size is called the *founder effect*. Whichever organisms found or start the new population will determine the new gene frequencies, regardless of whether or not their alleles were the most common ones in the original population.

An example of the founder effect can be seen in the gene frequency for the rare colour-blindness (non-X-linked) called *achromatopsia* on the Pingelapatoll. This atoll is in the pathway of the Pacific typhoons, and these storms have often nearly wiped out the population of the island. A typhoon in 1775 followed by a famine left only 30 survivors. Today, the population is about 1600, 5% of whom are homozygous for the *acromatopsia* allele. It is postulated that the chief in 1775, who had several wives, was a carrier for this allele. The resultant population may have been enriched for the allele by means of genetic drift after the typhoon, as well as experiencing the founder effect.

Another example can be seen in the Amish people. These people moved to very localized areas of Pennsylvania, Ohio, and Indiana from

Berne Canton, Switzerland in the 1700s. Subsequently, they kept themselves in virtually complete genetic isolation from other people of the areas. Their incidence of Ellis—van Creveld syndrome is quite high. This syndrome is an autosomal recessive disorder; 50% of those homozygous for it die in infancy from heart or respiratory difficulties. The survivors are disproportionately dwarfed, being short but with very short extremities. They often have extra fingers and sometimes extra toes (polydactyly). Of 43 such cases investigated in 26 families, the parents were descendants of Samuel King and his wife. One of these two members of the founding population must have carried the recessive allele for Ellis—van Creveld syndrome. The genetic isolation of the Amish people kept the allele among them, and the absence of selection against the heterozygote (which would have obscured the random trend seen) allowed the allele to build up to a high level by genetic drift.

A third example of this founder effect is observed in the Northern elephant seal, *Mirounga angustirostris*, which was reduced to N = 20 in the 1890s but now has a population size of about 30,000. This seal population was examined for electrophoretic variants by Bonnell and Selander, and of 24 genes examined, not one was polymorphic.

Population Diversification and Speciation

The change from a relatively homogeneous population of some organism into two or more different populations (often constituting different species) is called *speciation*. Speciation can be very rapid or can occur gradually. It is difficult to define a species perfectly satisfactorily, but one operational definition is a group of potentially interbreeding organisms that are reproductively isolated from all other kinds of organisms.

This definition has come under attack recently by most botanists and some zoologists as well. Ehrlich and Raven, and more recently, Levin, have pointed out that in many cases, individuals of a recognized species are potentially able (in the laboratory, for example) to interbreed with individuals of one or more other species. In addition, different populations of individuals of the same species are often actually reproductively isolated from each other, although the populations include morphologically identical members of their own species. Why don't such populations diverge from each other, if there is no gene flow between them?

Sewall Wright suggested long ago that populations achieve adaptive peaks that are separated from one another by poorly adapted valleys. Ehrlich and Raven, as well as Levin, argue that by occupying a particular niche, the organism obligates itself to a certain suite of adaptations, and thus selection retains similarity without the necessity for gene flow between populations. As you can see, these two concepts are strikingly similar and offer a satisfying explanation for these puzzling observations. In the light of such information, it is probably more appropriate to consider the population, rather than the species, as the unit of evolution. Species might then be defined as a convenient taxonomic category, rather than a reproductively isolated, interbreeding unit of organisms. The word "speciation" will be used in this chapter to imply diversification, without necessarily implying that the unit of speciation is the species itself.

In the classical sense, speciation is the process of subdividing a potentially interbreeding population such that the subdivisions are reproductively isolated from each other. An obvious change in structure, function, and ecological niche need not occur. Populations that can be differentiated reproductively but not morphologically, so called *sibling species*, are relatively common in nature. It has been possible to distinguish such populations via electrophoretic differences between the proteins of the sibling species.

The process of speciation can theoretically occur when the populations are in contact with each other through occupying the same geographical area (*sympatric populations*) or when there is geographical isolation (*allopatric populations*). The isolation required is of the gene pools rather than the organisms them selves. In practice, sympatric speciation is rare and results from large genomic changes such as polyploidization.

Rapid Reproductive Isolation in *Stephanomeria* Derivatives

An interesting case of two new species having arisen by a similar mechanism from the same parent species involves the plant *Stephanomeria exigua coronaria*, a composite. There are about 35,000 individuals in this subspecies. The subspecies is self-incompatible, causing it to cross with other individuals rather than self-pollinate. Two self-pollinating populations evidently derived from this subspecies are "Malheurensis" (a provisional name, not yet accepted by taxonomists) and *Stephanomeria paniculata*.

The *Malheurensis* taxon (group, not necessarily a species) occurs only in one location in the dry plains of eastern Oregon, having about 750 plants in the subspecies. It is sympatric with the parent species at this site. Reproductive isolation is maintained by means of the very effective self-pollination of Malheurensis, a poor ability to form seed from interspecific (between different species) crosses, and several chromosomal rearrangements, including a reciprocal translocation. Since the pollen produced by the surviving hybrids rarely gets a complete haploid genome, owing to the chromosomal rearrangements, the pollen viability is only 25% from hybrids. Gottlieb, who has studied the electrophoretic patterns produced by enzymes encoded by 25 structural genes, found that the Malheurensis gene pool contains a highly limited extraction of alleles from the gene pool of the parent species.

The parent species is polymorphic at 60% of its genes, with 45 different alleles being represented at all the polymorphic loci together. The *Malheurensis* species is polymorphic at only 12% of its genes, with a total of seven alleles .In this new species, only several dozen possible phenotypes can be found. In the *Stephanomeria exigera coronaria*, nearly every individual plant has a distinctive electrophoretic pattern. The new species is restricted in range and, as shown by laboratory tests, is less well adapted than its progenitor species under a wide range of environmental conditions.

Table 12.1. The genetic diversity of *Stephanomeria exigua* ssp. *coronaria* and its sympatric derivative species "*Malheurensis*"

Gene	*Malheurensis*	*S. exigua coronaria*
LAP-1	1	3
LAP-2	1	2
EST-1A	1	2
EST-1B	1	2-3
EST-2	2	3
FST-3	3	5
EST-4	1	5
GOT-1	1	3
GOT-3	1	2
GDH	1	2
POM-1	1	2
APH-1	1	4
APH-2	1	2
ADH-1	1	≈4
ADH-2	2	≈4
	19	45-46

A different type of new species originating from the same parent species is *S. paniculata*. It is also self-pollinating, but it is allopatric to the parent. It occurs in open, sandy patches near roads from north California to east Washington. This species was also extensively studied by Gottlieb, who examined 600 individual plants electrophoretically. It proved to have only one allele (be monomorphic) for each of the 25 genes except for one: a *glutamate-oxaloacetate transaminase* (GOT) gene. Even that GOT gene was fixed in all populations; it was merely that one population had become fixed with a different allele of the GOT. In spite of the very limited genetic variability, this species has much wider dispersal and total number of individuals than *Malheurensis*.

Clines

A cline is a change in gene frequency, generally gradual, in space. The cline of the dark (melanic form) and light (non-melanic form) phenotypes for the moth, *Amathes glaveosa* in the Shetland Islands. The north is primarily inhabited by the dark form and the south by the light form. A cline of populations with different ratios of the two forms lies in between. Why doesn't the entire species come to equilibrium for these alleles? Evidently, the species consists of a number of potentially interbreeding populations which, in fact, breed *assortatively* rather than randomly. There is a greater chance of a (local × local) mating than of a (local × non-local mating) in assortative mating systems, but the gene flow between populations is accomplished by migration. Genetic/reproductive barriers do not usually occur between the populations of a cline, although exceptions exist.

White clover, *Trifolium repens*, has potential for cyanide production. The cline of the cyanide-producing phenotype in Europe. Superimposed on the map is a record of the January cold temperatures. One can distinguish that cyanide is produced much more frequently where the January temperature is warmest. It is conjectured that cyanide production is a good protection against being eaten but may cause suicide where the plants freeze and thus rupture the membranes, releasing cyanide into their own cytosol.

A famous example of a dine is the chromosome polymorphism in *Drosophila pseudoobscura*. In the Sierra Nevada mountains of California, Dobzhansky and his collaborators showed that 50% of the flies had the Standard chromosome arrangement in the sea level populations, but less than 10% had the Standard arrangement at timberline (3000 m). Increases with altitude in Arrowhead and Chiricahua arrangements of the chromosomes were noted. The synapsis between the Arrowhead (AR) and Standard (ST) third chromosomes. The Arrow head is an inversion of bands 70-76. Crossing over within the inversion loop would

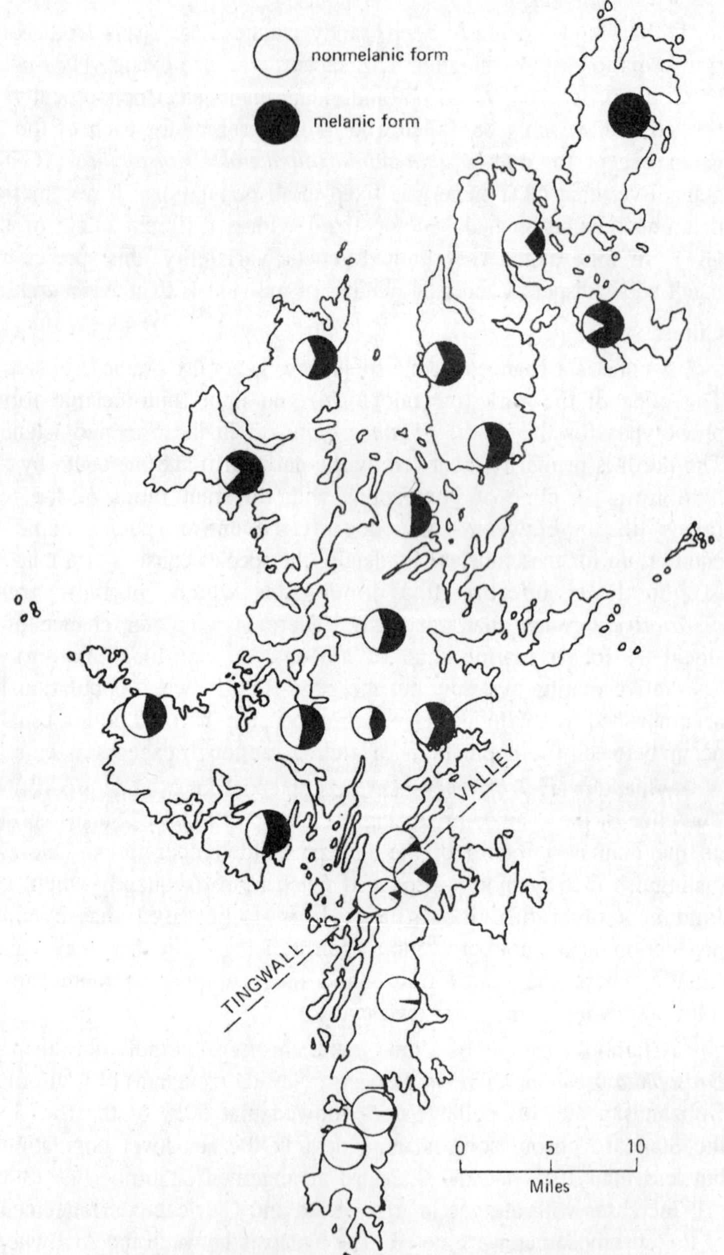

Fig. 12.2. The cline of pigmentation for the moth Amathes glaveosa in Shetland islands.

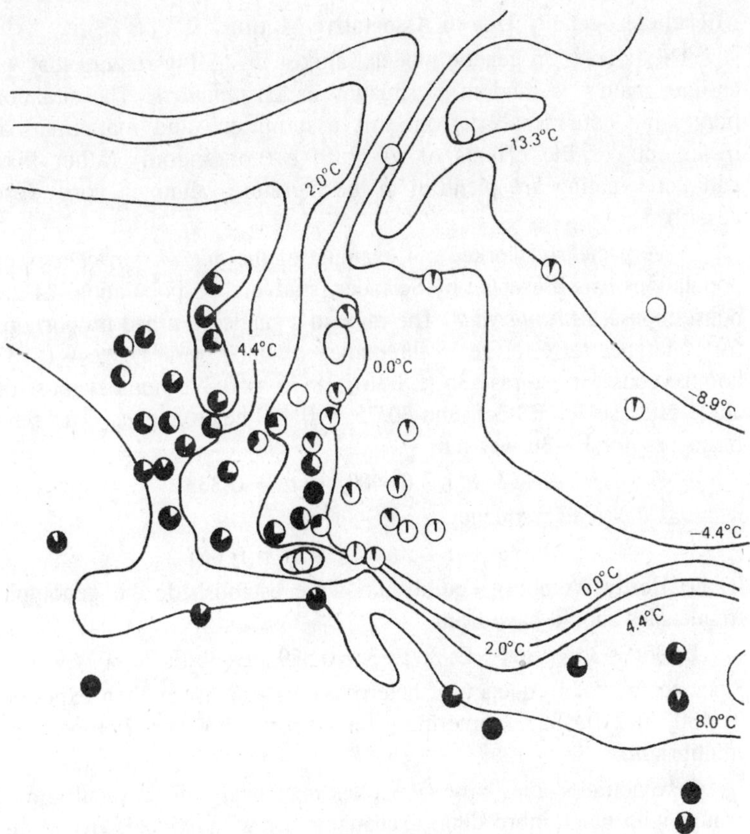

Fig. 12.3. Cline of cyanide production for white clover, Trifolium repens.

lead to nonviable gametes. *Drosophila* and other diptera are partially protected from these effects by the lack of crossing over in the male flies. In most cases, though, inversions help to isolate the populations from each other by making gene flow between individuals with different chromosome arrangements less likely to succeed.

An interesting aspect of the cline is that dines are often different for different characteristics of an organism. There may be a north-south cline for colour, but an east-west cline for shape, for example. The present view is that the individual populations along a dine are under somewhat different selective forces, so that different allelic proportions can be found in different locales. This idea raises the question: what is the phenotype geneticists call the wild type? Is it the phenotype found at the central point in each cline?

Microheterogeneity Due to Assortative Mating

The population genetic models, almost invariably require that we assume mating is random, a situation called *panmixis*. The situation may vary considerably from this assumption, and mathematical treatments of the effects of *assortative* (nonrandom) rather than panmictic mating are plentiful in the literature, although none were described.

A very clearly worked out example of the *microheterogeneity* of populations was presented by Selander, studying the populations of the house mouse, *Musmusculus*. The mice in a single barn had proportions for genotypes for esterase 3—three alleles, which were 17/75 = 0.133 homozygous for esterase-3b (ES-3b); 28/75 = 0.373 homozygous for allele esterase-3c (ES-3c); and 30/75 = 0.400 heterozygous. The gene frequency for ES-3b was thus

$$0.133 + 1/2\ (0.400) = p = 0.333$$

whereas the gene frequency for ES-3c was

$$0.373 + 1/2\ (0.400) = q = 0.667$$

If the Hardy-Weinberg equilibrium were established, the genotypic frequencies should have been

Es-3b/Es-3b=0.181 Es-3c/Es-3c=0.329 Es-3b/Es-3c=0.489.

Since the actual frequency of heterozygotes was lower than expected (0.400, not 0.489), something interfered with establishment of equilibrium.

Behavioural studies have shown that mice tend to form small family groups with much inbreeding, exchanging a few individuals from time to time with other family groups. The groups are quite territorial. We would be justified in attributing the low heterozygosity to the inbreeding, which can be obtained as follows

$$F = \frac{2pq - (\text{real genotype frequency for heterozygotes})}{2pq}$$

The value of F, the *coefficient of inbreeding*, is the proportion of the expected heterozygotes that are missing as a result of assortative rather than random mating. In this case

$$F = \frac{0.489 - 0.400}{0.489} = 0.18.$$

In a different barn, at the same farm, Selander collected mice in 870 traps set in a grid pattern on the east side of the barn floor. The 591 mice caught on two successive nights were tested for esterase alleles by electrophoresis, where Es-3b is of medium mobility but Es-

3c is slow. The lines were drawn around areas that may be family groups, based on their high (50% or 80%) probability of having the same gene frequencies throughout.

An additional feature of this microheterogeneity may be its stability. On a slightly higher level, the gene frequencies for mice on the east and west sides of the same barn floor were compared in August 1967, December 1967, and May 1968. The tabulated data show the stability of the Es-3 gene frequencies at these three samplings. The east side consistently had higher proportions of Es-3c than the west side of the same barn.

Table 12.2. Repeated sampling of gene frequencies for alleles of esterase-2 in two halves of barn in austin

	West side		*East side*	
Allele Electrophoresis:	*Es-3b (medium mobility)*	*Es-3c (slow mobility)*	*Es-3b (medium mobility)*	*Es-3c (slow mobility)*
Sampling date				
August	0.472	0.528	0.343	0.657
December	0.446	0.554	0.284	0.716
May	0.433	0.567	0.350	0.650
Averages	p = 0.450	q = 0.550	p = 0.326	q = 0.674

Hamrick and Allard have also shown, in a study of the oat *Avena* polymorphism and microheterogeneity, that isozyme changes are notable over a matter of a few yards.

We might wonder what possible advantage would offset the risk of homozygosity for one of the deleterious recessives as a result of inbreeding? This question has been discussed in the literature since Sewall Wright's early treatments, and it is still postulated that the high degree of inbreeding in small clusters has evolutionary advantages. Each population can be viewed as a partially independent evolutionary experiment, where failure results in extinction of that population but not of the whole species. Success may lead to a population that moves out to share its new, successful alleles with the rest of the species due to population pressure at its original locale.

Physical Isolation and Renewed Contact

Two populations from the same species can become physically isolated from one another so that no migration between them can occur. Subsequently, either genetic drift or selection pressures or both may modify these populations such that they become different in structure and function. If interbreeding is still possible, such populations may be considered subspecies. If reproductive barriers evolved, the populations would be considered separate species.

At sites where two populations come into contact after having been isolated, gene flow between them may be reinitiated. If they are already separate species, this process is called *introgressive hybridization*. Whether or not the populations have become reproductively isolated, their zone of contact is referred to as the *hybrid zone* or the *suture zone*. Such a zone exists for a number of species in northern Florida, where Remington has described an interface between continental populations and distinguishable peninsular subspecies. During the Pleistocene interglacial period, the peninsula was severed by a strait so that the populations there were physically isolated for long enough to become subspecies. When the peninsula was rejoined to the mainland, a suture zone was formed, along which gene flow between the morphologically distinguishable forms occurs.

Types of Reproductive Isolation Mechanisms

The previous section summarized some patterns of population diversity that might lead to speciation. The critical separating mechanisms must prevent or greatly diminish contact between the two gene pools in order to separate one parent species into two species. The mechanisms separating some species have been briefly mentioned earlier in this chapter. Isolation mechanisms for sympatric species can either completely prevent mating (*premating isolation*) or have an adverse effect on matings between individuals or different species after the mating occurs (*postmating isolation*).

Premating Isolation

Premating isolation often involves behavioural differences between possible mates. For example, fireflies of different species mate at different times of night. In addition, female fireflies respond to a light pattern emitted by males of their own species but not to light patterns emitted by males of other species. In plants, differences in flower colour and form may attract quite different pollinating animals to the different species. Often potentially compatible plants flower at different times of the year. If not, pollen can be released and/or received at different hours of the day. Sometimes, physiological incompatibilities form part of the premating isolating mechanism. The pistil of one plant species may be incompatible with the pollen of another. In animals, copulation may fail to produce progeny because sperm are not released in crosses between different species.

Postmating Isolation

Postmating mechanisms may impair the viability of the sperm and/or egg, the viability of the hybrid, or the fecundity of the hybrid. In crosses between the frog species *Rana pipiens* and *Rana sylvatica*, the development of zygotes is normal up to the early gastrula stage, where development ceases. This example illustrates an effect on viability of the hybrid. The inversions and other chromosomal rearrangements common among *Drosophila* have been mentioned. Since crossing over will produce some inviable eggs in the hybrids, these inversions, and so on, impair the fecundity of the hybrid flies. The chromosomes in were both found in *Drosophila pseudoobscura*. This species can be crossed with *D. persimilis*, resulting in sterile progeny. In this case, genes affecting the fertility were detected by Dobzhansky throughout the genome.

Mechanisms for Enhancing Outbreeding

It is interesting that some of the same mechanisms that can cause reproductive isolation in some species will lead instead to more outbreeding in other species. One example of such a reverse effect is in pistil/pollen incompatibility in plants. In the previous section, the role of such incompatibility in preventing crosses between species was discussed. In fact, inbreeding reduces variation and causes expression of deleterious alleles that might well have remained masked during outbreeding. The effects on vigor of corn that has been outbreed first, and then inbred again. There are genetic mechanisms in plants that cause self-incompatibility, thus increasing the likelihood of out breeding. One common mechanism of this type involves an incompatibility genes, with a large number of different alleles s-1 through s-N. An individual

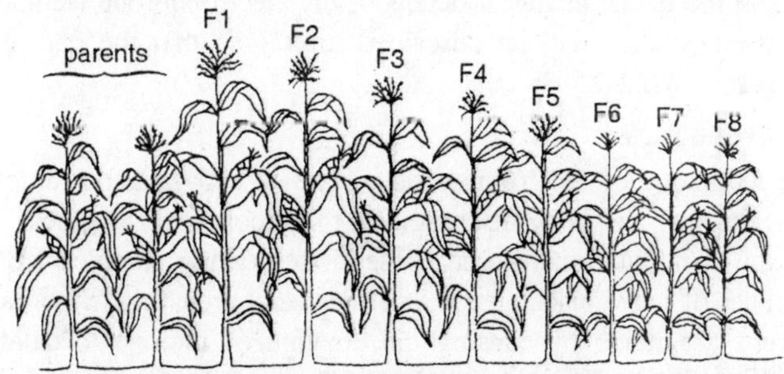

Fig. 12.4. Effects of outbreeding and subsequent inbreeding on vigour of corn.

plant can be fertilized only by pollen with a completely different incompatibility allele than either of its own. A plant with s-3/s-6 genotype could not be fertilized by pollen with either s-3 or s-6 incompatibility allele. Since its own pollen has to be of one of these two kinds, it cannot self-fertilize. This type of mechanism is found in the primrose, *Oenothera organensis*, for example, with many (20 or more) alleles. The primrose could be thought of as reproductively isolated from itself at the pre-mating level!

Behaviour may also work against inbreeding. An example is in the eel, *Anguilla rostrata*. Rather than mating in their localized populations along the Atlantic coast of North America, eels from the entire range of locations travel to the Sargasso Sea to mate, thus greatly increasing the chance of panmixis.

Stable Polymorphisms

In earlier chapters, the term wild type has been used in a manner that implied most of the individuals of a population or species were identical, and the one special individual being examined differed by having a mutant allele at one gene (or a few well-defined genes). This view of the species has not been substantiated by Molecular ecologists. In fact, various organisms that have been examined by electrophoresis appear to have approximately 30% of their genes *polymorphic*. The degree of heterozygosity is also rather high; in mice (*Mus musculas*) on a farm in California, 11% of the loci of a typical mouse were heterozygous. These number are rather similar to findings in *Drosophila* and in man. The results suggest typical individuals need not he anywhere near genetically identical.

In this section, it will he our task to try to account for some of the reasons why populations are so polymorphic that the concept of a wild type is virtually meaningless. Before trying to account for the variation, some of the technical details of the electrophoretic techniques used in these studies will be described briefly so that the data cited will be understandable.

Gel Electrophoresis

The technique of electrophoresis in gels to separate molecules of different charges has been discussed in this book several times. The application to populations merely necessitates taking samples from a large number of individuals, electrophoresing, and preparing zymograms. *Zymograms* are made by causing the enzyme of interest to catalyze the reaction of a modified reactant (or substrate) to form a coloured product. The coloured product is localized in the area to which the enzyme has migrated during electrophoresis. If one has an enzyme that does not form dimers, tetramers, and so on, interpretation of allelic differences is relatively simple. Gel electrophoresis of such an enzyme: phosphoglucomutase of *Drosophila pseudoobscura*. Enzymes encoded by three alleles are distinguished by their mobilities. Homozygous and heterozygous genotypes have been analyzed by electrophoresis followed by a modified phosphoglucomutase reaction to form stained hands where the enzyme molecules were found.

The situation is a little more complex when the enzyme is active as a dimer. Let a_1 be the allele coding for A1 isozyme and a_2 be the allele coding for A2. Then, each homozygote still has one band of enzyme that can be found by electrophoresis and staining for enzyme activity as described. The heterozygote has three bands rather than two. If the heterozygote is a_1a_2 then the active enzyme molecules, dimers, can be A1A1, or A1A2, or A2A2. Each of these classes has its own distinct mobility. Imagine that A1A1 had a mobility of 100 mm and A2A2 had a mobility of 110 mm. We would expect A1A2 to have a mobility of 105 mm. Three bands of activity would be seen, at 100, 105, and 110 mm. An enzyme that behaves in this way is malate dehydrogenase of *Drosophila equinoxialis*. Gel electrophoresis of individual flies, followed by staining for malate dehydrogenase activity. Each fly had one or three bands of activity, just as the model predicts for the dimer case.

What if the active enzyme is a tetramer? The tetramer of polypeptide chain types A and B would have five bands: A_4, A_3B, A_2B_2, A_1B_3 and B_4. An enzyme of this structure is lactate dehydrogenase.

A noteworthy complexity in these systems is the effect of post-translational modification. Variation in xanthine dehydrogenase (XDH) of *Drosophila melanogaster* results from variation at the rosy locus, the structural gene for XDH. In addition, the XDH structure is affected by alleles at maroonlike (*mal*), cinnamon (*cin*), and also xanthine dehydrogenase (*lxd*). The active enzyme XDH appears to be a dimer of identical polypeptide chains so that the *mal*, *cin*, and *Ixd* effects are thought to be post—translational modifications of amino acids in the completed amino acid chain. This idea is supported by the finding that the modifying genes *mal*, *cin*, and *lxd* also affect the activity of aldehyde oxidase (AO), encoded by the *aldox* gene.

Finnerty and Johnson have recently examined the implications of this system for electrophoretic analysis, gel-sieving analysis, and thermolability analysis of variants thought to be alleles. They conclude that modifications of XDH and AO under the influence of *mal* and *lxd* cause changes in enzyme shape, enzyme activity, and possibly subunit dissociation behaviour. Both *mal* and *lxd* alleles display simple dominance rather than codominance. The effects of the two genes are additive on AO and XDH, rather than alternative. The heterozygotes for genes for posttranslational modifications, because of the dominance of such genes, produce one XDH band rather than the three bands predicted for heterozygotes of actual alleles of XDH structural genes. The characteristics of the XDH are quite different depending on the modifications, however.

Euphausia Adaptive Strategies and Polymorphism

One hypothesis relating polymorphism to ecology was put forward by Bretsky and Lorenz. They proposed that more environmental stability could be correlated with less polymorphism. The extinction patterns observed in fossils of marine invertebrates often were among organisms from stable environments, where it seemed possible that less polymorphism would have led to less potential for adaptation if conditions changed. Ayala and Valentine tested this hypothesis using several organisms. One series of particular note was *Euphausia*, the krill. *Euphausia superba* is circumantarctic, *E. mucronata* is temperate, and *E. distinguenda* is tropical, but the three species are otherwise closely related.

The data show that the most diversity is found for the tropical species *E. distinguenda* and the least for the circumantarctic species. These two environments are both physically stable; however, the food supply of the krill varies from abundance to scarcity in the antarctic but varies from abundance to superabundance in the tropics. Ayala and Valentine postulate that the unstable (and insufficient) levels of food supply probably limit the genetic diversity of the *E. superba*. In this case, the level of food supply, rather than the stability of the physical environment, was required to explain the different levels of polymorphism.

Coarse and Fine grained Environments and Polymorphism

Levins proposed in 1968 that large, mobile species with good homeostatic mechanisms would perceive their environment as *fine grained* and that smaller, more stationary, less homeostatic organisms would find their environments more coarse grained; that is, the environmental variations would have a larger impact on the organism. Levins postulated that this difference could be reflected in a greater

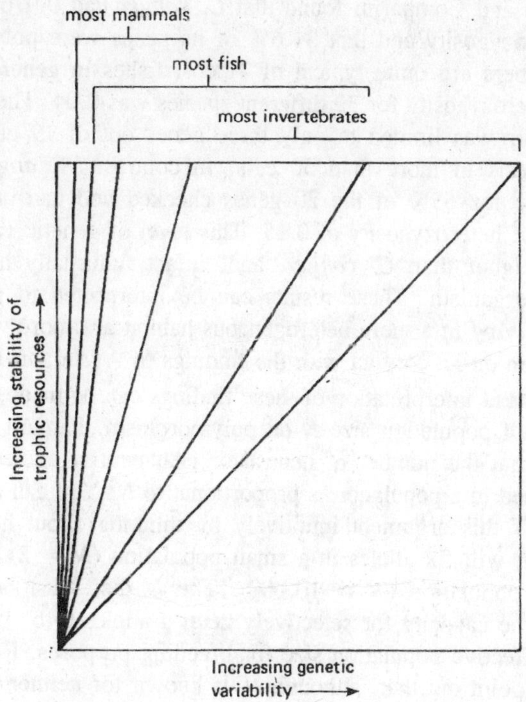

Fig. 12.5. Theoretical relationship between genetic variability and trophic resource stability in species of marine mammals, fishes and invertebrates, based on expectations of the hypothesis that generalized alleles are found in species with fine-grained strategies and specialized alleles in species with coarse-grained strategies.

genetic diversity among organisms forced to cope with a *coarse grained* environment. Invertebrates are more polymorphic than vertebrates in general. If one supposes that invertebrates have the coarse-grained experience and vertebrates the fine-grained experience, then the Levins model receives some support from this difference.

A comparison of related organisms that experience different environmental variations was undertaken by Rodino and Comparini. They compared the *Anguilla anguilla* L., which matures in fresh water rivers, then mates in the Sargasso Sea, with the conger eel, *Conger conger* L., which is strictly marine. *C. conger* mates in several areas of the deep ocean, and its spawn mature near the coasts where the water currents carry them. The two eels both migrate, and both are closely related members of the Teleost fishes, both belonging to the order Anguilliformes. Obviously, the *A. anguilla* experiences greater environmental variation than *C. conger*.

Rodino and Comparini found that *C. conger* had 0.076 observed mean heterozygosity and that 31.6% of its genes were polymorphic. These numbers are quite typical of Teleost fishes in general, whose average heterozygosity for 39 different species was 0.04. The variation in *C. conger* was limited to only three genes out of 19 checked by electrophoresis in more than 50 eels. In contrast, *A. anguilla* was polymorphic for 65% of the 20 genes checked and had an average frequency of heterozygosity of 0.15. This level of genetic variation is strikingly higher than *C. conger*, and, infact, unusually high for a vertebrate organism. These results can be interpreted to mean that organisms living in a more heterogeneous habitat are more variable, a conclusion in direct conflict with the findings of Ayala and Valentine.

A different interpretation of these findings can be made, based on the effects of population size N on polymorphism. Kimura and Ohta calculated that the number of neutral or near-neutral alleles that can be maintained in a population is proportional to N. You can appreciate the force of this argument intuitively by thinking about how easily genetic drift will fix alleles in a small population ($N = 2$) compared to a large population ($N = 10{,}000$). The N that must be used in predicting the capacity for selectively neutral alleles to be maintained is N_e the effective population size for breeding purposes. Rodino and Comparini point out that, although N_e is known for neither of the eel species studied, N_e is almost certainly much larger for *A. anguilla* than for *C. conger*. This is because *A. anguilla* mates panmictically with eels from throughout the North Atlantic in one large adult population at the Sargasso Sea, whereas *C. conger* has several mating areas instead. Thus, this finding of a large difference in polymorphism levels between rather closely related organisms is subject to at least two different interpretations.

Concluding Remark

Ecological genetics is the study of interrelationships between organisms and their environment, studied via genetic techniques. The study of alleles coding for proteins distinguishable by electrophoresis is particularly fruitful for ecologists studying population structure. Evolutionary genetics is attempting to see, via molecular genetics of the organisms existing today, what past relationships may have existed between organisms. The ecological niche is the temporal and spatial part of the environment occupied by a particular organism. In some cases the nature of the niche occupied by an organism can be shown to depend heavily on the genetically controlled colour patterns of the organism. The selective pressure against various alleles can favour the heterozygote in some niches, but favor one or the other homozygote in other niches. The environment of an organism may augment genetic drift, causing a bottle neck in population size or the opportunity to found a new population. These events can lead to rapid large changes in gene frequency, such as establishment of high frequencies of *achromatopsia* and Ellis-van Creveld syndrome in isolated human populations.

Speciation, separation of a relatively homogeneous population into two or more different populations, can occur when the populations are sympatric or allopatric. Rapid reproductive isolation can be achieved, for example, by suddenly arising self-fertility genes in a self-incompatible species. This type of new species is less genetically diverse than the parent species. Clines are changes in gene frequency in space. Genetic barriers usually do not exist along a dine; all members of the population can potentially interbreed. Clines are evidently often the result of different selective forces upon populations in different areas along the dine. Generally not all of the genetic differences between members of a population have the same chine. Instead, each character has a dine in response to one or more major features of the environment that may have no effect on another pair of alleles. Renewed contact between populations, after prolonged separation, can lead to an area of parallel dines called a *hybrid zone*.

Ecologists have shown that panmixis need not apply, even in small areas. Analysis of mice in a single barn floor showed that stable persisting genetic subgroups, evidently preferring to mate with their own group members, exist even within such a small locale.

Reproductive isolation can be achieved by premating or postmating isolation. Premating isolation often results from behavioral differences. Postmating isolation results from lowered fertility or fecundity of the hybrids. Behavioral mechanisms can also lead to outbreeding rather than reproductive isolation and inbreeding.

Stable polymorphisms in natural populations are generally high; one allele has not displaced its alternative allele(s), but the gene frequencies are not changing progressively with generations of mating. Many polymorphisms have been detected by analysis of proteins which move differently upon electrophoresis, but which are active in either electrophoretically distinguishable form. The factors responsible for the high degree of polymorphism have not been conclusively established, but the stability of the environment and the levels of food available have emerged as important factors.

13

EVOLUTIONARY GENETICS

Biological evolution is a genetically based, heritable change in one or more characteristics of a population or species over the course of many generations. Evolution can be viewed on a small scale as it relates to a single gene, or it can be viewed on a larger scale as it relates to the formation of new species. This process, also known as *microevolution*, concerns the changing composition of gene pools with regard to particular alleles. As we have seen, several evolutionary forces, such as mutation, inbreeding, genetic drift, migration, and natural selection, affect the allele and genotypic frequencies within natural populations.

In the first part of this chapter, we will be concerned with evolution on a larger scale, which leads to the origin of new species. This phenomenon was central to the development of evolutionary theory In more advanced courses, you may also learn about evolution on an even grander scale. The term *macroevolution* refers to evolutionary changes above the species level. It concerns the diversity of organisms established over long periods of time through the accumulated evolution and extinction of many species. Macroevolution involves relatively large changes in form and function that are sufficient to produce new species and higher taxa.

At the end of this chapter, we will tie together molecular genetics and the evolution of species. The development of techniques for analyzing chromosomes and DNA sequences has enhanced greatly our understanding of evolutionary processes. The term *molecular evolution* refers to the molecular changes in genetic material that underlie the phenotypic changes associated with evolution. The topic of molecular evolution is a fitting way to end our discussion of genetics, since it integrates the ongoing theme of this text—the relationship between molecular genetics and Traits in the grandest and most profound ways. Theodosius Dobzhansky, an influential evolutionary scientist, once said, "Nothing in biology makes sense except in the light of evolution." The extraordinarily diverse and seemingly bizarre array of species that exist on our planet are explained naturally within the context of evolution. An examination of molecular evolution allows us to make sense of the existence of these species at both the population and the molecular levels.

Origin of Species

The theory of evolution is attributed largely to Charles Darwin, a British naturalist, who was born in 1809. Like many great scientists, Darwin had a very broad background in science, which enabled him to see connections among different disciplines. His thinking was influenced greatly by theories of geology indicating that the Earth is very old, and that slow geological processes can lead eventually to substantial changes in the Earth's characteristics.

A second important influence on Darwin was his own experimental observations. His famous voyage on the *Beagle*, which lasted from 1832 to 1836, involved a careful examination of many different species. He observed the similarities among many discrete species, yet noted the differences that enabled them to be adapted to their environmental conditions. He was particularly struck by the distinctive adaptations of island species. For example, the finches found on the Galapagos Islands had unique phenotypic characteristics as compared with similar finches found on the mainland.

Finally, a third important impact on Darwin was a paper published in 1798, *Essay on the Principle of Population*, by Thomas Malthus, an English economist. Malthus asserted that the population size of humans can, at best, increase arithmetically due to increased land usage and improvements in agriculture, while the reproductive potential of humans can increase geometrically. He argued that famine, war, and disease will limit population growth, especially among the poor.

With these three ideas in mind, Darwin had largely formulated his theory of evolution by the mid-1840s. He then spent several years studying barnacles with out having published his ideas. The geologist Charles Lyell, who had greatly influenced Darwin's thinking, strongly encouraged Darwin to publish his theory of evolution. In 1856, Darwin began to write a long book to explain his ideas. In 1858, however,

Alfred Wallace, a naturalist working in the East Indies, sent Darwin an unpublished manuscript to read prior to its publication. In it, Wallace proposed the same ideas concerning evolution. Darwin therefore quickly excerpted some of his writings on this subject, and two papers, one by Darwin and one by Wallace, were published in the Linnaean Society of London. These papers were not widely recognized. A short time later, however, Darwin finished his book, The *Origin of Species*, which expounded his ideas in greater detail, and with experimental support. This book, which received high praise from many scientists and scorn from others, started a great debate concerning evolution. Although there were some flaws in his ideas, Darwin's work represents one of the most important contributions to our understanding of biology.

The theory of evolution was called "the theory of descent with modification through variation and natural selection" by Charles Darwin. As its name suggests, it is based on two fundamental principles: genetic variation and natural selection. A modern interpretation of evolution can view these principles with regard to speciation and microevolution:

1. *Genetic variation at the species level*: As we have seen, genetic variation is a consistent feature of most natural populations. Darwin ob served that many species exhibit a great amount of phenotypic variation. Although the theory of evolution preceded Mendel's pioneering work in genetics, Darwin observed that offspring resemble their parents more than they do unrelated individuals. Therefore, he assumed that some phenotypic variation is passed from parent to offspring.

   At the micro evolutionary level: Genetic variation results from differences in DNA sequences that create alleles, changes in chromosome structure, and alterations in chromosome number. These differences are caused by random mutations. Alternative alleles may affect the functions of the proteins they encode and thereby affect the phenotype of the organism. Likewise, changes in chromosome structure and number may affect gene expression and thereby influence the phenotype of the individual.
2. *Natural selection at the species level*: Darwin agreed with Maithus that most species produce many more offspring than survive and reproduce. This creates a "*struggle for existence*" that results in a "*survival of the fittest*?' Over the course of many generations, those individuals who happen to possess the most favourable traits will dominate the composition of the population. The ultimate

result of natural selection is to make a species better adapted to its environment and/or more efficient at reproduction.

At the microevolutionary level: Some alleles may encode proteins that provide the individual with a selective advantage. Over time, natural selection may change the allele frequencies of genes and thereby lead to the fixation of beneficial alleles and the elimination of detrimental alleles.

In this section, we will examine the features of evolution as it occurs in natural populations over time.

Biological Species is a Group of Reproductively Isolated Individuals

Perhaps it is now obvious that the study of evolution requires a dialogue among naturalists, taxonomists, and geneticists. However, after Darwin's death, and during the first three decades of the 20th century, naturalists and taxonomists hardly interacted with geneticists. This situation changed dramatically due to the efforts of Theodosius Dobzhansky.

Dobzhansky was a naturalist and a taxonomist (with an interest in insects) who starting working with Thomas Hunt Morgan in 1927 (a pioneer in studying genetic mutations in *Drosophila*). Dobzhansky's profound perception was that an understanding of evolution must be rooted in the discreteness and discontinuity among different species. A key contribution of Dobzhansky was the idea that discrete species exist due to *isolating mechanisms*. He proposed that species are reproductively isolated from other species. In 1937, he published a book, *Genetics and the Origin of Species*, which laid the foundation for our modern understanding of evolution. Unlike Darwin (who thought that environmental variation was somehow responsible for genetic variation), he argued that random mutations provide the basis of variation, but he agreed with Darwin that natural selection is responsible for creating adaptive changes in species. In addition, Dobzhansky acknowledged that forces such as genetic drift and migration could also play important roles.

Ernst Mayr at Harvard University expanded on the ideas of Dobzhansky to provide a biological definition of a species. According to Mayr's *biological species concept*, a *species* is defined as a group of individuals whose members have the potential to interbreed with one another in nature to produce viable, fertile offspring, but who cannot successfully interbreed with members of other species. There are several different ways to achieve reproductive isolation. These can be classified as prezygotic mechanisms that prevent the formation

Table 13.1. Types of reproductive isolation between different species

Mechanism	*Description*
Prezygotic mechanisms	
Habitat isolation	Species may occupy different habitats, so that they never come in contact with each other.
Temporal isolation	Species have different mating or flowering seasons or times of day, or become sexually mature at different times of the year.
Sexual isolation	Sexual attraction between males and females of different animal species is limited due to differences in behavior, physiology, or morphology.
Mechanical isolation	the anatomical structures of genitalia prevents mating between different species
Gametic isolation	Gametic transfer takes place, but the gametes fail to unite with each other. This can occur because the male and female gamete fail to attract, because they are unable to fuse, or because the male gametes are inviable in the female reproductive tract of another species.
Postzygotic mechanisms	
Hybrid inviability	The egg of one species is fertilized by the sperm from another species, but the fertilized egg fails to develop past early embryonic stages.
Hybrid sterility	The interspecies hybrid survives, but it is sterile. For example, the mule is a cross between a female horse (*Equus caballus*) and a male donkey (*Equus asinus*).
Hybrid breakdown	The F_1 interspecies hybrid is viable and fertile, but succeeding generations (i.e., F_2, etc.) become increasing inviable. This is usually due to formation of less-fit genotypes by genetic recombination.

of a zygote, and postzygotic mechanisms that prevent the development of a viable individual after fertilization has taken place. Reproductive isolation in nature may be circumvented when species are kept in captivity. For example, different species of the genus *Drosophila* rarely mate with each other in nature. In the laboratory, however, it is fairly easy to produce interspecies hybrids.

Hugh Paterson at the University of Witwatersrand in South Africa proposed an alternative way to view a species known as the *species recognition concept*. This idea agrees with the biological species concept in that it recognizes that species are reproductive communities that are isolated from other species. The difference in the species recognition concept concerns the mechanisms by which new species arise. According to Paterson, the sexual recognition of members of the same species acts as a positive force in evolution. There is variation in the ability of mates to locate each other and participate in a productive fertilization. Therefore, natural selection favours the survival of members who can recognize each other and mate successfully. In this regard, recognition can have a very broad meaning. It could involve factors such as visual recognition, behavioural recognition, or even cellular recognition (e.g., sperm and egg cells).

The species recognition concept is consistent with many examples of secondary sexual traits that lead to reproductive isolation. The term *sexual selection*, originally described by Darwin, refers to "the advantage which certain individuals have over others of the same sex and species solely in respect of reproduction." In many species of animals, sexual selection affects male characteristics more in tensely than it does female. Unlike females, who tend to be fairly uniform in their reproductive success, male success tends to be more variable, with some males mating with many females and others not mating at all. There are many examples of male traits that appear to be solely for the purpose of recognition by the female of the species. These differences allow the females to recognize and choose an appropriate mate. By comparison, the females of these species look rather similar to each other.

Speciation Usually Occurs via a Branching Process called Cladogenesis

By examining the fossil record, evolutionary biologists have found two different patterns of speciation. During *anagenesis* (from the Greek, ana, meaning "up' and genesis, meaning "origin"), a single species is transformed into a different species over the course of many generations. During this process, the characteristics of the species change due to both neutral evolutionary forces and adaptive forces promoted by natural selection. As a result of natural selection, the new species maybe better adapted to survive in its original environment, or the environment may have changed so that the new species is better adapted to the new surroundings.

By comparison, *cladogenesis* (from the Greek *clados*, meaning "branch") involves the division of a species into two or more species. This is the most common form of speciation. Although cladogenesis is usually thought of as a branching process, it commonly occurs as a budding process in which a single species divides into the original species plus a new species with different characteristics. If we view evolution as a tree, the new species buds from the original species and develops characteristics that prevent it from interbreeding with the original one.

Divergent Evolution can be Allopatric, Parapatric, or Sympatric

The divergence of one species into two or more discrete species is the most common form of speciation. Depending on the geographic locations of the evolving population(s), speciation is categorized as allopatric, parapatric, or sympatric. The following discussion will describe some examples.

Allopatric speciation (from the Greek *allas*, "other and Latin *patria*, "home land") is thought to be the most prevalent way for a species to diverge. It occurs when members of a species become geographically separated from the other members. This form of speciation can occur by the geographic subdivision of large populations by geological processes. For example, a mountain range may emerge and split a species that occupies the lowland regions. Or a creeping glacier may divide a population. Two species of antelope squirrels occupy opposite rims of the Grand Canyon. On the south rim is Harris's antelope squirrel (*Ammospermophilus harrisi*), while a closely related white-tailed antelope squirrel (*Ammospermophilus leucurus*) is found on the north rim. Presumably, these two species evolved from a common species that existed before the canyon was formed. Over time, the accumulation of genetic changes in the two populations led to the formation of two morphologically distinct species. Interestingly, birds that can easily fly across the canyon have not diverged into different species on the opposite rims.

Allopatric speciation can also occur via a second mechanism, known as the *founder effect*, which is thought to be more rapid and frequent than allopatric speciation caused by geological events. The founder effect, which was discussed in occurs when a small group migrates to a new location that is geographically separated from the main population. For example, a storm may force a small group of birds from the mainland to a distant island. In this case, migration between the island and the mainland populations is a very infrequent event. In a relatively short period of time, the founding population on the island may evolve in to a new species. Several evolutionary forces may contribute to this rapid evolution. First, genetic drift may quickly lead to the random fixation of certain alleles and the elimination of the other alleles from the population. Another factor is natural selection. The environment on an island may differ significantly from the mainland environment.

Table 13.2. Common genetic mechanisms that underlie allopatric, parapatric, and sympatric speciation.

Type of Speciation	*Common genetic mechanisms responsible for speciation*
Allopatric: two large Populations are separated by geographic barriers.	Many small genetic differences may accumulate over a long period of time, leading to reproductive isolation. Some of these genetic differences may be adaptive; others will be neutral.
Allopatric: a small founding population separates from the main population.	Genetic drift and shifts in adaptive peaks may lead to the rapid formation of a new species. If the group has moved to an environment that differs from its previous environment, natural selection is expected to favour beneficial alleles and eliminate harmful alleles.
Parapatric: two populations occupy overlapping ranges so that a limited amount of interbreeding occurs.	A new combination of alleles or chromosomal rearrangement may limit the amount of gene flow between neighbouring populations due to the very low fitness of hybrid offspring.
Sympatric: within a population occupying a single habitat in a continuous range, a small group evolves into a reproductively isolated species.	An abrupt genetic change leads to reproductive isolation. For example, a mutation may affect gamete recognition. In plants, the formation of a polyploidy often leads to the formation of new species, because the interspecies hybrid (e.g., diploid x tetraploid) is triploid and sterile.

To explain how natural selection may act on the phenotypic effects of many alleles, Sewall Wright developed the concept of *adaptive peaks* to describe combinations of many alleles that provide an optimal fitness for individuals in a stable environment. There are several different combinations of alleles that may lead to highly fit individuals. If a small change occurs in the frequency of alleles at one or a few loci, it will drive the population off an adaptive peak and into a (less fit) valley; natural selection, though, will tend to push the population back to the original peak.

When, however, a small group migrates to a new environment, the adaptive landscape is likely to be changed. Allele combinations

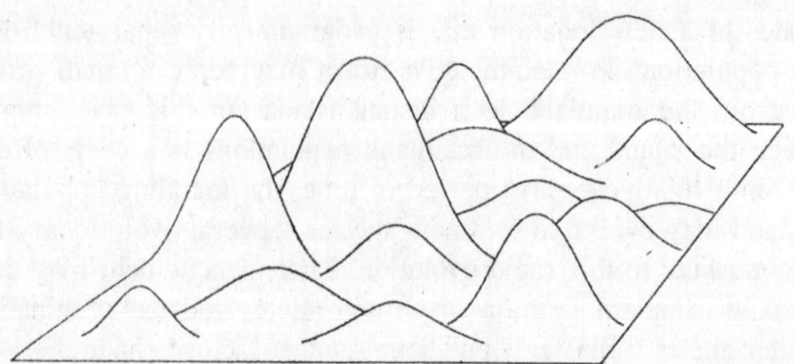

Fig. 13.1. The concept of adaptive peaks.

that were highly adaptive in the original environment may be much less so in the new environment. If this occurs, natural selection may lead to a shift toward a new adaptive peak in which the frequencies of many alleles have been changed. In other words, natural selection will act to change the morphological features of the organism to make it better adapted to its new environment. Similarly, in a species occupying a large geographic range, the adaptive of local population may differ to variation in the environmental condition with in the range.

Parapatric speciation (from the greek *para* "beside") occur when members of a species are separated only partially or when a species is very sedentary in these cases, the geographic separation is not complete. For example a mountain range may divide a species into two populations, but with breaks in the range where time two groups are connected physically. In these zones of contact, the members of two populations can interbreed, although this tends to occur in frequently. Likewise, parapatric speciation may occur among very sedentary species even though no geographic isolation exists. Certain organisms are so sedentary that 100 to 1000 meters may be sufficient to limit the interbreeding between neighbouring groups. Plants, terrestrial snails, rodents, grasshoppers, lizards, and many flightless insects may speciate in a parapatric manner.

Prior to parapatric speciation, the zones where two populations can interbreed are known as *hybrid zones*. For speciation to occur, the amount of gene flow within the hybrid zones must become very limited. In other words, there must be selection against the offspring produced in the hybrid zone. One way that this can happen is that each of the two parapatric populations may accumulate different neutral chromosomal rearrangements, such as inversions and balanced translocations. As was discussed earlier, if an individual has one chromosome with a large inversion and one that does not carry the inversion, crossingover during meiosis can lead to the production of grossly abnormal chromosomes. Therefore, such an individual is substantially less fertile. By comparison, an individual who is homozygous for two normal chromosomes or for two chromosomes carrying the same inversion will be fertile, because crossing over can proceed normally.

Finally, *sympatric speciation* (from the Greek *sym*, "together") occurs when members of a species initially occupy the same habitat within the same range. In plants, a common way for sympatry to occur is by the formation of polyploids. A complete non-disjunction of chromosomes during gamete formation can increase the number of chromosome sets within the same species (autopolyploidy) or between different species (allopolyploidy). Polyploidy is so frequent in plants that it is a major form of speciation. In ferns and flowering plants, about 30 to 35% of the species are polyploid. By comparison, polyploidy is much less common in animals, but it can occur. For example, roughly 30 species of reptiles and amphibians have been identified that are polyploids derived from diploid relatives.

The formation of a polyploid can lead abruptly to reproductive isolation. A an example, let's consider the probable formation of a natural species of common hemp nettle known as *Galeopsis tetrahit*. As described in this species is thought to be an allopolyploid between two diploid species, *Galeopsis pubescens* and *Galeopsis speciosa*. These two diploid species contain 16 chromosomes each (2n = 16), while *G. tetrahit* contains 32. What would happen in crosses between the allopolyploid and the diploid species? The allopolyploid crossed to another allopolyploid produces an allopolyploid. The allopolyploid is fertile, because all of its chromosomes occur in homologous pairs that can segregate evenly during meiosis. However, a cross between an allopolyploid and a diploid produces an offspring that is monoploid for one chromosome set and diploid for the other set. These offspring are expected to be sterile, because they will produce highly aneuploid gametes that have incomplete sets of chromosomes. This hybrid sterility causes the allopolyploid to be reproductively isolated from the diploid species.

Muntzing was Able to Re-create an Allopolyploid Species by Selective Breeding

In studying biological evolution, we cannot look into the past and know with complete certainty how events occurred. An alternative way to explore past events is called *retrospective testing*. In this procedure, a researcher formulates a hypothesis and then collects observations as a way to confirm or refute the hypothesis. Therefore, this approach relies on the observation and testing of already existing materials, which themselves are the results of past events. The careful study or manipulation of these materials may then be used to support or refute ideas concerning how evolution may have occurred in the past.

An interesting example of this approach is provided by a study in the 1930s by Arne Muntzing, a Swedish botanist. It was Muntzing who first proposed that *Galeopsis tetrahit* is an allopolyploid that arose from an interspecies cross between *G. pubescens* and *G. speciosa*. To confirm this hypothesis, he at tempted to "reenact" evolution by recreating an "artificial" *G. tetrahit* by crossing *G. pubescens* and *G. speciosa* in a way that produced an allopolyploid.

Hypothesis

G. tetrahit is an allopolyploid that contains a diploid chromosomal composition from *G. pubescens* and *G. speciosa*.

Testing the hypothesis

Starting material: Specimens of *G. pubescens*, *G. speciosa*, and *G. tetrahit*.

1. Cross *G. pubescens* to *G. speciosa*. Note: In most cases, this will produce alloploids that have 16 chromosomes.
2. Make an $F_1 \times F_1$ cross. The fertility of the alloploids is very low, because the *G. pubescens* chromosomes usually do not pair properly with the *G. speciosa* chromosomes during meiosis. For this reason, the F_2 offspring obtained are expected to have abnormal numbers of chromosomes.
3. Examine the chromosome compositions of 200 F_2 offspring.
4. Among 200 offspring examined, one of them had 16 chromosomes from *G. speciosa* and 8 chromosomes from *G. pubescens*. Presumably, this occurred because of nondisjunction of the *G. speciosa* chromosomes. Let's call this strain the "intermediate F_2 strain."
5. Mate the intermediate F_2 strain to *G. pubescens*.
6. Only one seed was obtained! It was diploid for the chromosomes of both species. Presumably, it arose from a normal *G. pubescens* gamete fusing with a gamete from the intermediate F_2 strain that had resulted from complete non disjunction. This allopolyploid strain was called "artificial" *G. tetrahit*.

7. Examine the morphological features of the "artificial" *G. tetrahit* and compare them with the natural species.
8. Explore the ability of the "artificial" *G. tetrahit* to be crossed with itself and with the natural *G. tetrahit* species.

Data

	Artificial *G. tetrahit*	*Natural* *G. tetrahit*	*G. pubescens*	*G. speciosa*
Chromosomal composition	32	32	16	16
Mean flower size (mm)	19	19	29	27
Calyx length (mm)	15	15	12	18
Degree of stem hairiness	Moderate	Moderate	Low	High
Fertility:	The artificial *G. tetrahit* could produce fertile offspring when it was crossed to another artificial *G. tetrahit* and it could produce fertile offspring when it was crossed to a natural strain of *G. tetrahit*. Microscopically, normal chromosome pairing was observed during meiosis in the artificial and natural *G. tetrahit* hybrid.			

Interpreting the data

Before we interpret the data, let's examine how the "artificial" *G. tetrahit* was produced. Muntzing knew that each of these two species occasionally produces diploid gametes (by abnormal non-disjunction), which could fuse together and produce an allopolyploid in a single step. However, he expected that this would be an extremely rare event. Therefore, rather than hoping for such a rare event to occur, he decided to make the allopolyploid in a two-step procedure.

Muntzing first produced a strain that was diploid for *G. speciosa* chromosomes and monoploid for the *G. pubescens* chromosomes. To develop such a strain, he first crossed *G. pubescens* to *G. speciosa*. Muntzing assumed that the F_1 alloploids would have a higher frequency of non disjunction because they would contain one set of chromosomes from two different species. He hoped that the $F_1 \times F_1$ cross would tend to produce gametes that exhibited nondisjunction. An occasional off spring might be perfectly diploid for the chromosomes of one species (e.g., *G. speciosa*) and perfectly monoploid for the other species (e.g., *G. pubescens*). Out of 200 F_2 off spring he examined, he found one F_2 strain that contained 16 chromosomes from *G. speciosa* and 8 chromosomes from *G. pubescens* (which we have named the inter mediate F_2 strain).

Due to its unusual chromosomal composition, Muntzing knew that the inter mediate F_2 strain would usually make highly aneuploid gametes that could not produce viable seeds. He reasoned that the intermediate F strain might occasionally produce a gamete that would be the result of complete non-disjunction. Such a gamete would be diploid for the *G. speciosa* chromosomes and monoploid for the *G. pubescens* chromosomes. If this gamete fused with a normal haploid gamete from *G. pubescens*, this would result in a viable polyploid offspring containing two sets of chromosomes from each species—the composition of *G. tetrahit*. Therefore, to obtain an offspring with the chromosomal composition of *G. tetrahit*, he crossed the intermediate F_2 strain to a normal *G. pubescens* strain. Since the intermediate F_2 strain usually produced aneuploid gametes, it was almost unable to make any viable seeds. However, Muntzing did obtain one seed that had the chromosomal composition he wanted: it had a diploid number of chromosomes from both species.

This seed was the artificial *G. tetrahit* Muntzing had hoped for. Based on morphological characteristics, the artificial *G. tetrahit* species confirmed his hypothesis. It had traits that were similar to the natural *G. tetrahit* species but different from *G. pubescens* and *G. speciosa*. Furthermore, the artificial and natural *G. tetrahit* strains could be mated to each other to produce fertile offspring. Overall, Muntzing came to the conclusion that: "not only the artificial tetraploid but probably also natural *G. tetrahit* represents a synthesis of *pubescens*- and *speciosa*-genomes. Consequently, this case might be regarded as the first instance where a species already existing in nature has been synthesized from the genomes of two other species."

Evolution can Proceed Gradually or be Punctuated by Periods of Rapid Change

As we have seen, there are many different genetic mechanisms that give rise to new species. For this reason, the rates of evolutionary change are not constant, although the degree of inconstancy has been debated since the time of Darwin. Even Darwin himself suggested that evolution can occur at fast and slow paces. The *gradualism* hypothesis suggests that each new species evolves continuously over long spans of time. The principal idea is that large phenotypic differences that produce the divergence of species are due to the accumulation of many small genetic changes.

In 1972, Miles Eldrege at the American Museum of Natural History in New York and Stephen Jay Gould at Harvard University proposed an alternative hypothesis, called *punctuated equilibrium*. According to this proposal, species exist relatively unchanged for many generations. During this period, the species is in equilibrium with its environment. These long periods of equilibrium are punctuated by relatively short periods (i.e., on a geological time scale) during which evolution occurs at a far more rapid rate.

The phenomenon of punctuated equilibrium is largely supported by the fossil record. Paleontologists rarely find a gradual transition of fossil forms. Instead, it is much more common to observe species appearing as new forms rather suddenly in a layer of rocks, persisting relatively unchanged for a very long period of time, and then suddenly becoming extinct. It is presumed that the transition period during which a previous species evolved into a new, morphologically distinct species was so short that few, if any, of the transitional members were preserved as fossils. Therefore, the fossil record primarily contains representatives from the long equilibrium periods.

As was discussed earlier, rapid evolutionary change can be explained by genetic phenomena. As we have seen throughout this text, single gene mutations can have dramatic effects on phenotypic characteristics. Therefore, only a small number of new mutations may be required to alter phenotypic characteristics, eventually producing a group of individuals that make up a new species. Likewise, genetic events such as changes in chromosome structure (e.g., inversions and translocations) or chromosome number may abruptly create individuals with new phenotypic traits. On an evolutionary time scale, these types of events can be rather rapid, since one or only a few genetic changes can have a major impact on the phenotype of the organism.

In conjunction with genetic changes, species may also be subjected to sudden environmental shifts that quickly drive the gene pool in a particular direction via natural selection. For example, a small group may migrate to a new environment in which different alleles provide better adaptation to the surroundings. Alternatively, a species may be subjected to a relatively sudden environmental event that has a major impact on survival. There may be a change in climate, or a new predator may have infiltrated the geographic range of the species. To ensure survival, these kinds of events may necessitate the rapid evolution of the gene pool by natural selection so that the new members of the population can survive the climatic change or have phenotypic characteristics that allow them to avoid the predator. Overall, the fossil record and known genetic phenomena tend to support the punctuated equilibrium model of evolution, although the debate is by no means re solved. Rapid evolutionary changes are quite common in many phylogenetic lines, leading to the rapid branching of evolutionary trees. However, gradual changes are sometimes observed in certain species over long periods of time. Furthermore, the gradual accumulation of mutations has been revealed by molecular analyses of DNA.

Molecular Evolution

Historically, taxonomists have categorized different species primarily on the basis of phenotypes. Initially, morphological differences were used to construct evolutionary trees in which species that are more similar in appearance tended to be placed closer together on the tree. In addition, species have been categorized based on differences in physiology, biochemistry, and even behaviour. However, since evolution is based on genetic changes, it makes sense to categorize species based on the properties of their genetic material (i.e., their genotypes). Those species that are closely related evolutionarily are expected to have greater similarities in their genetic material than are distantly related species.

The advent of molecular approaches for analyzing genetic material and gene products has revolutionized the field of evolution. In the past few decades, molecular genetics has taken on the dominant role in facilitating our understanding of speciation and evolution. Differences in nucleotide sequences are quantitative and can be analyzed using mathematical principles in conjunction with computer pro grams. Evolutionary changes at the DNA level can be objectively compared among different species to establish evolutionary relationships. Furthermore, this approach can be used to compare any two existing organisms, no matter how greatly they differ in their morphological traits. For example, we can compare DNA sequences between humans and bacteria, or between plants and fruit flies. Such comparisons would be very difficult at a morphological level.

There are many ways to compare the genetic material and gene products among different species. Table 13.3 lists the most common molecular approaches that geneticists use to evaluate evolutionary relationships. In this section, we will focus most of our attention on the evolution of gene sequences.

Homologous Genes are Derived from a Common Ancestral Gene

Two genes are said to be *homologous* if they are derived from the same ancestral gene. Genes can exhibit *interspecies homology* and

Table 13.3. Experimental approaches used to compare species at the molecular level

Technique	*Explanation*
DNA Level	
DNA sequencing	The degree of similarities and differences between DNA sequences are used to establish evolutionary relationships. This approach, along with the analysis of the deduced amino acid sequences of structural genes, is the most common and reliable method in use.
Hybridization	Genetic material from two different species is allowed to hybridize. The rate of hybridization will be faster for closely related species.
Restriction mapping	Segments of DNA are isolated from different species and subjected to restriction mapping. Closely related species will have more similar restriction maps.
Cytogenetic analysis	The chromosomes of different species are analyzed microscopically. Closely related species will have identical or similar chromosome numbers and banding patterns.
Protein Level	
Deduced amino acid sequence .	DNA sequencing is technically much easier to perform than amino acid sequencing. Therefore, to determine an amino acid sequence of a protein, the coding region of a structural gene is determined, and the amino acid sequence is then deduced using the genetic code. The amino acid sequences of a given protein will be more similar between closely related species.
Immunological methods	Antibodies that recognize specific macromolecules, usually on the cell surface, are tested on different species. Antibodies that recognize macromolecules from one species will often recognize closely related species, but not from distantly related species.

intraspecies homology. Hemoglobin is an oxygen-carrying protein found in all vertebrate species. Adult hemoglobin is composed of two different types of subunits, encoded by the α and β-globin genes. The sequences are homologous between humans and horses because of their evolutionary relationship. Therefore, the homology between human and horse globin genes is an example of interspecies homology.

Within a particular vertebrate species, several globin genes are homologous to each other. For example, the human α-globin gene is homologous to the human β-globin gene. On rare occasions, an unequal crossover event will add an extra copy of a gene to a chromosome. Over time, the two or more copies may accumulate distinct mutations that cause their sequences to diverge.

Variation in Gene Sequences can be Used to Construct Phylogenetic Trees

A *phylogenetic tree* is a diagram that describes the evolutionary relationships among different species. In 1963, Linus Pauling and Emile Zukerkandl were the first to suggest that molecular data should be used to establish evolutionary relationships. With advances in DNA sequencing technology, the amount of phylogenetic information in genetic sequences is immense. Molecular evolutionists want to analyze the information within genetic sequences to obtain a better understanding of evolutionary links.

Nucleotide and amino acid sequence data are particularly well suited to the construction of evolutionary trees, because genetic sequences change over the course of many generations due to the accumulation of mutations. Therefore, when comparing homologous genes in different organisms, the DNA sequences from closely related organisms are more similar to each other than are the sequences from distantly related species. In a sense, the relatively constant rate of neutral mutation acts as a "molecular clock" on which to measure evolutionary time. According to this idea, neutral mutations will become substituted at an average rate equal to the rate of mutation per generation. On this basis, the genetic divergence between species that is due to neutral mutations reflects the time elapsed since their latest common ancestor.

For molecular evolutionary studies, the DNA sequences of many genes have been obtained from a wide range of sources. Several different types of gene sequences have been used to construct evolutionary trees. One very commonly analyzed gene is that encoding 16S rRNA, a small subunit rRNA. This gene has been sequenced from thousands of different species. It is as reliable a molecular measure of phylogenetic relationships among organisms as is now available. Because

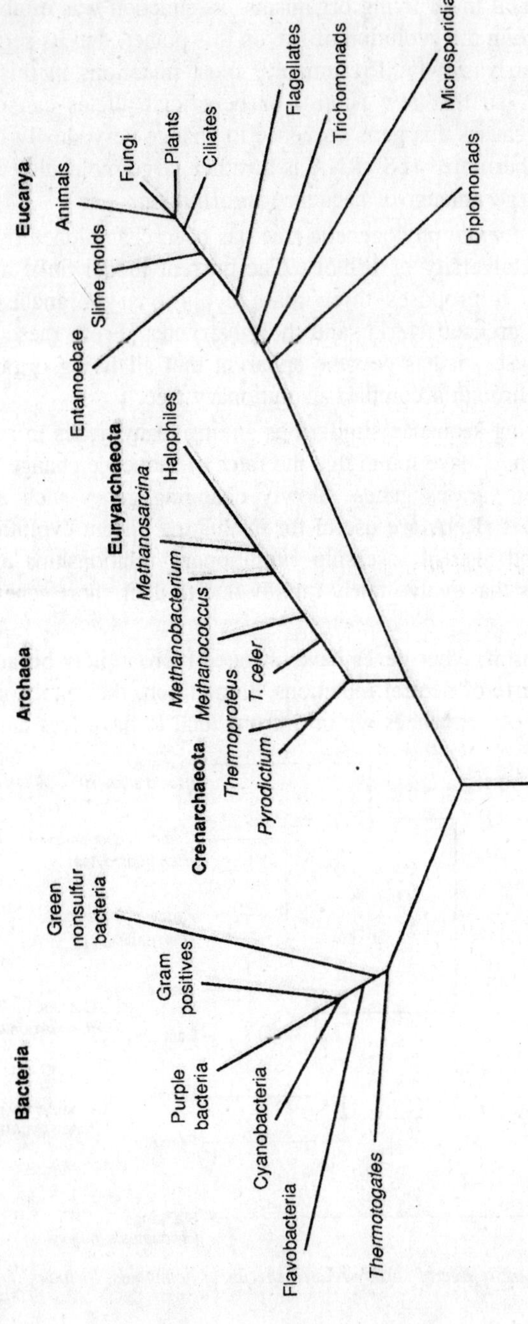

Fig. 13.2. A phylogenetic tree of all life on earth based on 16S rRNA data.

rRNA is universal in all living organisms, its function was established at an early stage in the evolution of life on this planet, and its sequence has changed fairly slowly. Presumably, most mutations in this gene are deleterious, so that few neutral or beneficial alleles can occur. This limitation causes this gene sequence to change very slowly during evolution. Furthermore, 16S rRNA is a rather large molecule, and so it contains a large amount of sequence information.

This *three-domain* phylogenetic tree has been championed by Carl Woese at the University of Illinois. The current model links all life on our planet. It proposes three main evolutionary branches: the eubacteria, the archaebacteria, and the eukaryotes. From these types of genetic analyses, it has become apparent that all living organisms are connected through a complex evolutionary tree.

By comparing sequence similarities among many genes in myriad species, researchers have found that the rates of sequence change differ dramatically for various genes. Slowly changing genes such as the gene that encodes rRNA are useful for evaluating distant evolutionary relationships and also in assessing evolutionary relationships among bacterial species that evolve fairly rapidly due to their short generation times.

By comparison, other genes have changed more rapidly because of a greater tolerance of neutral mutations. In addition, the mitochondrial genome and DNA sequences within introns tend to have less negative

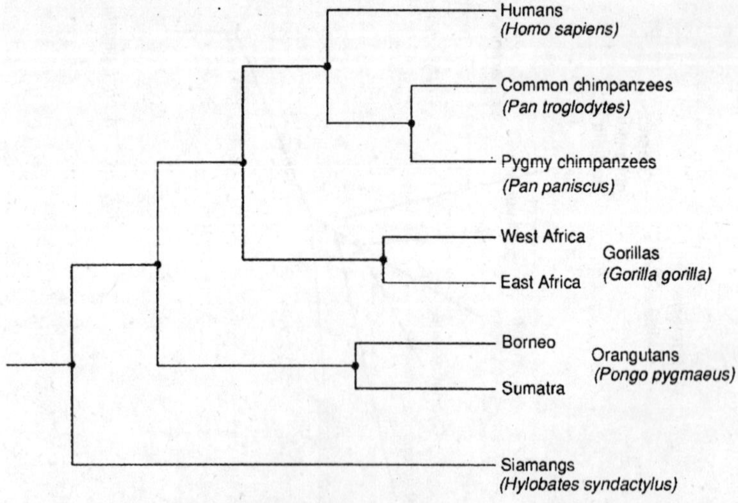

Fig. 13.3. A phylogenetic tree of closely related species of hominoids, including humans.

selective pressure, and so their sequences change fairly rapidly during evolution. More rapidly changing DNA sequences have been used to elucidate more recent evolutionary relationships, particularly among eukaryotic species that have long generation times and tend to evolve more slowly. In these cases, slowly evolving genes may not be very useful for establishing evolutionary relationships, because two closely related species likely have identical or nearly identical DNA sequences for such genes. Instead, it is easier to find sequence differences among closely related species when the DNA sequences are more rapidly changing. This tree was derived by comparing DNA sequences in a mitochondrial gene that encodes a protein called cytochrome oxidase subunit II.

Scientists can Examine the Relationships between Living and Extinct Flightless Birds by Analyzing Ancient DNA and then Comparing DNA Sequences

The vast majority of our knowledge concerning molecular evolution has come from the analysis of DNA samples collected from living species. Using this approach, we can infer the prehistoric changes that gave rise to present-day DNA sequences. As an alternative, scientists have discovered that it is occasionally possible to obtain DNA sequence information from species that have lived in the past. In 1984, the first successful attempt at determining DNA sequences from extinct species was accomplished by groups at the University of California at Berkeley and the San Diego Zoo, including Russell Higuchi, Barbara Bowman, Mary Freiberger, Oliver Rynder, and Allan Wilson. They obtained a sample of dried muscle from a museum specimen of the quagga (*Equus quagga*), a zebra-like species that became extinct in 1883. This piece of muscle tissue was obtained from an animal that had died 140 years ago. A sample of its skin and muscle had been preserved in salt in the Museum of Natural History at Mainz, Germany.

The researchers were able to extract DNA from the sample, clone pieces of it into vectors, and then sequence hybrid vectors containing the quagga DNA. At about the same time, Svante Paabo at the University of Munich obtained DNA sequences from an Egyptian mummy that was more than 2000 years old. These pioneering studies opened the field of *ancient DNA analysis*.

Since the mid-1980s, many researchers have become excited about the information that might be derived from sequencing DNA obtained from older specimens. Currently there is debate concerning how long

DNA can remain significantly intact after an organism has died. Over time, the structure of DNA is degraded by hydrolysis and the loss of purines. Nevertheless, under certain conditions (e.g., cold temperature, low oxygen, and so forth), DNA samples may be stable as long as 50,000 to 100,000 years.

In most studies involving prehistoric specimens (in particular, much older than the salt-preserved quagga sample), the ancient DNA is extracted from bone, dried muscle, or preserved skin. These samples are often obtained from museum specimens that have been gathered by archaeologists. However, it is unlikely that enough DNA will be extracted to enable a researcher to directly clone the DNA into a vector. Since 1985, however, the advent of PCR technology has made it possible to amplify the very small amounts of DNA using PCR primers that flank a region within the 12S rRNA gene, a slowly changing gene. In recent years, this approach has been used to elucidate the phylogenetic relationships between modern and extinct species.

In the experiment described here, published in 1992, Alan Cooper, Cecile Mourer-Chauvire, Geoffrey Chambers, Arndt von Haeseler, Allan Wilson, and Svante Paabo investigated the evolutionary relationships between some extinct and modern species of flightless birds. Two groups of flightless birds, the kiwis and the moas, existed in New Zealand during the Pleistocene. The moas are now extinct, although 11 species were formerly present. In this study, the researchers investigated the phylogenetic relationships between the kiwis and moas of New Zealand, and several other (non-extinct) species of flightless birds. These included the emu and the cassowary (found in Australia and New Guinea), the ostrich (found in Africa and formerly Asia), and two rheas (found in South America).

Hypothesis

Because DNA is a relatively stable molecule it can be PCR amplified from a preserved sample of a deceased organism and subject to DNA sequencing .A comparison of DNA sequence with modern species may help elucidate the phylogenetic relationship between extinct and modern species.

Testing the hypothesis

Starting material: Tissue samples from four extinct species of moas were obtained from museum specimens. Tissue samples were also obtained from three species of kiwis, one ostrich, one cassowary, one emu, and two species of rhea.

1. For soft tissue samples, treat with proteinase K (which digests protein) and a detergent that dissolves cell membranes. This releases the DNA from the cells.
2. Individually, mix the DNA samples with a pair of PCR primers that are complementary to the 12S rRNA gene. Note: Primers recognize the 12S rRNA gene.
3. Subject the samples to PCR.
4. Subject the amplified DNA fragments to DNA sequencing.
5. Align the DNA sequences to each other.

Interpreting the data

The data illustrate a multiple sequence alignment of the amplified DNA sequences. The first line shows the DNA sequence of one of the extinct moas species. Underneath it are the sequences of the other species. When the other sequences are identical to the first sequence, a dot is placed in the corresponding position. When the sequences are different, the nucleotide base (A, T, G, or C) is placed there. In a few regions, the genes are different lengths. In these cases, a dash is placed at the corresponding position.

As you can see from the large number of dots, the sequences among all of these flightless birds are very similar. To establish evolutionary relationships, we need to focus on the few differences that occur. Some surprising results were obtained. The sequences from the kiwis (a New Zealand species) are actually more similar to the sequence from the ostrich (an African species) than they are to those of the moas, which were once found in New Zealand. Likewise, the kiwis are more similar to the emu and cassowary (found in Australia and New Guinea) than they are to the moas. Contrary to their original expectations, the authors concluded that the kiwis are more closely related to Australian and African flightless birds than they are to the moas. They proposed that New Zealand was colonized twice by ancestors of flightless birds.

Genetic Variation at the Molecular Level Commonly is Associated with Neutral Changes in Gene Sequences; Non-neutral Changes are Acted on by Natural Selection

We learned earlier that genetic variation is prevalent among natural populations. During the last few decades, a great debate has occurred among population and evolutionary geneticists. The debate centers on the reason for genetic variation in natural populations. Is it due primarily to mutations that are favored by natural selection, or to random genetic events?

A *nonneutral mutation* is one that affects the phenotype of the organism and can be acted on by natural selection. A non neutral mutation may only subtly alter the phenotype of an organism, or it may have a major impact. According to Darwin, natural selection is the agent that leads to evolutionary change in populations. It selects for the survival of the fittest and thereby promotes the establishment of beneficial alleles and the elimination of deleterious ones. Therefore, many geneticists have assumed that natural selection is the dominant force in changing the genetic composition of natural populations, thereby leading to variation.

In 1968, Motoo Kimura, at the National Institute of Genetics in Mishima, Japan, proposed the *neutral theory of evolution*. According to this idea, most genetic variation observed in natural populations is due to the accumulation of neutral mutations. *Neutral mutations* do not affect the phenotype of the organism; neutral alleles are not acted on by natural selection. For example, a mutation within a structural gene that changes a glycine codon from GGG to GGC would not affect the amino acid sequence of the encoded protein. Since neutral mutations do not affect phenotype, they spread throughout a population according to their frequency of appearance and to genetic drift. This theory has been called the "survival of the luckiest" and also *non-Darwinian evolution* to contrast it with Darwin's "survival of the fittest." Kimura agreed with Darwin that natural selection is responsible for adaptive changes in a species during evolution. His main argument is that most modern variation in gene sequences is explained by neutral variation rather than adaptive variation. In 1974, Kimura, along with his colleague Tomoko Ohta, Suggested five principles that govern the evolution of genes at the molecular level:

1. For each protein, the rate of evolution, in terms of amino acid substitutions, is approximately constant with regard to neutral substitutions that do not affect protein structure or function.

   Evidence: As an example, the amount of genetic variation between the coding sequence of the human α-globin and β-globin genes is approximately the same as the difference between the α-globin and β-globin genes in carp. This type of comparison holds true among many different genes compared among many different species.

2. Proteins that are functionally less important for the survival of an organism, or parts of a protein that are less important for its function, tend to evolve faster than more important proteins or

regions of a protein. In other words, during evolution, less important proteins will accumulate amino acid substitutions more rapidly than important proteins.

Evidence: Certain proteins are critical for survival, and their structure is exquisitely tuned to their function. An example are the histone proteins necessary for nucleosome formation in eukaryotes. Histone genes tolerate very few mutations and have evolved extremely slowly. By comparison, fibrinopeptides, which bind to fibrinogen to form a blood clot, evolve very rapidly. Presumably, the sequence of amino acids in this polypeptide is not very important in allowing it to aggregate and form a clot. Another example concerns the amino acid sequences of enzymes. It is known that amino acid substitutions are very rare within the active site (which is critical for function), but are more frequent in other parts of the protein.

3. Those mutant amino acid substitutions that do not disrupt the existing structure and function of a protein (conservative substitutions) occur more frequently in evolution than disruptive amino acid changes.

   Evidence: When examining the rate of change of the coding sequence within structural genes, nucleotide substitutions are more likely to occur in the wobble base than in the first or second base within a codon. Mutations in the wobble base are often silent (i.e., do not change the amino acid sequence of the protein). In addition, conservative substitutions (i.e., a similar amino acid substitution, such as a non-polar amino acid for another non-polar amino acid) are fairly common. By comparison, non conservative substitutions are less frequent, although they do occur. Nonsense and frameshift mutations are very rare within the coding sequences of genes. Also, intron sequences evolve more rapidly than exon sequences.
4. Gene duplication must always precede the emergence of a gene having a new function Evidence: When a single copy of a gene exists in a species, it usually plays a functional role similar to that of the homologous gene found in an other species. Gene duplications have created gene families in which each family member can evolve somewhat different functional roles.
5. Selective elimination of definitely deleterious mutations and random fixation of selectively neutral or very slightly deleterious alleles occur far more frequently in evolution than Darwinian selection of definitely advantageous mutants.

Evidence: As mentioned in #3, silent and conservative mutations are much more common than non conservative substitutions. Presumably these nonconservative mutations usually have a negative effect on the phenotype of the organism, so that they are effectively eliminated from the population by natural selection. On rare occasions, however an amino acid substitution due to a mutation may have a beneficial effect on the phenotype. For example, a non conservative mutation in the β-globin gene produces H^s, which gives an individual resistance to malaria in the heterozygous condition.

In general, the DNA sequencing of hundreds of thousands of different genes from thousands of species has provided compelling support for these five principles of gene evolution at the molecular level. However, the argument is by no means resolved. There are some geneticists, called *selectionists*, who oppose the neutralist theory. They often can offer persuasive theoretical arguments in favour of natural selection as the primary factor promoting genetic variation. In any case, the argument is largely a quantitative rather than a qualitative one. Each school of thought accepts that genetic drift and natural selection both play key roles in evolution. The neutralists argue that most genetic variation arises from neutral genetic mutations and genetic drift, whereas the selectionists argue that beneficial mutations and natural selection are primarily responsible.

Evolution is Associated with Changes in Chromosome Structure and Number

In this section, we have emphasized mutations that alter the DNA sequences within genes. In addition to gene mutations, however, other types of changes, such as gene duplications, inversions, translocations, and changes in chromosome number, are important features of evolution.

As was discussed earlier in this chapter, changes in chromosome structure and/or number may not always be adaptive, but they can lead to reproductive isolation and the origin of new species. As an example of variation of chromosome structure among closely related species, compares the banding pattern of the three largest chromosomes in humans, and the corresponding chromosomes in chimpanzees, gorillas, and orangutans. The banding patterns are strikingly similar, because these species are closely related evolutionarily. However, there are some interesting differences. Humans have one large chromosome 2, but this chromosome is divided into two separate chromosomes in the other three species. This explains why humans have 23 types of chromosomes while the apes have 24. This may have occurred by a fusion of the two smaller chromosomes during the development of the human lineage. Another interesting change in chromosome structure is seen in chromosome 3. The banding patterns among humans, chimpanzees, and gorillas are very similar, but the orangutan has a large inversion that flips the arrangement of bands in the centromeric region.

Concluding Remark

Biological evolution involves heritable changes in one or more characteristics in a population or species over the course of many generations. In the first part of this chapter, we were concerned primarily with how evolution results in the formation of new species. To become a separate species, a population must be reproductively isolated from all other species. This enables their gene pool to evolve as a single unit. There are several ways that reproductive isolation can occur. The *biological species concept* emphasizes reproductive isolation as an important event that leads to speciation, whereas the *species recognition concept* emphasizes sexual selection as a positive force during speciation.

Speciation is usually a branching process (*cladogenesis*), although *anagenesis* (transformation of a single species) occasionally occurs. Depending on geographic barriers, divergent speciation may be *allopatric*, *parapatric*, or *sympatric*. Allopatric speciation, which involves geographic barriers, is thought to be the most widespread form of speciation. It can occur slowly, due to the gradual formation of a geographic barrier, or rapidly, due to the founder effect. *Parapatric speciation* is similar to allopatric, except that the geographic barriers are less complete, with ever-decreasing levels of interbreeding taking place in *hybrid zones*. *Sympatric speciation* does not re quire geographic barriers. Instead, an abrupt genetic event may lead to reproductive isolation. In plants, the formation of polyploids is a common form of sympatric speciation. Polyploids are reproductively isolated, because they produce sterile hybrids when crossed to non-polyploid species. In general, the fossil record suggests that speciation is often a rapid process that punctuates long periods during which the species is in equilibrium with its environment. This *punctuated equilibrium* hypothesis is contrasted with *gradualism*, which proposes a slower but steady rate of evolution due to the accumulation of many small genetic changes.

Molecular evolution is the study of the molecular changes in the genetic material during evolution. There are many ways to investigate the genetic material, but the analysis of DNA sequences and (the deduced) amino acid sequences are the most commonly used methods and perhaps the most informative. Homologous genes are derived from a common ancestral gene. *Interspecies homology* can be used to reconstruct *phylogenetic trees*. When two species are closely related evolutionarily, they will tend to have gene sequences that are more similar to each other.

Kimura and Ohta proposed five principles of molecular evolution that are consistent with our present knowledge of gene sequences. The *neutral theory of evolution* argues that most variation in gene sequences is neutral. In the case of structural genes, *neutral mutations* are not expected to significantly alter the structure and function of the encoded protein; many *nonneutral mutations* are highly deleterious and are thus eliminated quickly from the gene pool; other non neutral mutations are adaptive and are acted on by natural selection to change the characteristics of a species. According to the neutral theory of evolution, though, these adaptive changes represent a small proportion of the total number of genetic changes that occur during evolution. In addition to changes at the gene level, it is also common for changes in chromosome structure and number to occur. These events are often instrumental in leading to reproductive isolation.

In a broad sense, evolutionary biologists would like to know how genetic changes have led to the phenotypic characteristics of present-day species. At the heart of this question is speciation: How do we explain the existence of so many different species? To answer this question, biologists have tried to understand how two different, but closely related species, have become reproductively isolated. By observing species in nature, they have identified several prezygotic and postzygotic mechanisms that prevent the production of viable, interspecies offspring. Researchers have also correlated reproductive isolation with experimentally observable genetic changes. For example, parapatric speciation may be associated with the accumulation of chromosomal inversions that prevent the production of fertile offspring within hybrid zones of contact. Sympatric speciation can occur via the formation of allopolyploids. This speciation mechanism can be con firmed by producing "artificial" polyploids.

Evolutionary biologists are also interested in the rates of evolutionary change. Experimentally, the fossil record suggests that it is more common for evolution to follow a pattern of punctuated equilibrium, in which a species is well adapted to its environment for a long period of time (the equilibrium), which is punctuated by short periods of rapid evolutionary change. These short periods may occur due to abrupt genetic changes (e.g., allopolyploidism or new alleles that have a dramatic effect on phenotype), or there may be a short period of strong selective pressure (e.g., the founder effect, a new predator in the region, or a significant environmental change).

At the molecular level, evolution is due to alterations in DNA sequences, and changes in chromosome structure and number. Geneticists have found a great amount of variation within most species. There are several experimental methods for detecting genetic variation at the DNA and protein levels. The earliest methods involved the study of allozymes; DNA sequencing is now the most common way to study molecular evolution. By comparing homologous gene sequences within a species and among different species, evolutionary biologists have been able to probe the relationship between organismal evolution and changes in DNA sequences. They have also constructed evolutionary trees that describe the phylogenetic relationships among many species. More recently, it has even been possible to analyze DNA sequences from some extinct species.

With all of this sequence information available, researchers have debated the origin of genetic variation in modern species. The data suggest that most genetic variation occurs through neutral changes that accumulate due to genetic drift. Assuming that the rate of new mutation is essentially constant, the accumulation of neutral changes provides a biological clock to measure the time scale of evolution. However, not all variation is neutral. Adaptive changes, which are acted on by natural selection, must also occur and alter the phenotypic characteristics of species over time, thereby leading to the evolution of new, better adapted species.

Common ancestor, 252
Complementary gene action, 38
Complete nondisjunction, 111
Condensin, 72
Conger conger, 293
Conglomerate, 263
Constitutive heterochromatin, 78, 162
Constitutive puffs, 76
Correlation, 189
Correlation coefficient, 190, 212
Counted, 5
Covariance, 189
Cri-du-chat, 133
Cri-du-chat syndrome, 92
Crossing over, 51
Cytogeneticist, 89, 160
Cytoplasm, 44

D

D. melanogaster, 122, 246
D. persimilis, 289
D. pseudoobscura, 246
Darwinian fitness, 265, 275
Datura stramonium, 104
Deficiency, 91, 117, 122
Deletions, 91, 117
Deoxyribonucleic acid, 45
Development, 240
Developmental genetics, 214
Diakinesis, 174
Dicentric bridge, 99
Dicentric chromosome, 99
Different, 29
Differentiated, 235
Differentiation, 241
Dihybrid, 10
Diploid set, 49
Diplotene, 174
Disassortative mating, 252
Discontinuous traits, 186
DNA-DNA hybridization, 84
Dominance, 47
Dominance hypothesis, 210
Dominant, 6, 47
Dominant gene, 7
Dorsoventral, 241
Dorsoventral axis, 217
Double-bar, 94
Drosophila, 51, 53, 67, 69, 72, 75, 76, 86, 93, 94, 96, 107, 110, 112, 118, 123, 136, 146, 164, 168, 175, 176, 197, 215, 216, 217, 218, 220, 221, 224, 225, 226, 227, 228, 229, 233, 234, 235, 236, 237, 240, 241, 242, 285, 289, 290, 299, 301
Drosophila embryo, 225
Drosophila equinoxialis, 291
Drosophila melanogaster, 50, 102, 179, 196, 215, 291
Drosophila pseudoobscura, 246, 283, 289, 291
Duplicate genes, 39
Duplication, 91, 117

E

E proteins, 236
E. coli, 56, 59, 66, 67
E. distinguenda, 292
E. mucronata, 292
E. superba, 292
Ecological genetics, 278
Ecological niche, 278
Egg-laying defective, 230
Embryogenesis, 216
Endoplasmic reticulum, 44
Endopolyploid, 118
Endopolyploidy, 107
Engrailed, 226
Environmentally induced puffs, 76
Enzyme, 246

Epistatic, 36
Equal segregation, 8
Equus quagga, 315
Euchromatin, 78
Eukaryotes, 44
Eukaryotic, 44
Euoenothera, 132
Euphausia, 292
Euphausia superba, 292
Euploid, 102
Even-skipped, 218, 226
Expressivity, 40, 41

F

Facultative heterochromatin, 162
Familial down syndrome, 100
Fate map, 229, 241
Festuca arundinacea, 116
Fine grained, 292
First filial generation, 4
Fixation coefficient, 253
Founder effect, 262, 275, 279, 302
Frequency distribution, 186
Fusarium, 20

G

G banding, 160
G bands, 77, 90
G-tetraplex, 81
G. pubescens, 306, 307, 308
G. speciosa, 306, 307, 308
G. tetrahit, 305, 306, 307, 308
Gain of function mutation, 227
Galeopsis pubescens, 305
Galeopsis speciosa, 305
Galeopsis tetrahit, 305, 306
Gap genes, 218, 241
Gastrulation, 216
Gene amplification, 86
Gene duplication, 93
Gene family, 84, 96, 117
Gene flow, 263, 275
Gene knockout, 234, 241
Gene pair, 7
Gene pool, 244, 274
Genetic load, 273, 276
Genetic markers, 23
Genetic material, 45
Genetic variation, 88
Genome mutation, 88
Genotype, 47
Genotype frequencies, 247, 275
Giemsa stain, 77, 90
Glutamate-oxaloacetate transaminase, 283
Gradualism, 308, 321
Gurke, 239

H

H. niger, 114
Haploid, 49
Hardy—Weinberg equation, 249
Hardy—Weinberg equilibrium, 274
Helix-loop-helix, 236
Hemoglobins, 96
Hemophilia, 97
Heritability, 198, 212
Heritability in the broad sense, 212
Heterochronic mutations, 232, 241
Heterogametic, 50
Heterokaryon, 116
Heterosis, 116, 209
Heterozygote, 8, 47
Heterozygote advantage, 270
Hippotragus equinus, 114
Histone, 69
Histone acetyl transferases, 73
Homeobox, 228
Homeodomain, 228
Homeologous, 114
Homeotic, 227
Homeotic gene, 217, 241

Homogametic, 50
Homologous, 49, 96, 310
Homologous genes, 233
Homozygote, 8, 47
Homozygous dominant, 8
Homozygous recessive, 9
Hox complex, 234, 241
Hox gene, 234, 235
Hunchback, 225
Hybrid cell, 116
Hybrid vigor, 116, 209
Hybrid zone, 288, 295, 304, 321
Hybrids, 8

I

Imaginal disks, 216
In situ, 164, 165, 224
In vivo, 236
Inbred, 252
Inbreeding, 206, 252
Inbreeding path, 252
Incomplete, 47
Incomplete dominance, 26
Independent, 11
Independent assortment, 12
Induction, 221
Intercalary heterochromatin, 78
Interphase, 48
Interspecies homology, 310, 322
Interspecies matings, 110, 118
Interstitial deficiency, 92
Intraspecies homology, 311
Introgressive hybridization, 288
Inversion, 91, 117
Inversion heterozygote, 98
Inversion loop, 98
Isochromosome, 169
Isolating mechanisms, 299

K

Karyotype, 89, 168
Kinetochore, 79
Klinefelter's syndrome, 146, 167
Kruppel, 218

L

Lampbrush chromosomes, 76
Lampbrush loops, 165
Leptotene, 172
Linear order, 52
Linkage group, 53
Local populations, 245
Loci, 52
Lock-and-key model, 60
Locusta, 176
Lolium multiflorum, 116
Long interspersed elements, 85

M

Macroevolution, 296
MADS box, 240
Malheurensis, 281, 282, 283
Map distance, 52
Map units, 52
Mean fitness of the population, 267
Meiosis, 49
Meiotic nondisjunction, 110
Metabolism, 44
Metacentric, 90
Metamorphosis, 216
Metaphase II, 175
Microconvules, 171
Microevolution, 296
Microheterogeneity, 286
Mirounga angustirostris, 280
Mitochondria, 44
Mitosis, 48
Mitotic nondisjunction, 110
Modified Mendelian ratios, 35
Modifier gene, 37
Molecular evolution, 296, 321
Monohybrid, 10
Monomorphic, 246

Monoploid, 106
Monopterous, 239
Monosomy, 167
Morphogens, 220
Mosaic, 118
Mosaicism, 112
Multiple allelism, 27
Musmusculus, 286
Mutability, 60
Mutant, 50
Mutatlon, 50, 64
Mutation rate, 257, 275
Myogenic bHLH proteins, 241
Myoglobin, 96

N

N. tabacum, 155
Nanos, 222, 223
Narrow sense heritability, 201, 212
Natural selection, 264, 275
Neurospora, 40
Neutral forces, 275
Neutral mutations, 318, 322
Neutral processes, 255
Neutral theory of evolution, 318, 322
Nicotiana tabacum, 155
Non-Darwinian evolution, 318
Non-disjunction, 110, 118, 166
Nonhistone proteins, 73
Nonneutral mutation, 318, 322
Normal distribution, 211
Normal distribution curve, 186
Nuclear membrane, 44
Nuclease-hypersensitive sites, 72
Nucleosomes, 69
Nucleus, 44

O

O. argillicola, 132
O. biennis, 133
O. grandiflora, 132
O. hookeri, 132
O. irrigua, 133
O. parviflora, 133
O. strigosa, 133
Oenothera, 131, 132
Oenothera organensis, 290
Organelles, 44
Origin of species, 298
Outbreeding, 252
Over dominance Hypothesis, 210
Overdominance, 270
Oxytricha nova, 82

P

Pachytene, 174
Pair, 49
Pair-rule genes, 218, 241
Panmixis, 286
Paralithodes camtschatica, 137
Parapatric, 321
Parapatric speciation, 304, 321
Parasegment, 216, 241
Parental generation, 4
Parental types, 47
Parthenogenesis, 117
Partial, 47
Particulate inheritance, 1
Pedigree analysis, 17
Penetrance, 40, 41
Pericentric inversion, 97
Phenotype, 47
Phylogenetic tree, 312, 322
Physically, 52
Pistillata, 240
Pisum sativum, 2
Plasmodium falciparum, 270
Pleiotropic, 33
Pneumonoccus, 54
Point centromere, 79
Polycomb, 228

Polygenic, 212
Polygenic inheritance, 193
Polymorphic, 274, 290
Polymorphism, 246
Polyploid, 103, 118
Polytene chromosomes, 94, 118
Population, 245
Population genetics, 244
Position, 52
Position effect, 94, 118
Positional information, 218, 241
Postmating isolation, 288
Potential, 5
Premating isolation, 288
Primula, 38
Probability, 11
Prokaryotes, 44
Prokaryotic, 44
Prophase II, 175
Propositus, 17
Protoplasm, 44
Protoplast, 116, 154
Proximodistal axis, 217
Pseudo-overdominance, 211
Pseudodominance, 93, 117, 123
Pseudotsuga menziesii, 245
Pubescens, 308
Puffs, 163
Punctuated equilibrium, 309, 321
Pure line, 3

Q

Q banding, 160
Quantitative genetics, 184, 211
Quantitative trait loci, 193, 212
Quantitative traits, 186, 211

R

R banding, 160
R-bands, 78
Rana pipiens, 289
Rana sylvatica, 289
Random genetic drift, 258, 275
Raphanobrassica, 153
Raphanus, 115, 153
Realizator genes, 228, 241
Realized heritability, 208
Recessive, 6, 47
Recessive gene, 7
Recessive lethal, 33
Reciprocal, 47
Reciprocal cross, 4
Reciprocal translocation, 91, 118, 170
Recombinant types, 47
Regional centromeres, 79
Repetitive DNA, 84
Replication, 62
Replication error, 64
Replication-coupling assembly factor, 72
Restitution, 120
Retrospective testing, 305
Retrotransposons, 84
Reverse bands, 78
Reverse genetics, 234
Reversion, 50
Rhoeo, 131
Ribosomal transcription unit, 178
Ribosomes, 44
Right-left axis, 217
Root apical meristems, 238
Rophanus, 114

S

S. paniculata, 282
Saccharomyces cerevisiae, 79, 83
Satellite DNA, 77
Scaffold, 73, 172
Segment-polarity genes, 218, 241
Segmentation, 241
Segmentation genes, 218, 225

Segments, 216, 241
Selection coefficient, 266
Selectionists, 320
Selective breeding, 205
Self-incompatibility, 29
Self-maintenance, 45
Self-reproduction, 45
Sex chromosomes, 50
Sex linkage, 51
Sexual selection, 301
Shoot apical meristem, 238
Short interspersed elements, 85
Sibling species, 281
Simple translocation, 91
Sister chromatid, 49, 172
Sister chromatid exchanges, 180
SMC proteins, 72
Somatic cells, 48
Sordaria macrospora, 176
Speciation, 280
Species, 299
Species recognition concept, 301, 321
Specific group, 205
Specific set of conditions, 205
Speciosa, 308
Stage-specific puffs, 76
Standard deviation, 188, 212
Stephanomeria exigera coronaria, 282
Stephanomeria paniculata, 281
Streptococcus pneumoniae, 54
Struggle for existence, 298
Submetacentric, 90
Subpopulations, 245
Subvital, 34
Suppression, 38
Survival of the fittest, 298
Suture zone, 288
Sympatric, 321
Sympatric populations, 281
Sympatric speciation, 305, 321
Synapse, 49
Synaptinemal complex, 172
Synaptinemal complex node, 176
Syndrome, 133

T

T cell, 231
T-loop, 83
T. aestivum, 193
Telocentric, 90
Telomerase, 80
Telomeres, 79
Telophase I, 175
Telophase II, 175
Terminal deficiency, 92
Terminalize, 174
Testcross, 12
Tetrad, 174
Tetraploid, 103
Thalassemia, 87
Three-domain, 314
Threshold concentration, 220
Tissue-specific puffs, 76
Totipotent, 238
Tradescantia, 131
Transfer experiments, 58
Transformed, 54
Transforming principle, 55
Translocation, 51. 91, 117
Translocation cross, 101
Tricale, 154
Trifolium, 29
Trifolium repens, 283
Triplets, 61
Triploid, 102
Triradius, 203
Trisomy, 103, 167
Triticale, 153, 154
Triticum aestivum, 108, 154, 193
Triturus cristatus, 165

Triturus marmoratus, 177
Turner's syndrome, 168

U

Ultra-bar, 94, 95, 96
Uninemic, 67
Unique DNA, 84

V

Variation, 205
Verticillium, 20
Virions, 46
Viscoelastic, 68

W

Wild type, 50
Wingless, 226

X

Xenopus, 86, 233
Xenopus laevis, 233

Z

Zygotene, 172
Zygotic genes, 226
Zymograms, 290